Quantum

QUANTUM MECHANICS

500 Problems with Solutions

G. Aruldhas

Formerly Professor and Head of Physics
and Dean, Faculty of Science
University of Kerala

PHI Learning Private Limited

Delhi-110092

2016

₹ 325.00

QUANTUM MECHANICS: 500 Problems with Solutions
G. Aruldhas

ISBN-978-81-203-4069-5

Seventh Printing ... ... **July, 2016**

Published by Asoke K. Ghosh, PHI Learning Private Limited, Rimjhim House, 111, Patparganj Industrial Estate, Delhi-110092 and Printed by V.K. Batra at Pearl Offset Press Private Limited, New Delhi-110015.

To

my wife, **Myrtle**

Our children

Vinod & **Anitha, Manoj** & **Bini, Ann** & **Suresh**

and

Our grandchildren

Nithin, Cerene, Tina, Zaneta, Juana, Joshua, Tesiya, Lidiya, Ezekiel

for their unending encouragement and support

Contents

Preface

This comprehensive, in-depth treatment of quantum mechanics in the form of problems with solutions provides a thorough understanding of the subject and its application to various physical and chemical problems. Learning to solve problems is the basic purpose of a course since it helps in understanding the subject in a better way. Keeping this in mind, considerable attention is devoted to work out these problems. Typical problems illustrating important concepts in Quantum Mechanics have been included in all the chapters. Problems from the simple plugged-ins to increasing order of difficulty are included to strengthen the students' understanding of the subject.

Every effort has been taken to make the book explanatory, exhaustive, and user-friendly. Besides helping students to build a thorough conceptual understanding of Quantum Mechanics, the book will also be of considerable assistance to readers in developing a more positive and realistic impression of the subject.

It is with a deep sense of gratitude and pleasure that I acknowledge my indebtedness to my students for all the discussions and questions they have raised. I express my sincere thanks to the Publishers, PHI Learning, for their unfailing cooperation and for the meticulous processing of the manuscript. Finally, I acknowledge my gratitude to my wife, **Myrtle**, and our children for the encouragement, cooperation, and academic environment they have provided throughout my career.

Above all, I thank my Lord **Jesus Christ** who has given me wisdom, knowledge, and guidance throughout my life.

G. Aruldhas

CHAPTER 1

Quantum Theory

Quantum physics, which originated in the year 1900, spans the first quarter of the twentieth century. At the end of this important period, Quantum Mechanics emerged as the overruling principle in Physics.

1.1 Planck's Quantum Hypothesis

Quantum physics originated with Max Planck's explanation of the black body radiation curves. Planck assumed that the atoms of the walls of the black body behave like tiny electromagnetic oscillators, each with a characteristic frequency of oscillation. He then boldly put forth the following suggestions:

1. An oscillator can have energies given by

$$E_n = nh\nu, \qquad n = 0,\ 1,\ 2,\ \ldots \tag{1.1}$$

   where ν is the oscillator frequency and h is **Planck's constant** whose value is 6.626×10^{-34} J s.

2. Oscillators can absorb energy from the cavity or emit energy into the cavity only in discrete units called **quanta**, i.e.,

$$\Delta E_n = \Delta n h\nu = h\nu \tag{1.2}$$

   Based on these postulates, Planck derived the following equation for the spectral energy density u_ν of black body radiation:

$$u_\nu = \frac{8\pi h\nu^3}{c^3} \frac{d\nu}{\exp(h\nu/kT) - 1} \tag{1.3}$$

1.2 Photoelectric Effect

On the basis of quantum ideas, Einstein succeeded in explaining the photoelectric effect. He extended Planck's idea and suggested that light is not only absorbed or emitted in quanta but also propagates as quanta of energy $h\nu$, where ν is the frequency of radiation. The individual quanta of light are called **photons**. Einstein's photoelectric equation

$$h\nu = h\nu_0 + \frac{1}{2}mv^2 \tag{1.4}$$

explained all aspects of photoelectric effect. In Eq. (1.4), $h\nu$ is the energy of the incident photon, $h\nu_0$ is the **work function** of the metallic surface, and ν_0 is the threshold frequency. Since the rest mass of photon is zero,

$$E = cp \quad \text{or} \quad p = \frac{E}{c} = \frac{h\nu}{c} = \frac{h}{\lambda} \tag{1.5}$$

1.3 Compton Effect

Compton allowed x-rays of monochromatic wavelngth λ to fall on a graphite block and measured the intensity of scattered x-rays. In the scattered x-rays, he found two wavelengths—the original wavelength λ and another wavelength λ' which is larger than λ. Compton showed that

$$\lambda' - \lambda = \frac{h}{m_0 c}(1 - \cos\phi) \tag{1.6}$$

where m_0 is the rest mass of electron and ϕ is the scattering angle. The factor h/m_0c is called the **Compton wavelength**.

1.4 Bohr's Theory of Hydrogen Atom

Niels Bohr succeeded in explaining the observed hydrogen spectrum on the basis of the following two postulates:

(i) An electron moves only in certain allowed circular orbits which are stationary states in the sense that no radiation is emitted. The condition for such states is that the orbital angular momentum of the electron is given by

$$mvr = n\hbar, \qquad n = 1, 2, 3, \ldots \tag{1.7}$$

where $\hbar = h/2\pi$ is called the **modified Planck's constant**, v is the velocity of the electron in the orbit of radius r, and m is the electron mass.

(ii) Emission or absorption of radiation occurs only when the electron makes a transition from one stationary state to another. The radiation has a definite frequency ν_{mn} given by the condition

$$h\nu_{mn} = E_m - E_n \tag{1.8}$$

where E_m and E_n are the energies of the states m and n, respectively.

According to Bohr's theory, the radius of the nth orbit is

$$r_n = \frac{n^2\hbar^2}{kme^2}, \qquad k = \frac{1}{4\pi\varepsilon_0} \tag{1.9}$$

where ε_0 is the permittivity of vacuum and its experimental value is 8.854×10^{-12} C^2 N^{-1} m^{-2}.

The radius of the first orbit is called **Bohr radius** and is denoted by a_0, i.e.

$$a_0 = \frac{4\pi\varepsilon_0\hbar^2}{me^2} = 0.53 \text{ Å} \tag{1.10}$$

In terms of a_0, from Eq. (1.9), we have

$$r_n = n^2 a_0 \tag{1.11}$$

The total energy of the hydrogen atom in the nth state is

$$E_n = -\frac{me^4}{32\pi^2\varepsilon_0^2\hbar^2}\cdot\frac{1}{n^2} = -\frac{13.6}{n^2}\text{ eV}, \qquad n = 1, 2, 3, \ldots \tag{1.12}$$

When the electron drops from the mth to nth state, the frequency of the emitted line ν_{mn} is given by

$$h\nu_{mn} = \frac{me^4}{32\pi^2\varepsilon_0^2\hbar^2}\left(\frac{1}{n^2} - \frac{1}{m^2}\right), \qquad m > n \geq 1 \tag{1.13}$$

For hydrogen-like systems,

$$E_n = -\frac{Z^2 me^4}{32\pi^2\varepsilon_0^2\hbar^2}\frac{1}{n^2}, \qquad n = 1, 2, 3, \ldots \tag{1.14}$$

The parameters often used in numerical calculations include the **fine structure constant** α and the **Rydberg constant** R given by

$$\alpha = \frac{e^2}{4\pi\varepsilon_0 c\hbar} = \frac{1}{137} \tag{1.15}$$

$$R = \frac{me^4}{8\varepsilon_0^2 ch^3} = 10967757.6 \text{ m}^{-1} \tag{1.16}$$

The Rydberg constant for an atom with a nucleus of infinite mass is denoted by R_∞, which is the same as R in (1.16).

Different spectral series of hydrogen atom can be obtained by substituting different values for m and n in Eq. (1.13).

(i) The Lyman series

$$\frac{1}{\lambda} = R\left(\frac{1}{1^2} - \frac{1}{m^2}\right), \qquad m = 2, 3, 4, \ldots \tag{1.17}$$

(ii) The Balmer series

$$\frac{1}{\lambda} = R\left(\frac{1}{2^2} - \frac{1}{m^2}\right), \qquad m = 3, 4, 5, \ldots \tag{1.18}$$

(iii) The Paschen series

$$\frac{1}{\lambda} = R\left(\frac{1}{3^2} - \frac{1}{m^2}\right), \qquad m = 4, 5, 6, \ldots \tag{1.19}$$

(iv) The Brackett series

$$\frac{1}{\lambda} = R\left(\frac{1}{4^2} - \frac{1}{m^2}\right), \qquad m = 5, 6, 7, \ldots \tag{1.20}$$

(v) The Pfund series

$$\frac{1}{\lambda} = R\left(\frac{1}{5^2} - \frac{1}{m^2}\right), \qquad m = 6, 7, 8, \ldots \tag{1.21}$$

1.5 Wilson–Sommerfeld Quantization Rule

In 1915, Wilson and Sommerfeld proposed the general quantization rule

$$\oint p_i\, dq_i = n_i h, \qquad n_i = 0, 1, 2, 3, \ldots \tag{1.22}$$

where $\oint$ is over one cycle of motion. The q_i's and p_i's are the generalized coordinates and generalized momenta, respectively. In circular orbits, the angular momentum $L = mvr$ is a constant of motion. Hence, Eq. (1.22) reduces to

$$mvr = \frac{nh}{2\pi}, \qquad n = 1, 2, 3, \ldots \tag{1.23}$$

which is Bohr's quantization rule. The quantum number $n = 0$ is left out as it would correspond to the electron moving in a straight line through the nucleus.

PROBLEMS

1.1 The work function of barium and tungsten are 2.5 eV and 4.2 eV, respectively. Check whether these materials are useful in a photocell, which is to be used to detect visible light.

Solution. The wavelength λ of visible light is in the range 4000–7000 Å. Then,

$$\text{Energy of 4000 Å light} = \frac{hc}{\lambda} = \frac{(6.626 \times 10^{-34}\,\text{J s})\,(3 \times 10^{8}\ \text{m/s})}{(4000 \times 10^{-10}\text{m})(1.6 \times 10^{-19}\ \text{J/eV})} = 3.106\ \text{eV}$$

$$\text{Energy of 7000 Å light} = \frac{6.626 \times 10^{-34} \times 3 \times 10^{8}}{7000 \times 10^{-10} \times 1.6 \times 10^{-19}} = 1.77\ \text{eV}$$

The work function of tungsten is 4.2 eV, which is more than the energy range of visible light. Hence, barium is the only material useful for the purpose.

1.2 Light of wavelength 2000 Å falls on a metallic surface. If the work function of the surface is 4.2 eV, what is the kinetic energy of the fastest photoelectrons emitted? Also calculate the stopping potential and the threshold wavelength for the metal.

Solution. The energy of the radiation having wavelength 2000 Å is obtained as

$$\frac{hc}{\lambda} = \frac{(6.626 \times 10^{-34}\,\text{J s})\,(3 \times 10^{8}\ \text{m/s})}{(2000 \times 10^{-10}\text{m})(1.6 \times 10^{-19}\ \text{J/eV})} = 6.212\ \text{eV}$$

Work function = 4.2 eV

KE of fastest electron = 6.212 – 4.2 = 2.012 eV

Stopping potential = 2.012 V

$$\text{Threshold wavelength } \lambda_0 = \frac{hc}{\text{Work function}}$$

$$\lambda_0 = \frac{(6.626 \times 10^{-34}\,\text{J s})\,(3 \times 10^{8}\ \text{m/s})}{(4.2\ \text{eV})(1.6 \times 10^{-19}\ \text{J/eV})} = 2958\ \text{Å}$$

1.3 What is the work function of a metal if the threshold wavelength for it is 580 nm? If light of 475 nm wavelength falls on the metal, what is its stopping potential?

Solution.

$$\text{Work function} = \frac{hc}{\lambda_0} = \frac{(6.626 \times 10^{-34}\,\text{J s})\,(3 \times 10^{8}\ \text{m/s})}{(580 \times 10^{-9}\text{m})(1.6 \times 10^{-19}\ \text{J/eV})} = 2.14\ \text{eV}$$

$$\text{Energy of 475 nm radiation} = \frac{hc}{\lambda} = \frac{(6.626 \times 10^{-34}\,\text{J s})\,(3 \times 10^{8}\ \text{m/s})}{(475 \times 10^{-9}\text{m})(1.6 \times 10^{-19}\ \text{J/eV})} = 2.62\ \text{eV}$$

Stopping potential = 2.62 – 2.14 = 0.48 V

1.4 How much energy is required to remove an electron from the $n = 8$ state of a hydrogen atom?

Solution. Energy of the $n = 8$ state of hydrogen atom $= \dfrac{-13.6\ \text{eV}}{8^2} = -0.21\ \text{eV}$

The energy required to remove the electron from the $n = 8$ state is 0.21 eV.

1.5 Calculate the frequency of the radiation that just ionizes a normal hydrogen atom.

Solution. Energy of a normal hydrogen atom = –13.6 eV

Frequency of radiation that just ionizes is equal to

$$\frac{E}{h} = \frac{13.6 \text{ eV } (1.6 \times 10^{-19} \text{ J/eV})}{6.626 \times 10^{-34} \text{ J s}} = 3.284 \times 10^{15} \text{ Hz}$$

1.6 A photon of wavelength 4 Å strikes an electron at rest and is scattered at an angle of 150° to its original direction. Find the wavelength of the photon after collision.
Solution.

$$\Delta\lambda = \lambda' - \lambda = \frac{h}{m_0 c}(1 - \cos 150°)$$

$$= \frac{6.626 \times 10^{-34} \text{ J s} \times 1.866}{(9.11 \times 10^{-31} \text{ kg})(3 \times 10^8 \text{ m/s})} = 0.045 \text{ Å}$$

$$\lambda' = \lambda + 0.045 \text{ Å} = 4.045 \text{ Å}$$

1.7 When radiation of wavelength 1500 Å is incident on a photocell, electrons are emitted. If the stopping potential is 4.4 volts, calculate the work function, threshold frequency and threshold wavelength.

Solution. Energy of the incident photon $= \dfrac{hc}{\lambda}$

$$= \frac{(6.626 \times 10^{-34} \text{ J s})(3 \times 10^8 \text{ m/s})}{(1500 \times 10^{-10} \text{ m})(1.6 \times 10^{-19} \text{ J/eV})} = 8.28 \text{ eV}$$

Work function = 8.28 – 4.4 = 3.88 eV

$$\text{Threshold frequency } \nu_0 = \frac{3.88 \text{ eV } (1.6 \times 10^{-19} \text{ J/eV})}{6.626 \times 10^{-34} \text{ J s}} = 9.4 \times 10^{14} \text{ Hz}$$

$$\text{Threshold wavelength } \lambda_0 = \frac{c}{\nu_0} = \frac{3 \times 10^8 \text{ m/s}}{9.4 \times 10^{14} \text{ s}^{-1}} = 3191 \text{ Å}$$

1.8 If a photon has wavelength equal to the Compton wavelength of the particle, show that the photon's energy is equal to the rest energy of the particle.
Solution. Compton wavelength of a particle = h/m_0c

Wavelength of a photon having energy $E = \dfrac{hc}{E}$

Equating the above two equations, we get

$$\frac{h}{m_0 c} = \frac{hc}{E} \quad \text{or} \quad E = m_0 c^2$$

which is the rest energy of the particle.

1.9 x-rays of wavelength 1.4 Å are scattered from a block of carbon. What will be the wavelength of scattered x-rays at (i) 180°, (ii) 90°, and (iii) 0°?
Solution.

$$\lambda = \lambda + \frac{h}{m_0 c}(1 - \cos\phi), \qquad \lambda = 1.4 \text{ Å}$$

$$\frac{h}{m_0 c} = \frac{6.626 \times 10^{-34}\ \text{J s}}{9.1 \times 10^{-31}\ \text{kg}\ (3 \times 10^8\ \text{m/s})} = 0.024\ \text{Å}$$

(i) $\lambda' = \lambda + \frac{h}{m_0 c} \times 2 = 1.45\ \text{Å}$

(ii) $\lambda' = \lambda + \frac{h}{m_0 c} = 1.42\ \text{Å}$

(iii) $\lambda' = \lambda + \frac{h}{m_0 c}(1 - 1) = 1.4\ \text{Å}$

1.10 Determine the maximum wavelength that hydrogen in its ground state can absorb. What would be the next smallest wavelength that would work?

Solution. The maximum wavelength corresponds to minimum energy. Hence, transition from $n = 1$ to $n = 2$ gives the maximum wavelength. The next wavelength the ground state can absorb is the one for $n = 1$ to $n = 3$.

The energy of the ground state, $E_1 = -13.6$ eV

Energy of the $n = 2$ state, $E_2 = \frac{-13.6}{4}\ \text{eV} = -3.4\ \text{eV}$

Energy of the $n = 3$ state, $E_3 = \frac{-13.6}{9}\ \text{eV} = -1.5\ \text{eV}$

$$\text{Maximum wavelength} = \frac{hc}{E_2 - E_1}$$

$$= \frac{(6.626 \times 10^{-34}\ \text{J s})\ (3 \times 10^8\ \text{m/s})}{10.2\ \text{eV} \times 1.6 \times 10^{-19}\ \text{J/eV}}$$

$$= 122 \times 10^{-9}\ \text{m} = 122\ \text{nm}$$

$$\text{Next maximum wavelength} = \frac{hc}{E_3 - E_1} = 103\ \text{nm}$$

1.11 State the equation for the energy of the nth state of the electron in the hydrogen atom and express it in electron volts.

Solution. The energy of the nth state is

$$E_n = -\frac{me^4}{8\varepsilon_0^2 h^2} \frac{1}{n^2}$$

$$= \frac{-(9.11 \times 10^{-31}\ \text{kg})\ (1.6 \times 10^{-19}\ \text{C})^4}{8(8.85 \times 10^{-12}\ \text{C}^2\ \text{N}^{-1}\ \text{m}^{-2})^2 (6.626 \times 10^{-34}\ \text{J s})^2 n^2}$$

$$= \frac{-21.703 \times 10^{-19}}{n^2}\ \text{J} = \frac{21.703 \times 10^{-19}\ \text{J}}{1.6 \times 10^{-19} n^2\ \text{J/eV}}$$

$$= -\frac{13.56}{n^2}\ \text{eV}$$

1.12 Calculate the maximum wavelength that hydrogen in its ground state can absorb. What would be the next maximum wavelength?

Solution. Maximum wavelength correspond to minimum energy. Hence the jump from ground state to first excited state gives the maximum λ.

Energy of the ground state = –13.6 eV
Energy of the first excited state = –13.6/4 = –3.4 eV
Energy of the n = 3 state = –13.6/9 = –1.5 eV
Maximum wavelength corresponds to the energy 13.6 – 3.4 = 10.2 eV

$$\text{Maximum wavelength} = \frac{c}{\nu} = \frac{hc}{E_2 - E_1}$$

$$= \frac{(6.626 \times 10^{-34}\,\text{J s}) \times (3.0 \times 10^{8}\,\text{m/s})}{10.2 \times 1.6 \times 10^{-19}\,\text{J}}$$

$$= 122 \times 10^{-9}\,\text{m} = 122\,\text{nm}$$

The next maximum wavelength corresponds to a jump from ground state to the second excited state. This requires an energy 13.6 eV – 1.5 eV = 12.1 eV, which corresponds to the wavelength

$$\lambda = \frac{hc}{E_3 - E_1}$$

$$= \frac{(6.626 \times 10^{-34}\,\text{J s}) \times (3.0 \times 10^{8}\,\text{m/s})}{12.1 \times 1.6 \times 10^{-19}\,\text{J}}$$

$$= 103 \times 10^{-9}\,\text{m} = 103\,\text{nm}$$

1.13 A hydrogen atom in a state having binding energy of 0.85 eV makes a transition to a state with an excitation energy of 10.2 eV. Calculate the energy of the emitted photon.

Solution. Excitation energy of a state is the energy difference between that state and the ground state.

Excitation energy of the given state = 10.2 eV
Energy of the state having excitation energy 10.2 eV = –13.6 + 10.2 = – 3.4 eV
Energy of the emitted photon during transition from – 0.85 eV to –3.4 eV
= –0.85 – (–3.4) = 2.55 eV

Let the quantum number of –0.85 eV state be n and that of –3.4 eV state be m. Then,

$$\frac{13.6}{n^2} = 0.85 \quad \text{or} \quad n^2 = 16 \quad \text{or} \quad n = 4$$

$$\frac{13.6}{m^2} = 3.4 \quad \text{or} \quad m^2 = 4 \quad \text{or} \quad m = 2$$

The transition is from n = 4 to n = 2 state.

1.14 Determine the ionization energy of the He^+ ion. Also calculate the minimum frequency a photon must have to cause ionization.

Solution. Energy of a hydrogen-like atom in the ground state = $-Z^2 \times 13.6$ eV

Ground state energy of He^+ ion = – 4 × 13.6 eV = – 54.4 eV

Ionization energy of He^+ ion = 54.4 eV

The minimum frequency of a photon that can cause ionization is

$$\nu = \frac{E}{h} = \frac{54.4 \text{ eV} (1.6 \times 10^{-19} \text{ J/eV})}{6.626 \times 10^{-34} \text{ J s}} = 13.136 \times 10^{15} \text{ Hz}$$

1.15 Calculate the velocity and frequency of revolution of the electron of the Bohr hydrogen atom in its ground state.

Solution. The necessary centripetal force is provided by the coulombic attraction, i.e.

$$\frac{mv^2}{r} = \frac{ke^2}{r^2}, \qquad k = \frac{1}{4\pi\varepsilon_0}$$

Substituting the value of r from Eq. (1.9), the velocity of the electron of a hydrogen atom in its ground state is obtained as

$$v_1 = \frac{e^2}{2\varepsilon_0 h} = \frac{(1.6 \times 10^{-19} \text{C})^2}{2(8.85 \times 10^{-12} \text{ C}^2\text{N}^{-1} \text{ m}^{-2})(6.626 \times 10^{-34} \text{ J s})}$$

$$= 2.18 \times 10^6 \text{ ms}^{-1}$$

$$\text{Period } T = \frac{2\pi r}{v_1}$$

Substituting the value of r and v_1, we obtain the frequency of revolution of the electron in the ground state as

$$\nu_1 = \frac{me^4}{4\varepsilon_0^2 h^3} = \frac{(9.11 \times 10^{-31} \text{ kg})(1.6 \times 10^{-19} \text{ C})^4}{4(8.85 \times 10^{-12} \text{ C}^2\text{N}^{-1} \text{ m}^{-2})(6.626 \times 10^{-34} \text{ J s})^3}$$

$$= 6.55 \times 10^{15} \text{ Hz}$$

1.16 What is the potential difference that must be applied to stop the fastest photoelectrons emitted by a surface when electromagnetic radiation of frequency 1.5×10^{15} Hz is allowed to fall on it? The work function of the surface is 5 eV.

Solution. The energy of the photon is given by

$$h\nu = (6.626 \times 10^{-34} \text{ J s})(1.5 \times 10^{15} \text{ s}^{-1})$$

$$= \frac{(6.626 \times 10^{-34} \text{ J s})(1.5 \times 10^{15} \text{ s}^{-1})}{1.6 \times 10^{-19} \text{ J/eV}} = 6.212 \text{ eV}$$

Energy of the fastest electron = 6.212 – 5.0 = 1.212 eV

Thus, the potential difference required to stop the fastest electron is 1.212 V

1.17 x-rays with λ = 1.0 Å are scattered from a metal block. The scattered radiation is viewed at 90° to the incident direction. Evaluate the Compton shift.

Solution. The compton shift

$$\Delta\lambda = \frac{h}{m_0 c}(1 - \cos\phi) = \frac{(6.626 \times 10^{-34} \text{ J s})(1 - \cos 90°)}{(9.11 \times 10^{-31} \text{ kg})(3 \times 10^8 \text{m s}^{-1})}$$

$$= 2.42 \times 10^{-12} \text{ m} = 0.024 \text{ Å}$$

1.18 From a sodium surface, light of wavelength 3125 and 3650 Å causes emission of electrons whose maximum kinetic energy is 2.128 and 1.595 eV, respectively. Estimate Planck's constant and the work function of sodium.

Solution. Einstein's photoelectric equation is

$$\frac{hc}{\lambda} = \frac{hc}{\lambda_0} + \text{kinetic energy}$$

$$\frac{hc}{3125 \times 10^{-10}\ \text{m}} = \frac{hc}{\lambda_0} + 2.128\ \text{eV} \times (1.6 \times 10^{-19}\ \text{J/eV})$$

$$\frac{hc}{3650 \times 10^{-10}\ \text{m}} = \frac{hc}{\lambda_0} + 1.595\ \text{eV}\ (1.6 \times 10^{-19}\ \text{J/eV})$$

$$\frac{hc}{10^{-10}}\left(\frac{1}{3125} - \frac{1}{3650}\right) = 0.533 \times 1.6 \times 10^{-19}\ \text{J}$$

$$h = \frac{0.533 \times 1.6 \times 10^{-19} \times 10^{-10} \times 3125 \times 3650}{525 \times 3 \times 10^{8}}\ \text{J s}$$

$$= 6.176 \times 10^{-34}\ \text{Js}$$

From the first equation, the work function

$$\frac{hc}{\lambda_0} = \frac{(6.176 \times 10^{-34}\ \text{J s})(3 \times 10^{8}\ \text{m/s})}{3125 \times 10^{-10}\ \text{m}} - 2.128 \times 1.6 \times 10^{-19}\ \text{J}$$

$$= 2.524 \times 1.6 \times 10^{-19}\ \text{J} = 2.524\ \text{eV}$$

1.19 Construct the energy-level diagram for doubly ionized lithium.

Solution.

$$E_n = -\frac{Z^2 \times 13.6}{n^2}\ \text{eV} = -\frac{9 \times 13.6}{n^2}\ \text{eV}$$

$$= -\frac{122.4}{n^2}\ \text{eV}$$

$E_1 = -122.4$ eV $\quad E_2 = -30.6$ eV

$E_3 = -13.6$ eV $\quad E_4 = -7.65$ eV

These energies are represented in Fig. 1.1.

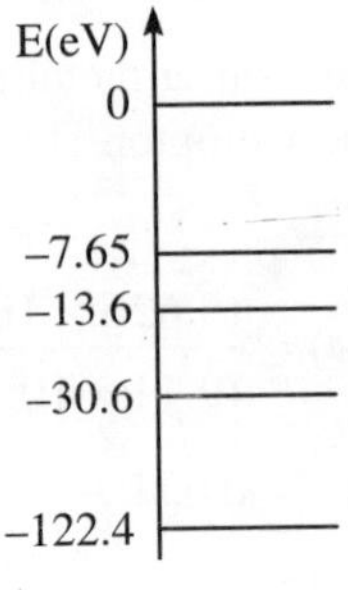

Fig. 1.1 Energy level diagram for doubly ionized lithium (not to scale).

1.20 What are the potential and kinetic energies of the electron in the ground state of the hydrogen atom?

Solution.

$$\text{Potential energy} = -\frac{1}{4\pi\varepsilon_0}\frac{e^2}{r}$$

Substituting the value of r from Eq. (1.9), we get

$$\text{Potential energy} = -\frac{me^4}{16\pi^2\varepsilon_0^2\hbar^2} = -2E_1 = -27.2 \text{ eV}$$

$$\text{Kinetic energy} = \text{total energy} - \text{potential energy}$$

$$= -13.6 \text{ eV} + 27.2 \text{ eV} = 13.6 \text{ eV}$$

1.21 Show that the magnitude of the potential energy of an electron in any Bohr orbit of the hydrogen atom is twice the magnitude of its kinetic energy in that orbit. What is the kinetic energy of the electron in the $n = 3$ orbit? What is its potential energy in the $n = 4$ orbit?

Solution.

$$\text{Radius of the Bohr orbit } r_n = n^2 a_0$$

$$\text{Potential energy} = -\frac{1}{4\pi\varepsilon_0}\frac{e^2}{r_n} = -\frac{1}{4\pi\varepsilon_0}\frac{e^2}{n^2 a_0} = -\frac{27.2}{n^2} \text{ eV}$$

$$\text{Kinetic energy} = \text{Total energy} - \text{Potential energy}$$

$$= -\frac{13.6}{n^2} \text{ eV} + \frac{27.2}{n^2} \text{ eV} = \frac{13.6}{n^2} \text{ eV}$$

$$\text{KE in the } n = 3 \text{ orbit} = \frac{13.6}{9} = 1.51 \text{ eV}$$

$$\text{Potential energy in the } n = 4 \text{ orbit} = -\frac{27.2}{16} = -1.7 \text{ eV}$$

1.22 Calculate the momentum of the photon of largest energy in the hydrogen spectrum. Also evaluate the velocity of the recoiling atom when it emits this photon. The mass of the atom = 1.67×10^{-27} kg.

Solution. The photon of the largest energy in the hydrogen spectrum occurs at the Lyman series limit, that is, when the quantum number n changes from ∞ to 1. For Lyman series, we have

$$\frac{1}{\lambda} = R\left(\frac{1}{1^2} - \frac{1}{m^2}\right), \qquad m = 2, 3, 4, \ldots$$

For the largest energy, $m = \infty$. Hence,

$$\frac{1}{\lambda} = R$$

$$\text{Momentum of the photon} = \frac{h\nu}{c} = \frac{h}{\lambda} = hR$$

$$= (6.626 \times 10^{-34} \text{ J s})\,(1.0967 \times 10^{7} \text{ m}^{-1})$$

$$= 7.267 \times 10^{-27} \text{ kg m s}^{-1}$$

$$\text{Velocity of recoil of the atom} = \frac{\text{momentum}}{\text{mass}}$$

$$= \frac{7.266 \times 10^{-27}\,\text{kg m s}^{-1}}{1.67 \times 10^{-27}\,\text{kg}} = 4.35\,\text{m s}^{-1}$$

1.23 Show that the electron in the Bohr orbits of hydrogen atom has quantized speeds $\mathbf{v}_n = c\alpha/n$, where α is the fine structure constant. Use this result to evaluate the kinetic energy of hydrogen atom in the ground state in eV.

Solution. According to the Bohr postulate,

$$m\mathbf{v}r = n\hbar, \qquad n = 1, 2, 3, \ldots$$

The coulombic attraction between the electron and the proton provides the necessary centripetal force, i.e.,

$$\frac{m\mathbf{v}^2}{r} = \frac{ke^2}{r^2}, \qquad k = \frac{1}{4\pi\varepsilon_0}$$

$$m\mathbf{v}r = \frac{ke^2}{\mathbf{v}}$$

Combining the two equations for $m\mathbf{v}r$, we obtain

$$\frac{ke^2}{\mathbf{v}} = n\hbar \quad \text{or} \quad \mathbf{v} = \frac{ke^2}{n\hbar}$$

$$\mathbf{v} = \frac{ke^2}{c\hbar}\frac{c}{n} = \frac{\alpha c}{n} \qquad \text{since } \alpha = \frac{ke^2}{c\hbar}$$

$$\text{Kinetic energy} = \frac{1}{2}m\mathbf{v}^2 = \frac{1}{2}m\frac{c^2\alpha^2}{n^2}$$

$$= \frac{1}{2}\frac{(9.1 \times 10^{-31}\,\text{kg})(3 \times 10^8\,\text{m s}^{-1})^2}{137^2}\frac{1}{n^2}$$

$$= \frac{21.8179 \times 10^{-19}\,\text{J}}{n^2} = \frac{21.8179 \times 10^{-19}\,\text{J}}{n^2(1.6 \times 10^{-19}\,\text{J/eV})}$$

$$= 13.636\frac{1}{n^2}\,\text{eV}$$

Kinetic energy in the ground state = 13.636 eV

1.24 In Moseley's data, the K_α wavelengths for two elements are found at 0.8364 and 0.1798 nm. Identify the elements.

Solution. The K_α x-ray is emitted when a vacancy in the K-shell is filled by an electron from the L-shell. Inside the orbit of L-electron, there are z-protons and the one electron left in the K-shell. Hence the effective charge experienced by the L-electron is approximately $(Z - 1)e$. Consequently, the energy of such an electron is given by

$$E_n = \frac{(Z - 1)^2 13.6\,\text{eV}}{n^2}$$

Then, the frequency of the K_α line is

$$\nu_{K\alpha} = \frac{(Z-1)^2\, 13.6\,\text{eV}}{h}\left(\frac{1}{1^2} - \frac{1}{2^2}\right)$$

$$= \frac{3}{4}\,\frac{(Z-1)^2\, 13.6\,\text{eV}}{h}$$

$$= \frac{3}{4}\,\frac{(Z-1)^2\,(13.6\,\text{eV})(1.6\times 10^{-19}\ \text{J/eV})}{6.626\times 10^{-34}\ \text{J s}}$$

$$= 2.463 \times 10^{15}\,(Z-1)^2\ \text{s}^{-1}$$

Since $\nu = c/\lambda$, we have

$$\frac{3\times 10^8\ \text{m s}^{-1}}{0.8364\times 10^{-9}\ \text{m}} = 2.463\times 10^{15}(Z-1)^2\ \text{s}^{-1}$$

$$Z - 1 = 12.06 \quad \text{or} \quad Z = 13$$

Hence the element is aluminium. For the other one

$$\frac{3\times 10^8\ \text{m s}^{-1}}{0.1798\times 10^{-9}\ \text{m}} = 2.463\times 10^{15}(Z-1)^2\ \text{s}^{-1}$$

$$Z - 1 = 26, \qquad Z = 27$$

The element is cobalt.

1.25 Using the Wilson-Sommerfeld quantization rule, show that the possible energies of a linear harmonic oscillator are integral multiples of $h\nu_0$, where ν_0 is the oscillator frequency.

Solution. The displacement x with time t of a harmonic oscillator of frequency ν_0 is given by

$$x = x_0 \sin(2\pi\nu_0 t) \tag{i}$$

The force constant k and frequency ν_0 are related by the equation

$$\nu_0 = \frac{1}{2\pi}\sqrt{\frac{k}{m}} \quad \text{or} \quad k = 4\pi^2 m\nu_0^2 \tag{ii}$$

$$\text{Potential energy } V = \frac{1}{2}kx^2 = 2\pi^2 m\nu_0^2 x_0^2 \sin^2(2\pi\nu_0 t) \tag{iii}$$

$$\text{Kinetic energy } T = \frac{1}{2}m\dot{x}^2 = 2\pi^2 m\nu_0^2 x_0^2 \cos^2(2\pi\nu_0 t) \tag{iv}$$

$$\text{Total energy } E = T + V = 2\pi^2 m\nu_0^2 x_0^2 \tag{v}$$

According to the quantization rule,

$$\oint p_x\, dx = nh \quad \text{or} \quad m\oint \dot{x}\, dx = nh \tag{vi}$$

When x completes one cycle, t changes by period $T = 1/\nu_0$. Hence, substituting the values of x and dx, we obtain

$$4\pi^2 m\nu_0^2 x_0^2 \int_0^{1/\nu_0} \cos^2(2\pi\nu_0 t)\, dt = nh, \qquad n = 0, 1, 2, \ldots$$

$$2\pi^2 m\nu_0 x_0^2 = nh \quad \text{or} \quad x_0 = \left(\frac{nh}{2\pi^2 m\nu_0}\right)^{1/2}$$

Substituting the value of x_0 in Eq. (v), we get

$$E_n = nh\nu_0 = n\hbar\omega, \qquad n = 0, 1, 2, \ldots$$

That is, according to old quantum theory, the energies of a linear harmonic oscillator are integral multiples of $h\nu_0 = \hbar\omega$.

1.26 A rigid rotator restricted to move in a plane is described by the angle coordinate θ. Show that the momentum conjugate to θ is an integral multiple of $\hbar$. Use this result to derive an equation for its energy.

Solution. Let the momentum conjugate to the angle coordinate be p_θ which is a constant of motion. Then,

$$\int_0^{2\pi} p_\theta \, d\theta = p_\theta \int_0^{2\pi} d\theta = 2\pi p_\theta$$

Applying the Wilson-Sommerfeld quantization rule, we get

$$2\pi p_\theta = nh \quad \text{or} \quad p_\theta = n\hbar, \qquad n = 0, 1, 2, \ldots$$

Since $p_\theta = I\omega$, $I\omega = n\hbar$. Hence, the energy of a rotator is

$$E = \frac{1}{2} I\omega^2 = \frac{1}{2I}(I\omega)^2$$

$$E_n = \frac{n^2\hbar^2}{2I}, \qquad n = 0, 1, 2, \ldots$$

1.27 The lifetime of the $n = 2$ state of hydrogen atom is 10^{-8} s. How many revolutions does an electron in the $n = 2$ Bohr orbit make during this time?

Solution. The number of revolutions the electron makes in one second in the $n = 2$ Bohr orbit is

$$\nu_2 = \frac{E_2}{h} = \frac{(13.6\,\text{eV})(1.6 \times 10^{-19}\ \text{J/eV})}{4(6.626 \times 10^{-34}\ \text{J s})}$$

$$= 0.821 \times 10^{15}\ \text{s}^{-1}$$

No. of revolutions the electron makes in 10^{-8} s $= (0.821 \times 10^{15}\ \text{s}^{-1})(10^{-8}\ \text{s})$

$= 8.21 \times 10^6$

1.28 In a hydrogen atom, the nth orbit has a radius 10^{-5} m. Find the value of n. Write a note on atoms with such high quantum numbers.

Solution. In a hydrogen atom, the radius of the nth orbit r_n is

$$r_n = n^2 a_0$$

$$n^2 = \frac{10^{-5}\,\text{m}}{0.53 \times 10^{-10}\ \text{m}} = 1.887 \times 10^5$$

$$n = 434.37 \cong 434$$

Atoms having an outermost electron in an excited state with a very high principal quantum number n are called **Rydberg atoms**. They have exaggerated properties. In such atoms, the valence electron is in a large loosely bound orbit. The probability that the outer electron spends its time outside the $Z - 1$ other electrons is fairly high. Consequently, Z_{eff} is that due to Z-protons and $(Z - 1)$ electrons, which is 1. That is, $Z_{eff} = 1$ which gives an ionization energy of 13.6 eV/n^2 for all Rydberg atoms.

1.29 When an excited atom in a state E_i emits a photon and comes to a state E_f, the frequency of the emitted radiation is given by Bohr's frequency condition. To balance the recoil of the atom, a part of the emitted energy is used up. How does Bohr's frequency condition get modified?

Solution. Let the energy of the emitted radiation be $E_\gamma = h\nu$ and E_{re} be the recoil energy. Hence,

$$E_i - E_f = h\nu + E_{re}$$

By the law of conservation of momentum,

Recoil momentum of atom = momentum of the emitted γ-ray

$$p_{re} = \frac{h\nu}{c}$$

where c is the velocity of light,

$$E_{re} = \frac{p_{re}^2}{2M} = \frac{h^2\nu^2}{2Mc^2}$$

where M is the mass of recoil atom

Substituting the value of E_{re}, the Bohr frequency condition takes the form

$$E_i - E_f = h\nu + \frac{h^2\nu^2}{2Mc^2}$$

where ν is the frequency of the radiation emitted and M is the mass of the recoil nucleus.

1.30 Hydrogen atom at rest in the $n = 2$ state makes transition to the $n = 1$ state.

(i) Compute the recoil kinetic energy of the atom.

(ii) What fraction of the excitation energy of the $n = 2$ state is carried by the recoiling atom?

Solution. Energy of the $n = 2 \rightarrow n = 1$ transition is given by

$$E_2 - E_1 = \left(-\frac{13.6\,\text{eV}}{2^2}\right) - \left(-\frac{13.6\,\text{eV}}{1^2}\right) = 10.2\ \text{eV}$$

$$= 10.2 \times 1.6 \times 10^{-19}\ \text{J}$$

(i) From Problem 1.29, the recoil energy

$$E_{re} = \frac{h^2\nu^2}{2Mc^2} \quad (M\text{-mass of the nucleus})$$

$$= \frac{(E_2 - E_1)^2}{2Mc^2}$$

$$= \frac{(10.2 \times 1.6 \times 10^{-19}\ \text{J})^2}{2(9.1 \times 10^{-31}\ \text{kg})1836(3 \times 10^8\ \text{m/s})^2}$$

$$= 8.856 \times 10^{-27}\ \text{J}$$

$$= 5.535 \times 10^{-8}\ \text{eV}$$

(ii) Excitation energy of the $n = 2$ state is 10.2 eV. Then,

$$\frac{\text{Recoil energy}}{\text{Excitation energy}} = \frac{5.535 \times 10^{-8}\ \text{eV}}{10.2\ \text{eV}} = 5.4 \times 10^{-9}$$

1.31 In the lithium atom (Z = 3), the energy of the outer electron is approximated as

$$E = -\frac{(Z - \sigma)^2\, 13.6\,\text{eV}}{n^2}$$

where σ is the screening constant. If the measured ionization energy is 5.39 eV, what is the value of screening constant?

Solution. The electronic configuration for lithium is $1\,s^2\,2\,s^1$. For the outer electron, $n = 2$. Since the ionization energy is 5.39 eV, the energy of the outer electron $E = -5.39$ eV. Given

$$E = -\frac{(Z - \sigma)^2\, 13.6\,\text{eV}}{n^2}$$

Equating the two energy relations, we get

$$-\frac{(Z - \sigma)^2\, 13.6\,\text{eV}}{2^2} = -\,5.39\,\text{eV}$$

$$(Z - \sigma)^2 = \frac{4 \times 5.39\,\text{eV}}{13.6\,\text{eV}} = 1.5853$$

$$Z - \sigma = 1.259$$

$$\sigma = 3 - 1.259 = 1.741$$

1.32 The wavelength of the L_α line for an element has a wavelength of 0.3617 nm. What is the element? Use (Z – 7.4) for the effective nuclear charge.

Solution. The L_α transition is from $n = 3$ to $n = 2$. The frequency of the L_α transition is given by

$$\frac{c}{\lambda} = \frac{(Z - 7.4)^2\, 13.6\,\text{eV}}{h}\left(\frac{1}{2^2} - \frac{1}{3^2}\right)$$

$$\frac{3 \times 10^8\ \text{m/s}}{0.3617 \times 10^{-9}\ \text{m}} = \frac{(Z - 7.4)^2\,(13.6\,\text{eV} \times 1.6 \times 10^{-19}\ \text{J/eV})}{6.626 \times 10^{-34}\ \text{J s}} \times \frac{5}{36}$$

$$8.294 \times 10^{17}\ \text{s}^{-1} = (Z - 7.4)^2\,(0.456 \times 10^{15}\ \text{s}^{-1})$$

$$Z - 7.4 = 42.64 \quad \text{or} \quad Z = 50.04$$

The element is tin.

CHAPTER 2

Wave Mechanical Concepts

2.1 Wave Nature of Particles

Classical physics considered particles and waves as distinct entities. Quantum ideas firmly established that radiation has both wave and particle nature. This dual nature was extended to material particles by Louis de Broglie in 1924. The wave associated with a particle in motion, called **matter wave**, has the wavelength λ given by the de Broglie equation

$$\lambda = \frac{h}{p} = \frac{h}{m\mathbf{v}} \tag{2.1}$$

where p is the momentum of the particle. Electron diffraction experiments conclusively proved the dual nature of material particles in motion.

2.2 Uncertainty Principle

When waves are associated with particles, some kind of indeterminacy is bound to be present. Heisenberg critically analyzed this and proposed the uncertainty principle:

$$\Delta x \cdot \Delta p_x \simeq h \tag{2.2}$$

where Δx is the uncertainty in the measurement of position and Δp_x is the uncertainty in the measurement of the x-component of momentum. A more rigorous derivation leads to

$$\Delta x \cdot \Delta p_x \geq \frac{\hbar}{2} \tag{2.3}$$

Two other equally useful forms are the energy time and angular momentum-polar angle relations given respecting by

$$\Delta E \cdot \Delta t \geq \frac{\hbar}{2} \tag{2.4}$$

$$\Delta L_z \cdot \Delta \phi \geq \frac{\hbar}{2} \tag{2.5}$$

2.3 Wave Packet

The linear superposition principle, which is valid for wave motion, is also valid for material particles. To describe matter waves associtated with particles in motion, we requires a quantity which varies in space and time. This quantity, called the **wave function** $\Psi(r, t)$, is confined to a small region in space and is called the **wave packet** or **wave group**. Mathematically, a wave packet can be constructed by the superposition of an infinite number of plane waves with slightly differing k-values, as

$$\Psi(x, t) = \int A(k) \exp [ikx - i\omega(k)t] \, dk \tag{2.6}$$

where k is the wave vector and ω is the angular frequency. Since the wave packet is localized, the limit of the integral is restricted to a small range of k-values, say, $(k_o - \Delta k) < k < (k_o + \Delta k)$. The speed with which the component waves of the wave packet move is called the **phase velocity** $\mathbf{v}_p$ which is defined as

$$\mathbf{v}_p = \frac{\omega}{k} \tag{2.7}$$

The speed with which the envelope of the wave packet moves is called the **group velocity** $\mathbf{v}_g$ given by

$$\mathbf{v}_g = \frac{d\omega}{dk} \tag{2.8}$$

2.4 Time-dependent Schrödinger Equation

For a detailed study of systems, Schrödinger formulated an equation of motion for $\Psi(\boldsymbol{r}, t)$:

$$i\hbar \frac{\partial}{\partial t} \Psi(\boldsymbol{r},t) = \left[-\frac{\hbar}{2m} \nabla^2 + V(r) \right] \Psi(\boldsymbol{r},t) \tag{2.9}$$

The quantity in the square brackets is called the **Hamiltonian operator** of the system. Schrödinger realized that, in the new mechanics, the energy E, the momentum $\boldsymbol{p}$, the coordinate r, and time t have to be considered as operators operating on functions. An analysis leads to the following operators for the different dynamical variables:

$$E \to i\hbar \frac{\partial}{\partial t}, \qquad p \to -i\hbar\nabla, \qquad \boldsymbol{r} \to \boldsymbol{r}, \qquad t \to t \tag{2.10}$$

2.5 Physical Interpretation of $\Psi(\boldsymbol{r}, t)$

2.5.1 Probability Interpretation

A universally accepted interpretation of $\Psi(\boldsymbol{r}, t)$ was suggested by Born in 1926. He interpreted $\Psi^*\Psi$ as the position probability density $P(\boldsymbol{r}, t)$:

$$P(\boldsymbol{r}, t) = \Psi^*(\boldsymbol{r}, t)\, \Psi(\boldsymbol{r}, t) = |\Psi(\boldsymbol{r}, t)|^2 \tag{2.11}$$

The quantity $|\Psi(\boldsymbol{r}, t)|^2 \, d\tau$ is the probability of finding the system at time t in the elementary volume $d\tau$ surrounding the point $\boldsymbol{r}$. Since the total probability is 1, we have

$$\int_{\infty}^{\infty} |\Psi(\boldsymbol{r}, t)|^2 \, d\tau = 1 \tag{2.12}$$

If Ψ is not satisfying this condition, one can multiply Ψ by a constant, say N, so that $N\Psi$ satisfies Eq. (2.12). Then,

$$|N|^2 \int_{\infty}^{\infty} |\Psi(\boldsymbol{r}, t)|^2 \, d\tau = 1 \tag{2.13}$$

The constant N is called the **normalization constant**.

2.5.2 Probability Current Density

The probability current density $\boldsymbol{j}(\boldsymbol{r}, t)$ is defined as

$$\boldsymbol{j}(\boldsymbol{r}, t) = \frac{i\hbar}{2m} (\Psi \nabla \Psi^* - \Psi^* \nabla \Psi) \tag{2.14}$$

It may be noted that, if Ψ is real, the vector $\boldsymbol{j}(\boldsymbol{r}, t)$ vanishes. The function $\boldsymbol{j}(\boldsymbol{r}, t)$ satisfies the equation of continuity

$$\frac{\partial}{\partial t} P(\boldsymbol{r}, t) + \nabla \cdot \boldsymbol{j}(\boldsymbol{r}, t) = 0 \tag{2.15}$$

Equation (2.15) is a quantum mechanical probability conservation equation. That is, if the probability of finding the system in some region increases with time, the probability of finding the system outside decreases by the same amount.

2.6 Time-independent Schrödinger Equation

If the Hamiltonian operator does not depend on time, the variables $\boldsymbol{r}$ and t of the wave function $\Psi(\boldsymbol{r}, t)$ can be separated into two functions $\psi(\boldsymbol{r})$ and $\phi(t)$ as

$$\Psi(\boldsymbol{r}, t) = \psi(\boldsymbol{r}) \, \phi(t) \tag{2.16}$$

Simplifying, the time-dependent Schrödinger equation, Eq. (2.9), splits into the following two equations:

$$\frac{1}{\phi(t)} \frac{d\phi}{dt} = -\frac{iE}{\hbar} \tag{2.17}$$

$$\left[-\frac{\hbar}{2m} \nabla^2 + V(r) \right] \psi(r) = E\psi(r) \tag{2.18}$$

The separation constant E is the energy of the system. Equation (2.18) is the time-independent Schrödinger equation. The solution of Eq. (2.17) gives

$$\phi(t) = Ce^{-iEt/\hbar} \tag{2.19}$$

where C is a constant.

$\Psi(\mathbf{r}, t)$ now takes the form

$$\Psi(\mathbf{r}, t) = \psi(\mathbf{r})e^{-iEt/\hbar} \tag{2.20}$$

The states for which the probability density is constant in time are called **stationary states**, i.e.,

$$P(\mathbf{r}, t) = |\Psi(\mathbf{r}, t)|^2 = \text{constant in time} \tag{2.21}$$

Admissibility conditions on the wave functions

(i) The wave function $\Psi(\mathbf{r}, t)$ must be finite and single valued at every point in space.
(ii) The functions Ψ and $\nabla\psi$ must be continuous, finite and single valued.

PROBLEMS

2.1 Calculate the de Broglie wavelength of an electron having a kinetic energy of 1000 eV. Compare the result with the wavelength of x-rays having the same energy.

Solution. The kinetic energy

$$T = \frac{p^2}{2m} = 1000 \text{ eV} = 1.6 \times 10^{-16} \text{ J}$$

$$\lambda = \frac{h}{p} = \frac{6.626 \times 10^{-34} \text{ js}}{[2 \times (9.11 \times 10^{-31} \text{ kg}) \times (1.6 \times 10^{-16} \text{ J}]^{1/2}}$$

$$= 0.39 \times 10^{-10} \text{ m} = 0.39 \text{ Å}$$

For x-rays,

$$\text{Energy} = \frac{hc}{\lambda}$$

$$\lambda = \frac{(6.626 \times 10^{-34} \text{ J s}) \times (3 \times 10^8 \text{ m/s})}{1.6 \times 10^{-16} \text{ J}} = 12.42 \times 10^{-10} \text{ m} = 12.42 \text{ Å}$$

$$\frac{\text{Wavelength of x-rays}}{\text{de Broglie wavelength of electron}} = \frac{12.42 \text{ Å}}{0.39 \text{ Å}} = 31.85$$

2.2 Determine the de Broglie wavelength of an electron that has been accelerated through a potential difference of (i) 100 V, (ii) 200 V.

Solution.

(i) The energy gained by the electron = 100 eV. Then,

$$\frac{p^2}{2m} = 100 \text{ eV} = (100 \text{ eV})(1.6 \times 10^{-19} \text{ J/eV}) = 1.6 \times 10^{-17} \text{ J}$$

$$p = [2\,(9.1 \times 10^{-13} \text{ kg})(1.6 \times 10^{-17} \text{ J})]^{1/2}$$

$$= 5.396 \times 10^{-24} \text{ kg ms}^{-1}$$

$$\lambda = \frac{h}{p} = \frac{6.626 \times 10^{-34} \text{ J s}}{5.396 \times 10^{-24} \text{ kg ms}^{-1}}$$

$$= 1.228 \times 10^{-10} \text{ m} = 1.128 \text{ Å}$$

(ii)

$$\frac{p^2}{2m} = 200 \text{ eV} = 3.2 \times 10^{-17} \text{ J}$$

$$p = [2\,(9.1 \times 10^{-31} \text{ kg})(3.2 \times 10^{-17} \text{ J})]^{1/2}$$

$$= 7.632 \times 10^{-24} \text{ kg ms}^{-1}$$

$$\lambda = \frac{h}{p} = \frac{6.626 \times 10^{-34} \text{ J s}}{7.632 \times 10^{-24} \text{ kg ms}^{-1}}$$

$$= 0.868 \times 10^{-10} \text{ m} = 0.868 \text{ Å}$$

2.3 The electron scattering experiment gives a value of 2×10^{-15} m for the radius of a nucleus. Estimate the order of energies of electrons used for the experiment. Use relativistic expressions.

Solution. For electron scattering experiment, the de Broglie wavelength of electrons used must be of the order of 4×10^{-15} m, the diameter of the atom. The kinetic energy

$$T = E - m_0^2 c^2 = \sqrt{c^2 p^2 + m_0^2 c^4} - m_0 c^2$$

$$(T + m_0 c^2)^2 = c^2 p^2 + m_0^2 c^4$$

$$m_0^2 c^4 \left(1 + \frac{T}{m_0 c^2}\right)^2 = c^2 p^2 + m_0^2 c^4$$

$$c^2 p^2 = m_0^2 c^4 \left[\left(1 + \frac{T}{m_0 c^2}\right)^2 - 1\right]$$

$$p = m_0 c \left[\left(1 + \frac{T}{m_0 c^2}\right)^2 - 1\right]^{1/2}$$

$$\frac{h^2}{\lambda^2} = m_0^2 c^2 \left[\left(1 + \frac{T}{m_0 c^2}\right)^2 - 1\right]$$

$$\left(1 + \frac{T}{m_0 c^2}\right)^2 = \frac{h^2}{\lambda^2 m_0^2 c^2} + 1$$

$$= \frac{(6.626 \times 10^{-34} \text{ J s})^2}{(16 \times 10^{-30} \text{ m}^2) \times (9.11 \times 10^{-31} \text{ kg})^2 \times (3 \times 10^8 \text{ m/s})^2} + 1$$

$$= 3.6737 \times 10^5$$

$$T = 605.1 m_0 c^2$$

$$= 605.1 \times (9.11 \times 10^{-31} \text{ kg}) \times (3 \times 10^8 \text{ m/s})^2$$

$$= 496.12 \times 10^{-13} \text{ J} = \frac{496.12 \times 10^{-13} \text{ J}}{1.6 \times 10^{-19} \text{ J/eV}}$$

$$= 310 \times 10^6 \text{ eV} = 310 \text{ MeV}$$

2.4 Evaluate the ratio of the de Broglie wavelength of electron to that of proton when (i) both have the same kinetic energy, and (ii) the electron kinetic energy is 1000 eV and the proton KE is 100 eV.

Solution.

$$\text{(i) } \lambda_1 = \frac{h}{\sqrt{2m_1T_1}}; \quad \lambda_2 = \frac{h}{\sqrt{2m_2T_2}}; \quad \frac{\lambda_1}{\lambda_2} = \sqrt{\frac{m_2T_2}{m_1T_1}}$$

$$\frac{\lambda \text{ of electron}}{\lambda \text{ of proton}} = \sqrt{\frac{1836\ m_eT}{m_eT}} = \sqrt{1836} = 42.85$$

(ii) $T_1 = 1000$ eV; $T_2 = 100$ eV

$$\frac{\lambda \text{ of electron}}{\lambda \text{ of proton}} = \sqrt{\frac{1836 \times 100}{1000}} = 13.55$$

2.5 Proton beam is used to obtain information about the size and shape of atomic nuclei. If the diameter of nuclei is of the order of 10^{-15} m, what is the approximate kinetic energy to which protons are to be accelerated? Use relativistic expressions.

Solution. When fast moving protons are used to investigate a nucleus, its de Broglie wavelength must be comparable to nuclear dimensions, i.e., the de Broglie wavelength of protons must be of the order of 10^{-15} m. In terms of the kinetic energy T, the relativistic momentum p is given by (refer Problem 2.3)

$$p = m_0c\sqrt{\left(1 + \frac{T}{m_0c^2}\right) - 1}, \qquad \lambda = \frac{h}{p} \cong 10^{-15} \text{ m}$$

$$\frac{h^2}{\lambda^2} = m_0^2c^2\left[\left(1 + \frac{T}{m_0c^2}\right)^2 - 1\right]$$

Substitution of λ, m_0, h and c gives

$$T = 9.8912 \times 10^{-11} \text{ J} = 618.2 \text{ MeV}$$

2.6 Estimate the velocity of neutrons needed for the study of neutron diffraction of crystal structures if the interatomic spacing in the crystal is of the order of 2 Å. Also estimate the kinetic energy of the neutrons corresponding to this velocity. Mass of neutron = 1.6749×10^{-27} kg.

Solution. de Broglie wavelength

$$\lambda \cong 2 \times 10^{-10} \text{ m}$$

$$\lambda = \frac{h}{m\mathbf{v}} \quad \text{or} \quad \mathbf{v} = \frac{h}{m\lambda}$$

$$\mathbf{v} = \frac{6.626 \times 10^{-34} \text{ J s}}{(1.6749 \times 10^{-27} \text{ kg})(2 \times 10^{-10} \text{ m/s})} = 1.978 \times 10^3 \text{ ms}^{-1}$$

$$\text{Kinetic energy } T = \frac{1}{2}m\mathbf{v}^2 = \frac{1}{2}(1.6749 \times 10^{-27} \text{ kg})(1.978 \times 10^3 \text{ ms}^{-1})^2$$

$$= 3.2765 \times 10^{-21} \text{ J} = 20.478 \times 10^{-3} \text{ eV}$$

2.7 Estimate the energy of electrons needed for the study of electron diffraction of crystal structures if the interatomic spacing in the crystal is of the order of 2 Å.

Solution. de Broglie wavelength of electrons $\cong$ 2 Å = 2×10^{-10} m

Kinetic energy $$T = \frac{p^2}{2m} = \frac{(h/\lambda)^2}{2m}$$

$$T = \frac{(6.626 \times 10^{-34}\ \text{J s})^2}{2 \times (2 \times 10^{-10}\ \text{m})^2\,(9.11 \times 10^{-31}\ \text{kg})}$$

$$= 60.24 \times 10^{-19}\ \text{J} = 37.65\ \text{eV}$$

2.8 What is the ratio of the kinetic energy of an electron to that of a proton if their de Broglie wavelengths are equal?

Solution.

m_1 = mass of electron, m_2 = mass of proton,

$\mathbf{v}_1$ = velocity of electron, $\mathbf{v}_2$ = velocity of proton.

$$\lambda = \frac{h}{m_1\mathbf{v}_1} = \frac{h}{m_2\mathbf{v}_2} \quad \text{or} \quad m_1\mathbf{v}_1 = m_2\mathbf{v}_2$$

$$m_1\left(\frac{1}{2}m_1\mathbf{v}_1^2\right) = m_2\left(\frac{1}{2}m_2\mathbf{v}_2^2\right)$$

$$\frac{\text{Kinetic energy of electron}}{\text{Kinetic energy of proton}} = \frac{m_2}{m_1} = 1836$$

2.9 An electron has a speed of 500 m/s with an accuracy of 0.004%. Calculate the certainty with which we can locate the position of the electron.

Solution.

$$\text{Momentum } p = m\mathbf{v} = (9.11 \times 10^{-31}\ \text{kg}) \times (500\ \text{m/s})$$

$$\frac{\Delta p}{p} \times 100 = 0.004$$

$$\Delta p = \frac{0.004\,(9.11 \times 10^{-31}\ \text{kg})\,(500\ \text{m/s})}{100}$$

$$= 182.2 \times 10^{-34}\ \text{kg m s}^{-1}$$

$$\Delta x \cong \frac{h}{\Delta p} = \frac{6.626 \times 10^{-34}\ \text{J s}}{182.2 \times 10^{-34}\ \text{kg m s}^{-1}} = 0.0364\ \text{m}$$

The position of the electron cannot be measured to accuracy less than 0.036 m.

2.10 The average lifetime of an excited atomic state is 10^{-9} s. If the spectral line associated with the decay of this state is 6000 Å, estimate the width of the line.

Solution.

$$\Delta t = 10^{-9}\ \text{s}, \quad \lambda = 6000 \times 10^{-10}\ \text{m} = 6 \times 10^{-7}\ \text{m}$$

$$E = \frac{hc}{\lambda} \quad \text{or} \quad \Delta E = \frac{hc}{\lambda^2}\Delta\lambda$$

$$\Delta E \cdot \Delta t = \frac{hc}{\lambda^2}\Delta\lambda \cdot \Delta t \approx \frac{\hbar}{2} = \frac{h}{4\pi}$$

$$\Delta\lambda = \frac{\lambda^2}{4\pi c\Delta t} = \frac{36 \times 10^{-14}\ \text{m}^2}{4\pi\,(3 \times 10^8\ \text{m/s}) \times (10^{-9}\ \text{s})} = 9.5 \times 10^{-14}\ \text{m}$$

2.11 An electron in the $n = 2$ state of hydrogen remains there on the average of about 10^{-8} s, before making a transition to $n = 1$ state.

(i) Estimate the uncertainty in the energy of the $n = 2$ state.
(ii) What fraction of the transition energy is this?
(iii) What is the wavelength and width of this line in the spectrum of hydrogen atom?

Solution. From Eq. (2.4),

(i) $\Delta E \geq \dfrac{h}{4\pi\Delta t} = \dfrac{6.626 \times 10^{-34}\ \text{J s}}{4\pi \times 10^{-8}\ \text{s}}$

$= 0.527 \times 10^{-26}$ J $= 3.29 \times 10^{-8}$ eV

(ii) Energy of $n = 2 \to n = 1$ transition

$$= -13.6\ \text{eV}\left(\frac{1}{2^2} - \frac{1}{1^2}\right) = 10.2\ \text{eV}$$

$$\text{Fraction } \frac{\Delta E}{E} = \frac{3.29 \times 10^{-8}\ \text{eV}}{10.2\ \text{eV}} = 3.23 \times 10^{-9}$$

(iii) $\lambda = \dfrac{hc}{E} = \dfrac{(6.626 \times 10^{-34}\ \text{J s}) \times (3 \times 10^8\ \text{m/s})}{(10.2 \times 1.6 \times 10^{-19}\ \text{J})}$

$= 1.218 \times 10^{-7}$ m $= 122$ nm

$$\frac{\Delta E}{E} = \frac{\Delta\lambda}{\lambda} \quad \text{or} \quad \Delta\lambda = \frac{\Delta E}{E} \times \lambda$$

$$\Delta\lambda = (3.23 \times 10^{-9})\ (1.218 \times 10^{-7}\ \text{m})$$

$$= 3.93 \times 10^{-16}\ \text{m} = 3.93 \times 10^{-7}\ \text{nm}$$

2.12 An electron of rest mass m_0 is accelerated by an extremely high potential of V volts. Show that its wavelength

$$\lambda = \frac{hc}{[\text{eV}\,(\text{eV} + 2m_0c^2)]^{1/2}}$$

Solution. The energy gained by the electron in the potential is Ve. The relativistic expression for kinetic energy $= \dfrac{m_0c^2}{(1 - \mathbf{v}^2/c^2)^{1/2}} - m_0c^2$. Equating the two and rearranging, we get

$$\frac{m_0c^2}{(1 - \mathbf{v}^2/c^2)^{1/2}} - m_0c^2 = Ve$$

$$(1 - \mathbf{v}^2/c^2)^{1/2} = \frac{m_0c^2}{Ve + m_0c^2}$$

$$1 - \frac{\mathbf{v}^2}{c^2} = \frac{m_0^2 c^4}{(Ve + m_0c^2)^2}$$

$$\frac{\mathbf{v}^2}{c^2} = \frac{(Ve + m_0c^2)^2 - m_0^2c^4}{(Ve + m_0c^2)^2} = \frac{Ve(Ve + 2m_0c^2)}{(Ve + m_0c^2)^2}$$

$$\mathbf{v} = \frac{c[Ve(Ve + 2m_0c^2)]^{1/2}}{Ve + m_0c^2}$$

de Broglie Wavelength $\quad \lambda = \dfrac{h}{m\mathbf{v}} = \dfrac{h(1 - \mathbf{v}^2/c^2)^{1/2}}{m_0\mathbf{v}}$

$$\lambda = \frac{h}{m_0} \frac{m_0c^2}{Ve + m_0c^2} \frac{Ve + m_0c^2}{c[Ve(Ve + 2m_0c^2)]^{1/2}}$$

$$= \frac{hc}{[Ve(Ve + 2m_0c^2)]^{1/2}}$$

2.13 A subatomic particle produced in a nuclear collision is found to have a mass such that Mc^2 = 1228 MeV, with an uncertainty of ± 56 MeV. Estimate the lifetime of this state. Assuming that, when the particle is produced in the collision, it travels with a speed of 10^8 m/s, how far can it travel before it disintegrates?

Solution.

$$\text{Uncertainty in energy } \Delta E = (56 \times 10^6 \text{ eV})\,(1.6 \times 10^{-19} \text{ J/eV})$$

$$\Delta t = \frac{\hbar}{2}\frac{1}{\Delta E} = \frac{(1.05 \times 10^{-34} \text{ J s})}{2} \frac{1}{(56 \times 1.6 \times 10^{-13} \text{ J})}$$

$$= 5.86 \times 10^{-24} \text{ s}$$

Its lifetime is about 5.86×10^{-24} s, which is in the laboratory frame.

$$\text{Distance travelled before disintegration} = (5.86 \times 10\text{–}24 \text{ s})(10^8 \text{ m/s})$$

$$= 5.86 \times 10^{-16} \text{ m}$$

2.14 A bullet of mass 0.03 kg is moving with a velocity of 500 m s^{-1}. The speed is measured up to an accuracy of 0.02%. Calculate the uncertainty in x. Also comment on the result.

Solution.

$$\text{Momentum } p = 0.03 \times 500 = 15 \text{ kg m s}^{-1}$$

$$\frac{\Delta p}{p} \times 100 = 0.02$$

$$\Delta p = \frac{0.02 \times 15}{100} = 3 \times 10^{-3} \text{ kg m s}^{-1}$$

$$\Delta x \approx \frac{h}{2\Delta p} = \frac{6.626 \times 10^{-34} \text{ J s}}{4\pi \times 3 \times 10^{-3} \text{ km/s}} = 1.76 \times 10^{-31} \text{ m}$$

As uncertainty in the position coordinate x is almost zero, it can be measured very accurately. In other words, the particle aspect is more predominant.

2.15 Wavelength can be determined with an accuracy of 1 in 10^8. What is the uncertainty in the position of a 10 Å photon when its wavelength is simultaneously measured?

Solution.

$$\Delta\lambda \approx 10^{-8} \text{ m}, \quad \lambda = 10 \times 10^{-10} \text{ m} = 10^{-9} \text{ m}$$

$$p = \frac{h}{\lambda} \quad \text{or} \quad \Delta p = \frac{h}{\lambda^2}\Delta\lambda$$

$$\Delta x \cdot \Delta p \cong \frac{\Delta x \times \Delta\lambda \times h}{\lambda^2}$$

From Eq. (2.3), this product is equal to $\hbar/2$. Hence,

$$\frac{(\Delta x)(\Delta\lambda)\,h}{\lambda^2} = \frac{h}{4\pi}$$

$$\Delta x = \frac{\lambda^2}{4\pi\,\Delta\lambda} = \frac{10^{-18} \text{ m}^2}{4\pi \times 10^{-8} \text{ m}} = 7.95 \times 10^{-12} \text{ m}$$

2.16 If the position of a 5 k eV electron is located within 2 Å, what is the percentage uncertainty in its momentum?

Solution.

$$\Delta x = 2 \times 10^{-10} \text{ m}; \quad \Delta p \cdot \Delta x \cong \frac{h}{4\pi}$$

$$\Delta p \cong \frac{h}{4\pi\Delta x} = \frac{(6.626 \times 10^{-34} \text{ J s})}{4\pi\,(2 \times 10^{-10} \text{ m})} = 2.635 \times 10^{-25} \text{ kg m s}^{-1}$$

$$p = \sqrt{2mT} = (2 \times 9.11 \times 10^{-31} \times 5000 \times 1.6 \times 10^{-19})^{1/2}$$

$$= 3.818 \times 10^{-23} \text{ kg m s}^{-1}$$

$$\text{Percentage of uncertainty} = \frac{\Delta p}{p} \times 100 = \frac{2.635 \times 10^{-25}}{3.818 \times 10^{-23}} \times 100 = 0.69$$

2.17 The uncertainty in the velocity of a particle is equal to its velocity. If $\Delta p \cdot \Delta x \cong h$, show that the uncertainty in its location is its de Broglie wavelength.

Solution. Given $\Delta\mathbf{v} = \mathbf{v}$. Then,

$$\Delta p = m\Delta\mathbf{v} = m\mathbf{v} = p$$

$$\Delta x \times \Delta p \cong h \quad \text{or} \quad \Delta x \cdot p \cong h$$

$$\Delta x \cong \frac{h}{p} = \lambda$$

2.18 Normalize the wave function $\psi(x) = A\exp(-ax^2)$, A and a are constants, over the domain $-\infty \le x \le \infty$.

Solution. Taking A as the normalization constant, we get

$$A^2 \int_{-\infty}^{\infty} \psi^* \psi\, dx = A^2 \int_{-\infty}^{\infty} \exp(-2ax^2)\, dx = 1$$

Using the result (see the Appendix), we get

$$\int_{-\infty}^{\infty} \exp(-2ax^2)\, dx = \sqrt{\frac{\pi}{2a}}$$

$$A = \left(\frac{2a}{\pi}\right)^{1/4}$$

$$\psi(x) = \left(\frac{2a}{\pi}\right)^{1/4} \exp(-ax^2)$$

2.19 A particle constrained to move along the x-axis in the domain $0 \le x \le L$ has a wave function $\psi(x) = \sin(n\pi x/L)$, where n is an integer. Normalize the wave function and evaluate the expectation value of its momentum.

Solution. The normalization condition gives

$$N^2 \int_0^L \sin^2 \frac{n\pi x}{L}\, dx = 1$$

$$N^2 \int_0^L \frac{1}{2}\left(1 - \cos\frac{2n\pi x}{L}\right) dx = 1$$

$$N^2 \frac{L}{2} = 1 \quad \text{or} \quad N = \sqrt{\frac{2}{L}}$$

The normalized wave function is $\sqrt{2/L}\, \sin[(n\pi x)/L]$. So,

$$\langle p_x \rangle = \int_0^L \psi^* \left(-i\hbar \frac{d}{dx}\right) \psi\, dx$$

$$= -i\hbar \frac{2}{L}\frac{n\pi}{L} \int_0^L \sin\frac{n\pi x}{L} \cos\frac{n\pi x}{L}\, dx$$

$$= -i\hbar \frac{n\pi}{L^2} \int_0^L \sin\frac{2n\pi x}{L}\, dx = 0$$

2.20 Give the mathematical representation of a spherical wave travelling outward from a point and evaluate its probability current density.

Solution. The mathematical representation of a spherical wave travelling outwards from a point is given by

$$\psi(r) = \frac{A}{r} \exp(ikr)$$

where A is a constant and k is the wave vector. The probability current density

$$j = \frac{i\hbar}{2m}(\psi\nabla\psi^* - \psi^*\nabla\psi)$$

$$= \frac{i\hbar}{2m}|A|^2\left[\frac{e^{ikr}}{r}\nabla\left(\frac{e^{-ikr}}{r}\right) - \frac{e^{-ikr}}{r}\nabla\left(\frac{e^{ikr}}{r}\right)\right]$$

$$= \frac{i\hbar}{2m}|A|^2\left[\frac{e^{ikr}}{r}\left(-\frac{ik}{r}e^{-ikr} - \frac{e^{-ikr}}{r^2}\right) - \frac{e^{-ikr}}{r}\left(\frac{ik}{r}e^{ikr} - \frac{e^{ikr}}{r^2}\right)\right]$$

$$= \frac{i\hbar}{2m}|A|^2\left(\frac{-2ik}{r^2}\right) = \frac{\hbar k}{mr^2}|A|^2$$

2.21 The wave function of a particle of mass m moving in a potential $V(x)$ is $\Psi(x, t) = A\exp\left(-ikt - \frac{km}{\hbar}x^2\right)$, where A and k are constants. Find the explicit form of the potential $V(x)$.

Solution.

$$\Psi(x, t) = A\exp\left(-ikt - \frac{kmx^2}{\hbar}\right)$$

$$\frac{\partial\Psi}{\partial x} = -\frac{2kmx}{\hbar}\Psi$$

$$\frac{\partial^2\Psi}{\partial x^2} = \left(-\frac{2km}{\hbar} + \frac{4k^2m^2x^2}{\hbar^2}\right)\Psi$$

$$i\hbar\frac{\partial\Psi}{\partial t} = k\hbar\Psi$$

Substituting these values in the time dependendent Schrödinger equation, we have

$$k\hbar = -\frac{\hbar^2}{2m}\left(-\frac{2km}{\hbar} + \frac{4k^2m^2x^2}{\hbar^2}\right) + V(x)$$

$$k\hbar = k\hbar - 2mk^2x^2 + V(x)$$

$$V(x) = 2mk^2x^2$$

2.22 The time-independent wave function of a system is $\psi(x) = A\exp(ikx)$, where k is a constant. Check whether it is normalizable in the domain $-\infty < x < \infty$. Calculate the probability current density for this function.

Solution. Substitution of $\psi(x)$ in the normalization condition gives

$$|N|^2\int_{-\infty}^{\infty}|\psi|^2\,dx = |N|^2\int_{-\infty}^{\infty}dx = 1$$

As this integral is not finite, the given wave function is not normalizable in the usual sense. The probability current density

$$j = \frac{i\hbar}{2m}(\psi \nabla \psi^* - \psi^* \nabla \psi)$$

$$= \frac{i\hbar}{2m}|A|^2 [e^{ikx}(-ik)e^{-ikx} - e^{-ikx}(ik)e^{ikx}]$$

$$= \frac{i\hbar}{2m}|A|^2 (-ik - ik) = \frac{k\hbar}{m}|A|^2$$

2.23 Show that the phase velocity $\mathbf{v}_p$ for a particle with rest mass m_0 is always greater than the velocity of light and that $\mathbf{v}_p$ is a function of wavelength.

Solution.

$$\text{Phase velocity } \mathbf{v}_p = \frac{\omega}{k} = \nu\lambda; \quad \lambda = \frac{h}{p}$$

Combining the two, we get

$$p\mathbf{v}_p = h\nu = E = (c^2p^2 + m_0^2c^4)^{1/2}$$

$$p\mathbf{v}_p = cp\left(1 + \frac{m_0^2c^4}{c^2p^2}\right)^{1/2} = cp\left(1 + \frac{m_0^2c^2}{p^2}\right)^{1/2}$$

$$\mathbf{v}_p = c\left(1 + \frac{m_0^2c^2}{p^2}\right)^{1/2} \quad \text{or} \quad \mathbf{v}_p > c$$

$$\mathbf{v}_p = c\left(1 + \frac{m_0^2c^2\lambda^2}{h^2}\right)^{1/2}$$

Hence $\mathbf{v}_p$ is a function of λ.

2.24 Show that the wavelength of a particle of rest mass m_0, with kintic energy T given by the relativistic formula

$$\lambda = \frac{hc}{\sqrt{T^2 + 2m_0c^2T}}$$

Solution. For a relativistic particle, we have

$$E^2 = c^2p^2 + m_0^2c^4$$

Now, since

$$E = T + m_0c^2$$

$$(T + m_0c^2)^2 = c^2p^2 + m_0^2c^4$$

$$T^2 + 2m_0c^2T + m_0^2c^4 = c^2p^2 + m_0^2c^4$$

$$cp = \sqrt{T^2 + 2m_0c^2T}$$

$$\text{de Broglie wavelength } \lambda = \frac{h}{p} = \frac{hc}{\sqrt{T^2 + 2m_0c^2T}}$$

2.25 An electron moves with a constant velocity 1.1×10^6 m/s. If the velocity is measured to a precision of 0.1 per cent, what is the maximum precision with which its position could be simultaneously measured?

Solution. The momentum of the electron is given by

$$p = (9.1 \times 10^{-31}\ \text{kg})\ (1.1 \times 10^6\ \text{m/s})$$

$$= 1 \times 10^{-24}\ \text{kg m/s}$$

$$\frac{\Delta \mathbf{v}}{\mathbf{v}} = \frac{\Delta p}{p} = \frac{0.1}{100}$$

$$\Delta p = p \times 10^{-3} = 10^{-27}\ \text{kg m/s}$$

$$\Delta x \cong \frac{h}{4\pi \Delta p} = \frac{6.626 \times 10^{-34}\ \text{J s}}{4\pi \times 10^{-27}\ \text{kg m/s}} = 6.6 \times 10^{-7}\ \text{m}$$

2.26 Calculate the probability current density $j(x)$ for the wave function.

$$\psi(x) = u(x) \exp\,[i\phi(x)],$$

where u, ϕ are real.

Solution.

$$\psi(x) = u(x) \exp\,(i\phi); \quad \psi^*(x) = u(x) \exp\,(-i\phi)$$

$$\frac{\partial \psi}{\partial x} = \frac{\partial u}{\partial x} \exp\,(i\phi) + iu\frac{\partial \phi}{\partial x} \exp\,(i\phi)$$

$$\frac{\partial \psi^*}{\partial x} = \frac{\partial u}{\partial x} \exp\,(-i\phi) - iu\frac{\partial \phi}{\partial x} \exp\,(i\phi)$$

$$j(x) = \frac{i\hbar}{2m}\left(\psi\frac{\partial \psi^*}{\partial x} - \psi^*\frac{\partial \psi}{\partial x}\right)$$

$$= \frac{i\hbar}{2m}\left[ue^{i\phi}\left(\frac{\partial u}{\partial x}e^{-i\phi} - iu\frac{\partial \phi}{\partial x}e^{-i\phi}\right) - ue^{i\phi}\left(\frac{\partial u}{\partial x}e^{i\phi} + iu\frac{\partial \phi}{\partial x}e^{i\phi}\right)\right]$$

$$= \frac{i\hbar}{2m}\left[u\frac{\partial u}{\partial x} - iu^2\frac{\partial \phi}{\partial x} - u\frac{\partial u}{\partial x} - iu^2\frac{\partial \phi}{\partial x}\right]$$

$$= \frac{i\hbar}{2m}\left[-2iu^2\frac{\partial \phi}{\partial x}\right] = \frac{\hbar}{m}u^2\frac{\partial \phi}{\partial x}$$

2.27 The time-independent wave function of a particle of mass m moving in a potential $V(x) \doteq \alpha^2 x^2$ is

$$\psi(x) = \exp\left(-\sqrt{\frac{m\alpha^2}{2\hbar^2}}x^2\right), \quad \alpha \text{ being a constant.}$$

Find the energy of the system.

Solution. We have

$$\psi(x) = \exp\left(-\sqrt{\frac{m\alpha^2}{2\hbar^2}}\,x^2\right)$$

$$\frac{d\psi}{dx} = -\sqrt{\frac{2m\alpha^2}{\hbar^2}} \times \exp\left(-\sqrt{\frac{m\alpha^2}{2\hbar^2}}\, x^2\right)$$

$$\frac{d^2\psi}{dx^2} = -\sqrt{\frac{2m\alpha^2}{\hbar^2}}\left[1 - \sqrt{\frac{2m\alpha^2}{\hbar^2}}x^2\right]\exp\left(-\sqrt{\frac{m\alpha^2}{2\hbar^2}}\, x^2\right)$$

Substituting these in the time-independent Schrödinger equation and dropping the exponential term, we obtain

$$-\frac{\hbar^2}{2m}\left[-\sqrt{\frac{2m\alpha^2}{\hbar^2}} + \frac{2m\alpha^2}{\hbar^2}x^2\right] + a^2x^2 = E$$

$$\hbar\sqrt{\frac{\alpha^2}{2m}} - a^2x^2 + a^2x^2 = E$$

$$E = \frac{\hbar\alpha}{\sqrt{2m}}$$

2.28 For a particle of mass m, Schrödinger initially arrived at the wave equation

$$\frac{1}{c^2}\frac{\partial^2\Psi}{\partial t^2} = \frac{\partial^2\Psi}{\partial x^2} - \frac{m^2c^2}{\hbar^2}\Psi$$

Show that a plane wave solution of this equation is consistent with the relativistic energy momentum relationship.

Solution. For plane waves,

$$\Psi(x, t) = A \exp\,[i\,(kx - \omega t)]$$

Substituting this solution in the given wave equation, we obtain

$$\frac{(-i\omega)^2}{c^2}\Psi = (ik)^2\Psi - \frac{m^2c^2}{\hbar^2}\Psi$$

$$\frac{-\omega^2}{c^2} = -k^2 - \frac{m^2c^2}{\hbar^2}$$

Multiplying by $c^2\hbar^2$ and writing $\hbar\omega = E$ and $k\hbar = p$, we get

$$E^2 = c^2p^2 + m^2c^4$$

which is the relativistic energy-momentum relationship.

2.29 Using the time-independent Schrödinger equation, find the potential $V(x)$ and energy E for which the wave function

$$\psi(x) = \left(\frac{x}{x_0}\right)^n e^{-x/x_0},$$

where n, x_0 are constants, is an eigenfunction. Assume that $V(x) \to 0$ as $x \to \infty$.

Solution. Differentiating the wave function with respect to x, we get

$$\frac{d\psi}{dx} = \frac{n}{x_0}\left(\frac{x}{x_0}\right)^{n-1} e^{-x/x_0} - \frac{1}{x_0}\left(\frac{x}{x_0}\right)^{n} e^{-x/x_0}$$

$$\frac{d^2\psi}{dx^2} = \frac{n(n-1)}{x_0^2}\left(\frac{x}{x_0}\right)^{n-2} e^{-x/x_0} - \frac{2n}{x_0^2}\left(\frac{x}{x_0}\right)^{n-1} e^{-x/x_0} + \frac{1}{x_0^2}\left(\frac{x}{x_0}\right)^{n} e^{-x/x_0}$$

$$= \left[\frac{n(n-1)}{x^2} - \frac{2n}{x_0 x} + \frac{1}{x_0^2}\right]\left(\frac{x}{x_0}\right)^{n} e^{-x/x_0}$$

$$= \left[\frac{n(n-1)}{x^2} - \frac{2n}{x_0 x} + \frac{1}{x_0^2}\right]\psi(x)$$

Substituting in the Schrödinger equation, we get

$$-\frac{\hbar^2}{2m}\left[\frac{n(n-1)}{x^2} - \frac{2n}{x_0 x} + \frac{1}{x_0^2}\right]\psi + V\psi = E\psi$$

which gives the operator equation

$$E - V(x) = -\frac{\hbar^2}{2m}\left[\frac{n(n-1)}{x^2} - \frac{2n}{x_0 x} + \frac{1}{x_0^2}\right]$$

When $x \to \infty$, $V(x) \to 0$. Hence,

$$E = -\frac{\hbar^2}{2mx_0^2}$$

$$V(x) = \frac{\hbar^2}{2m}\left[\frac{n(n-1)}{x^2} - \frac{2n}{x_0 x}\right]$$

2.30 Find that the form of the potential, for which $\psi(r)$ is constant, is a solution of the Schrödinger equation. What happens to probability current density in such a case?

Solution. Since $\psi(r)$ is constant,

$$\nabla^2\psi = 0.$$

Hence the Schrödinger equation reduces to

$$V\psi = E\psi \quad \text{or} \quad V = E$$

The potential is of the form V which is a constant. Since $\psi(r)$ is constant, $\nabla\psi = \nabla\psi^* = 0$. Consequently, the probability current density is zero.

2.31 Obtain the form of the equation of continuity for probability if the potential in the Schrödinger equation is of the form $V(r) = V_1(r) + iV_2(\boldsymbol{r})$, where V_1 and V_2 are real.

Solution. The probability density $P(\boldsymbol{r}, t) = \Psi^*\Psi$. Then,

$$i\hbar\frac{\partial P}{\partial t} = i\hbar\frac{\partial}{\partial t}(\Psi\Psi^*) = \Psi\left(i\hbar\frac{\partial}{\partial t}\Psi^*\right) + \Psi^*\left(i\hbar\frac{\partial\Psi}{\partial t}\right)$$

The Schrödinger equation with the given potential is given by

$$i\hbar\frac{\partial\Psi}{\partial t} = \frac{-\hbar^2}{2m}\nabla^2\Psi + (V_1 + iV_2)\,\Psi$$

Substituting the values of $i\hbar\dfrac{\partial\Psi}{\partial t}$ and $i\hbar\dfrac{\partial\Psi^*}{\partial t}$, we have

$$i\hbar\frac{\partial P}{\partial t} = \frac{\hbar^2}{2m}(\Psi\nabla^2\Psi^* - \Psi^*\nabla^2\Psi) + 2iV_2P$$

$$i\hbar\frac{\partial P}{\partial t} = \frac{\hbar^2}{2m}[\nabla\cdot(\Psi\nabla\Psi^* - \Psi^*\nabla\Psi) + 2iV_2P]$$

$$\frac{\partial P}{\partial t} = \nabla\cdot\left(-\frac{i\hbar}{2m}\right)(\Psi\nabla\Psi^* - \Psi^*\nabla\Psi) + 2\frac{V_2}{\hbar}P$$

$$\frac{\partial P}{\partial t} + \nabla\cdot \boldsymbol{j}(\boldsymbol{r},t) = \frac{2V_2}{\hbar}P(\boldsymbol{r},t)$$

2.32 For a one-dimensional wave function of the form

$$\Psi(x,\, t) = A\,\exp\,[i\phi\,(x,\, t)]$$

show that the probability current density can be written as

$$j = \frac{\hbar}{m}|A|^2\frac{\partial\phi}{\partial x}$$

Solution. The probability current density $\boldsymbol{j}(\boldsymbol{r},\, t)$ is given by

$$j(\boldsymbol{r},\, t) = \frac{i\hbar}{2m}(\Psi\nabla\Psi^* - \Psi^*\nabla\Psi)$$

$$\Psi(x,\, t) = A\,\exp\,[i\phi\,(x,\, t)]$$

$$\Psi^*(x,\, t) = A^*\,\exp\,[-i\phi\,(x,\, t)]$$

$$\nabla\Psi = \frac{\partial\Psi}{\partial x} = iAe^{i\phi}\frac{\partial\phi}{\partial x}$$

$$\nabla\Psi^* = \frac{\partial\Psi^*}{\partial x} = -iA^*e^{-i\phi}\frac{\partial\phi}{\partial x}$$

Substituting these values, we get

$$j = \frac{i\hbar}{2m}\left[Ae^{i\phi}\left(-iA^*e^{-i\phi}\frac{\partial\phi}{\partial x}\right) - A^*e^{-i\phi}\left(iAe^{i\phi}\frac{\partial\phi}{\partial x}\right)\right]$$

$$= \frac{i\hbar}{2m}\left[-i|A|^2 - i|A|^2\right]\frac{\partial\phi}{\partial x} = \frac{\hbar}{m}|A|^2\frac{\partial\phi}{\partial x}$$

2.33 Let $\psi_0(x)$ and $\psi_1(x)$ be the normalized ground and first excited state energy eigenfunctions of a linear harmonic oscillator. At some instants of time, $A\psi_0 + B\psi_1$, where A and B are constants, is the wave function of the oscillator. Show that $\langle x\rangle$ is in general different from zero.

Solution. The normalization condition gives

$$\langle (A\psi_0 + B\psi_1) \,|\, (A\psi_0 + B\psi_1) \rangle = 1$$

$$A^2 \langle \psi_0 \,|\, \psi_0 \rangle + B^2 \langle \psi_1 \,|\, \psi_1 \rangle = 1 \quad \text{or} \quad A_2 + B_2 = 1$$

Generally, the constants A and B are not zero. The average value of x is given by

$$\langle x \rangle = \langle (A\psi_0 + B\psi_1) \,|\, x \,|\, (A\psi_0 + B\psi_1) \rangle$$

$$= A^2 \langle \psi_0 \,|\, x \,|\, \psi_0 \rangle + B^2 \langle \psi_1 \,|\, x \,|\, \psi_1 \rangle + 2AB \langle \psi_0 \,|\, x \,|\, \psi_1 \rangle$$

since A and B are real and $\langle \psi_0 \,|\, x \,|\, \psi_1 \rangle = \langle \psi_1 \,|\, x \,|\, \psi_0 \rangle$. As the integrands involved is odd,

$$\langle \psi_0 \,|\, x \,|\, \psi_0 \rangle = \langle \psi_1 \,|\, x \,|\, \psi_1 \rangle = 0$$

$$\langle x \rangle = 2AB \langle \psi_0 \,|\, x \,|\, \psi_1 \rangle$$

which is not equal to zero.

2.34 (i) The waves on the surface of water travel with a phase velocity $\mathbf{v}_p = \sqrt{g\lambda/2\pi}$, where g is the acceleration due to gravity and λ is the wavelength of the wave. Show that the group velocity of a wave packet comprised of these waves is $\mathbf{v}_p/2$. (ii) For a relativistic particle, show that the velocity of the particle and the group velocity of the corresponding wave packet are the same.

Solution.

(i) The phase velocity

$$\mathbf{v}_p = \sqrt{\frac{g\lambda}{2\pi}} = \sqrt{\frac{g}{k}}$$

where k is the wave vector.

By definition, $\mathbf{v}_p = \omega/k$, and hence

$$\frac{\omega}{k} = \sqrt{\frac{g}{k}} \quad \text{or} \quad \omega = \sqrt{gk}$$

The group velocity

$$\mathbf{v}_g = \frac{d\omega}{dk} = \frac{1}{2}\sqrt{\frac{g}{k}} = \frac{\mathbf{v}_p}{2}$$

(ii) Group velocity $\mathbf{v}_g = \dfrac{d\omega}{dk} = \dfrac{dE}{dp}$

For relativistic particle, $E^2 = c^2p^2 + m_0^2c^4$, and therefore,

$$\mathbf{v}_g = \frac{dE}{dp} = \frac{c^2 p}{E} = \frac{c^2 m_0 \mathbf{v}\sqrt{1 - \mathbf{v}^2/c^2}}{m_0 c^2 \sqrt{1 - \mathbf{v}^2/c^2}} = v$$

2.35 Show that, if a particle is in a stationary state at a given time, it will always remain in a stationary state.

Solution. Let the particle be in the stationary state $\Psi(x, 0)$ with energy E. Then we have

$$H\Psi(x, 0) = E\Psi(x, 0)$$

where H is the Hamiltonian of the particle which is assumed to be real. At a later time, let the wave function be $\Psi(x, t)$, i.e.,

$$\Psi(x, t) = \Psi(x, 0)\, e^{-iEt/\hbar}$$

At time t,

$$H\Psi(x, t) = H\Psi(x, 0)\, e^{-iEt/\hbar}$$
$$= E\Psi(x, 0)\, e^{-iEt/\hbar}$$
$$= E\Psi(x, t)$$

Thus, $\Psi(x, t)$ is a stationary state which is the required result.

2.36 Find the condition at which de Broglie wavelength equals the Compton wavelength

Solution.

$$\text{Compton wavelength } \lambda_C = \frac{h}{m_0 c}$$

where m_0 is the rest mass of electron and c is the velocity of light

$$\text{de Broglie wave length } \lambda = \frac{h}{m\mathbf{v}}$$

where m is the mass of electron when its velocity is $\mathbf{v}$. Since

$$m = \frac{m_0}{\sqrt{1 - \mathbf{v}^2/c^2}}$$

$$\lambda = \frac{h\sqrt{1 - \mathbf{v}^2/c^2}}{m_0 \mathbf{v}} = \frac{h\sqrt{(c^2 - \mathbf{v}^2)}}{m_0 c\mathbf{v}}$$

$$= \frac{h\mathbf{v}\sqrt{c^2/\mathbf{v}^2 - 1}}{m_0 c\mathbf{v}} = \frac{h}{m_0 c}\sqrt{c^2/\mathbf{v}^2 - 1}$$

$$= \lambda_c \sqrt{\frac{c^2}{\mathbf{v}^2} - 1}$$

When $\lambda = \lambda_C$,

$$\sqrt{\frac{c^2}{\mathbf{v}^2} - 1} = 1 \quad \text{or} \quad \frac{c^2}{\mathbf{v}^2} - 1 = 1$$

$$\frac{c^2}{\mathbf{v}^2} = 2 \quad \text{or} \quad \mathbf{v} = \frac{c}{\sqrt{2}}$$

2.37 The wave function of a one-dimensional system is

$$\psi(x) = Ax^n e^{-x/a},\ A, \text{ a and } n \text{ are constants}$$

If $\psi(x)$ is an eigenfunction of the Schrödinger equation, find the condition on $V(x)$ for the energy eigenvalue $E = -\hbar^2/(2ma^2)$. Also find the value of $V(x)$.

Solution.

$$\psi(x) = Ax^n e^{-x/a}$$

$$\frac{d\psi}{dx} = Anx^{n-1}e^{-x/a} - \frac{A}{a}x^n e^{-x/a}$$

$$\frac{d^2\psi}{dx^2} = Ae^{-x/a}\left[n(n-1)x^{n-2} - \frac{2n}{a}x^{n-1} + \frac{x^n}{a^2}\right]$$

With these values, the Schrödinger equation takes the form

$$-\frac{\hbar^2}{2m}Ae^{-x/a}\left[n(n-1)x^{n-2} - \frac{2n}{a}x^{n-1} + \frac{x^n}{a^2}\right] + V(x)\,Ax^n e^{-x/a} = E\,Ax^n e^{-x/a}$$

$$-\frac{\hbar^2}{2m}\left[\frac{n(n-1)}{x^2} - \frac{2n}{ax} + \frac{1}{a^2}\right] = E - V(x)$$

From this equation, it is obvious that for the energy $E = -\hbar^2/2ma^2$, $V(x)$ must tend to zero as $x \to \infty$. Then,

$$V(x) = -\frac{\hbar^2}{2ma^2} - \frac{\hbar^2}{2m}\left[\frac{n(n-1)}{x^2} - \frac{2n}{a} + \frac{1}{a^2}\right]$$

$$= \frac{\hbar^2}{2m}\left[\frac{n(n-1)}{x^2} - \frac{2n}{ax}\right]$$

2.38 An electron has a de Broglie wavelength of 1.5×10^{-12} m. Find its (i) kinetic energy and (ii) group and phase velocities of its matter waves.

Solution.

(i) The total energy E of the electron is given by

$$E = \sqrt{c^2p^2 + m_0^2c^4}$$

$$\text{Kinetic energy } T = E - m_0c^2 = \sqrt{c^2p^2 + m_0^2c^4} - m_0c^2$$

$$\text{de Broglie wavelength } \lambda = \frac{h}{p} \quad \text{or} \quad cp = \frac{hc}{\lambda}$$

$$cp = \frac{(6.626 \times 10^{-34}\ \text{J s})\ (3 \times 10^8\ \text{m s}^{-1})}{1.5 \times 10^{-12}\ \text{m}}$$

$$= 13.252 \times 10^{-14}\ \text{J}$$

$$E_0 = m_0c^2 = (9.1 \times 10^{-31}\ \text{kg})\ (3 \times 10^8\ \text{m s}^{-1})$$

$$= 8.19 \times 10^{-14}\ \text{J}$$

$$T = \sqrt{(13.252)^2 + (8.19)^2} \times 10^{-14}\ \text{J} - 8.19 \times 10^{-14}\ \text{J}$$

$$= 7.389 \times 10^{-14}\ \text{J} = 4.62 \times 10^5\ \text{eV}$$

(ii) $E = \sqrt{(13.252)^2 + (8.19)^2} \times 10^{-14}$ J $= 15.579 \times 10^{-14}$ J

$$E = \frac{E_0}{\sqrt{1 - \mathbf{v}^2/c^2}} \quad \text{or} \quad 1 - \frac{\mathbf{v}^2}{c^2} = \left(\frac{E_0}{E}\right)^2$$

$$\mathbf{v} = \left[1 - \left(\frac{E_0}{E}\right)^2\right]^{1/2} c = \left[1 - \left(\frac{8.19}{15.579}\right)^2\right]^{1/2} (3 \times 10^8 \text{ m s}^{-1})$$

$$= 0.851c$$

The group velocity will be the same as the particle velocity. Hence,

$$\mathbf{v}_g = 0.851c$$

$$\text{Phase velocity } \mathbf{v}_p = \frac{c^2}{\mathbf{v}} = \frac{c}{0.851} = 1.175c$$

2.39 The position of an electron is measured with an accuracy of 10^{-6} m. Find the uncertainty in the electron's position after 1 s. Comment on the result.

Solution. When $t = 0$, the uncertainty in the electron's momentum is

$$\Delta p \geq \frac{\hbar}{2\Delta x}$$

Since $p = m\mathbf{v}$, $\Delta p = m\,\Delta\mathbf{v}$. Hence,

$$\Delta\mathbf{v} \geq \frac{\hbar}{2m\Delta x}$$

The uncertainty in the position of the electron at time t cannot be more than

$$(\Delta x)_t = t\Delta\mathbf{v} \geq \frac{\hbar t}{2m\Delta x}$$

$$= \frac{(1.054 \times 10^{-34} \text{ J s})\, 1 \text{ s}}{2(9.1 \times 10^{-31} \text{ kg})\, 10^{-6} \text{ m}} = 57.9 \text{ m}$$

The original wave packet has spread out to a much wider one. A large range of wave numbers must have been present to produce the narrow original wave group. The phase velocity of the component waves has varied with the wave number.

2.40 If the total energy of a moving particle greatly exceeds its rest energy, show that its de Broglie wavelength is nearly the same as the wavelength of a photon with the same total energy.

Solution. Let the total energy be E. Then,

$$E_2 = c^2p^2 + m_0^2c^4 \cong c^2p^2$$

$$p = \frac{E}{c}$$

$$\text{de Broglie wavelength } \lambda = \frac{h}{p} = \frac{hc}{E}$$

For a photon having the same energy,

$$E = h\nu = \frac{hc}{\lambda} \quad \text{or} \quad \lambda = \frac{hc}{E}$$

which is the required result.

2.41 From scattering experiments, it is found that the nuclear diameter is of the order of 10^{-15} m. The energy of an electron in β-decay experiment is of the order of a few MeV. Use these data and the uncertainty principle to show that the electron is not a constituent of the nucleus.

Solution. If an electron exists inside the nucleus, the uncertainty in its position $\Delta x \cong 10^{-15}$ m. From the uncertainty principle,

$$(10^{-15}\ \text{m})\ \Delta p \geq \frac{\hbar}{2}$$

$$\Delta p \geq \frac{1.05 \times 10^{-34}\ \text{J s}}{2(10^{-15}\ \text{m})} = 5.25 \times 10^{-20}\ \text{kg m s}^{-1}$$

The momentum of the electron p must at least be of this order.

$$p \cong 5.25 \times 10^{-20}\ \text{kg m s}^{-1}$$

When the energy of the electron is very large compared to its rest energy,

$$E \cong cp = (3 \times 10^8\ \text{ms}^{-1})(5.25 \times 10^{-20}\ \text{kg m s}^{-1})$$

$$= \frac{15.75 \times 10^{-12}\ \text{J}}{1.6 \times 10^{-19}\ \text{J/eV}} = 9.84 \times 10^7\ \text{eV}$$

$$= 98.4\ \text{MeV}$$

This is very large compared to the energy of the electron in β-decay. Thus, electron is not a constituent of the nucleus.

2.42 An electron microscope operates with a beam of electrons, each of which has an energy 60 keV. What is the smallest size that such a device could resolve? What must be the energy of each neutron in a beam of neutrons be in order to resolve the same size of object?

Solution. The momentum of the electron is given by

$$p^2 = 2mE = 2\,(9.1 \times 10^{-31}\ \text{kg})(60 \times 1000 \times 1.6 \times 10^{-19}\ \text{J})$$

$$p = 13.218 \times 10^{-23}\ \text{kg m s}^{-1}$$

The de Broglie wavelength

$$\lambda = \frac{h}{p} = \frac{6.626 \times 10^{-34}\ \text{J s}}{13.216 \times 10^{-23}\ \text{kg m s}^{-1}}$$

$$= 5.01 \times 10^{-12}\ \text{m}$$

The smallest size an elecron microscope can resolve is of the order of the de Broglie wavelength of electron. Hence the smallest size that can be resolved is 5.01×10^{-12} m.

The de Broglie wavelength of the neutron must be of the order of 5.01×10^{-12} m. Hence, the momentum of the neutron must be the same as that of electron. Then,

$$\text{Momentum of neutron} = 13.216 \times 10^{-23}\ \text{kg m s}^{-1}$$

$$\text{Energy} = \frac{p^2}{2M} \quad (M \text{ is mass of neutron})$$

$$= \frac{(13.216 \times 10^{-23}\ \text{kg ms}^{-1})^2}{2 \times 1836(9.1 \times 10^{-31}\ \text{kg})} = 5.227 \times 10^{-18}\ \text{J}$$

$$= \frac{5.227 \times 10^{-18}\ \text{J}}{1.6 \times 10^{-19}\ \text{J/eV}} = 32.67\ \text{eV}$$

2.43 What is the minimum energy needed for a photon to turn into an electron-positron pair? Calculate how long a virtual electron-positron pair can exist.

Solution. The Mass of an electron-positron pair is $2m_ec^2$. Hence the minimum energy needed to make an electron-positron pair is 2 m_ec^2, i.e., this much of energy needs to be borrowed to make the electron-positron pair. By the uncertainty relation, the minimum time for which this can happen is

$$\Delta t = \frac{\hbar}{2 \times 2m_ec^2}$$

$$= \frac{1.05 \times 10^{-34}\ \text{J s}}{4(9.1 \times 10^{-31}\ \text{kg})(3 \times 10^8\ \text{m/s})^2}$$

$$= 3.3 \times 10^{-22}\ \text{s}$$

which is the length of time for which such a pair exists.

2.44 A pair of virtual particles is created for a short time. During the time of their existence, a distance of 0.35*fm* is covered with a speed very close to the speed of light. What is the value of mc^2 (in eV) for each of the virtual particle?

Solution. According to Problem 2.43, the pair exists for a time Δt given by

$$\Delta t = \frac{\hbar}{4mc^2}$$

The time of existence is also given by

$$\Delta t = \frac{0.35 \times 10^{-15}\ \text{m}}{3 \times 10^8\ \text{m/s}} = 1.167 \times 10^{-24}\ \text{s}$$

Equating the two expressions for Δt, we get

$$\frac{\hbar}{4mc^2} = 1.167 \times 10^{-24}\ \text{s}$$

$$mc^2 = \frac{1.05 \times 10^{-34}\ \text{J s}}{4 \times 1.167 \times 10^{-24}\ \text{s}} = 2.249 \times 10^{-11}\ \text{J}$$

$$= \frac{2.249 \times 10^{-11}\ \text{J}}{1.6 \times 10^{-19}\ \text{J/eV}} = 140.56 \times 10^6\ \text{eV}$$

$$= 140.56\ \text{MeV}$$

2.45 The uncertainty in energy of a state is responsible for the natural line width of spectral lines. Substantiate.

Solution. The equation

$$(\Delta E)(\Delta t) \geq \frac{\hbar}{2} \tag{i}$$

implies that the energy of a state cannot be measured exactly unless an infinite amount of time is available for the measurement. If an atom is in an excited state, it does not remain there indefinitely, but makes a transition to a lower state. We can take the mean time for decay τ, called the lifetime, as a measure of the time available to determine the energy. Hence the uncertainty in time is of the order of τ. For transitions to the ground state, which has a definite energy E_0 because of its finite lifetime, the spread in wavelength can be calculated from

$$E - E_0 = \frac{hc}{\lambda}$$

$$|\Delta E| = \frac{hc\,|\Delta\lambda|}{\lambda^2}$$

$$\frac{\Delta\lambda}{\lambda} = \frac{\Delta E}{E - E_0} \tag{ii}$$

Using Eq. (i) and identifying $\Delta t \cong \tau$, we get

$$\frac{\Delta\lambda}{\lambda} = \frac{\hbar}{2\tau(E - E_0)} \tag{iii}$$

The energy width $\hbar/\tau$ is often referred to as the **natural line width**.

2.46 Consider the electron in the hydrogen atom. Using $(\Delta x)(\Delta p) \simeq \hbar$, show that the radius of the electron orbit in the ground state is equal to the Bohr radius.

Solution. The energy of the electron in the hydrogen atom is the given by

$$E = \frac{p^2}{2m} - \frac{ke^2}{r}, \qquad k = \frac{1}{4\pi\varepsilon_0}$$

where p is the momentum of the electron. For the order of magnitude of the position uncertainty, if we take $\Delta x \cong r$, then

$$\Delta p \cong \frac{\hbar}{r} \quad \text{or} \quad (\Delta p)^2 = \frac{\hbar^2}{r^2}$$

Taking the order of momentum p as equal to the uncertainty in momentum, we get

$$(\Delta p)^2 = \langle p^2 \rangle = \frac{\hbar^2}{r^2}$$

Hence, the total energy

$$E = \frac{\hbar}{2mr^2} - \frac{ke^2}{r}$$

For E to be minimum, $(dE/dr) = 0$. Then,

$$\frac{dE}{dr} = -\frac{\hbar^2}{mr^3} + \frac{ke^2}{r^2} = 0$$

$$r = \frac{\hbar^2}{kme^3} = a_0$$

which is the required result.

2.47 Consider a particle described by the wave function $\Psi(x, t) = e^{i(kx - \omega t)}$.

(i) Is this wave function an eigenfunction corresponding to any dynamical variable or variables? If so, name them.

(ii) Does this represent a ground state?

(iii) Obtain the probability current density of this function.

Solution.

(i) Allowing the momentum operator $-i\hbar\,(d/dx)$ to operate on the function, we have

$$-i\hbar\frac{d}{dx}\,e^{i(kx - \omega t)} = i\hbar(ik)\,e^{i(kx - \omega t)}$$
$$= \hbar k\,e^{i(kx - \omega t)}$$

Hence, the given function is an eigenfunction of the momentum operator. Allowing the energy operator $-i\hbar\,(d/dt)$ to operate on the function, we have

$$i\hbar\frac{d}{dt}\,e^{i(kx - \omega t)} = i\hbar(-i\omega)\,e^{i(kx - \omega t)}$$
$$= \hbar\omega\,e^{i(kx - \omega t)}$$

Hence, the given function is also an eigenfunction of the energy operator with an eigenvalue $\hbar\omega$.

(ii) Energy of a bound state is negative. Here, the energy eigenvalue is $\hbar\omega$, which is positive. Hence, the function does not represent a bound state.

(iii) The probability current density

$$j = \frac{i\hbar}{2m}\,(\psi\nabla\psi^* - \psi^*\nabla\psi)$$
$$= \frac{i\hbar}{2m}\,(-ik - ik) = \frac{\hbar k}{m}$$

2.48 Show that the average kinetic energy of a particle of mass m with a wave function $\psi(x)$ can be written in the form

$$T = \frac{\hbar^2}{2m}\int_{-\infty}^{\infty}\left|\frac{d\psi}{dx}\right|^2 dx$$

Solution. The average kinetic energy

$$\langle T\rangle = \frac{\langle p^2\rangle}{2m} = -\frac{\hbar^2}{2m}\int_{-\infty}^{\infty}\psi^*\frac{d^2\psi}{dx^2}\,dx$$

Integrating by parts, we obtain

$$\langle T\rangle = -\frac{\hbar^2}{2m}\left[\psi^*\frac{d\psi}{dx}\right]_{-\infty}^{\infty} + \frac{\hbar^2}{2m}\int_{-\infty}^{\infty}\frac{d\psi^*}{dx}\frac{d\psi}{dx}\,dx$$

As the wave function and derivatives are finite, the integrated term vanishes, and so

$$\langle T\rangle = \frac{\hbar^2}{2m}\int_{-\infty}^{\infty}\left|\frac{d\psi}{dx}\right|^2 dx$$

2.49 The energy eigenvalue and the corresponding eigenfunction for a particle of mass m in a one-dimensional potential $V(x)$ are

$$E = 0, \qquad \psi(x) = \frac{A}{x^2 + a^2}$$

Deduce the potential $V(x)$.

Solution. The Schrödinger equation for the particle with energy eigenvalue $E = 0$ is

$$-\frac{\hbar^2}{2m}\frac{d^2\psi}{dx^2} + V(x)\psi = 0, \qquad \psi = \frac{A}{x^2 + a^2}$$

$$\frac{d\psi}{dx^2} = -\frac{2Ax}{(x^2 + a^2)^2}$$

$$\frac{d^2\psi}{dx^2} = -2A\left[\frac{1}{(x^2 + a^2)^2} - \frac{4x^2}{(x^2 + a^2)^3}\right]$$

$$= \frac{2A(3x^2 - a^2)}{(x^2 + a^2)^3}$$

Substituting the value of $d^2\psi/dx^2$, we get

$$-\frac{\hbar^2}{2m}\frac{2A(3x^2 - a^2)}{(x^2 + a^2)^3} + \frac{V(x)A}{x^2 + a^2} = 0$$

$$V(x) = \frac{\hbar^2(3x^2 - a^2)}{m(x^2 + a^2)^2}$$

CHAPTER 3

General Formalism of Quantum Mechanics

In this chapter, we provide an approach to a systematic the mathematical formalism of quantum mechanics along with a set of basic postulates.

3.1 Mathematical Preliminaries

(i) The scalar product of two functions $F(x)$ and $G(x)$ defined in the interval $a \le x \le b$, denoted as (F, G), is

$$(F, G) = \int_a^b F^*(x)\, G(x)\, dx \tag{3.1}$$

(ii) The functions are orthogonal if

$$(F, G) = \int_a^b F^*(x)\, G(x)\, dx = 0 \tag{3.2}$$

(iii) The norm of a function N is defined as

$$N = (F, F)^{1/2} = \left[\int_a^b |F(x)|^2\, dx\right]^{1/2} \tag{3.3}$$

(iv) A function is normalized if the norm is unity, i.e.,

$$(F, F) = \int_a^b F^*(x) F(x)\, dx = 1 \tag{3.4}$$

(v) Two functions are orthonormal if

$$(F_i, F_j) = \delta_{ij}, \qquad i, j = 1, 2, 3, \ldots \tag{3.5}$$

where δ_{ij} is the Kronecker delta defined by

$$\delta_{ij} = \begin{cases} 1 & \text{if } i = j \\ 0 & \text{if } i \neq j \end{cases} \tag{3.6}$$

(vi) A set of functions $F_1(x)$, $F_2(x)$, ... is linearly dependent if a relation of the type

$$\sum_i c_i F_i(x) = 0 \tag{3.7}$$

exists, where c_i's are constants. Otherwise, they are linearly independent.

3.2 Linear Operator

An operator can be defined as the rule by which a different function is obtained from any given function. An operator A is said to be linear if it satisfies the relation

$$A\,[c_1 f_1(x) + c_2 f_2(x)] = c_1 A f_1(x) + c_2 A f_2(x) \tag{3.8}$$

The commutator of operators A and B, denoted by $[A, B]$, is defined as

$$[A, B] = AB - BA \tag{3.9}$$

It follows that

$$[A, B] = -[B, A] \tag{3.10}$$

If $[A, B] = 0$, A and B are said to **commute**. If $AB + BA = 0$, A and B are said to **anticommute**. The inverse operator A^{-1} is defined by the relation

$$AA^{-1} = A^{-1}A = I \tag{3.11}$$

3.3 Eigenfunctions and Eigenvalues

Often, an operator A operating on a function multiplies the function by a consant, i.e.,

$$A\psi(x) = a\psi(x) \tag{3.12}$$

where a is a constant with respect to x. The function $\psi(x)$ is called the **eigenfunction** of the operator A corresponding to the **eigenvalue** a. If a given eigenvalue is associated with a large number of eigenfunctions, the eigenvalue is said to be degenerate.

3.4 Hermitian Operator

Consider two arbitrary functions $\psi_m(x)$ and $\psi_n(x)$. An operator A is said to be Hermitian if

$$\int_{-\infty}^{\infty} \psi_m^* A\psi_n \, dx = \int_{-\infty}^{\infty} (A\psi_m)^* \psi_n \, dx \tag{3.13}$$

An operator A is said to be anti-Hermitian if

$$\int_{-\infty}^{\infty} \psi_m^* A\psi_n \, dx = -\int_{-\infty}^{\infty} (A\psi_m)^* \psi_n \, dx \tag{3.14}$$

Two important theorems regarding Hermitian operators are:

(i) The eigenvalues of Hermitian operators are real.

(ii) The eigenfunctions of a Hermitian operator that belong to different eigenvalues are orthogonal.

3.5 Postulates of Quantum Mechanics

There are different ways of stating the basic postulates of quantum mechanics, but the following formulation seems to be satisfactory.

3.5.1 Postulate 1—Wave Function

The state of a system having n degrees of freedom can be completely specified by a function Ψ of coordinates $q_1, q_2, \ldots, q_n$ and time t which is called the **wave function** or **state function** or **state vector** of the system. Ψ, and its derivatives must be continuous, finite and single valued over the domain of the variables of Ψ.

The representation in which the wave function is a function of coordinates and time is called the **coordinate representation**. In the **momentum representation**, the wave function is a function of momentum components and time.

3.5.2 Postulate 2—Operators

To every observable physical quantity, there corresponds a Hermitian operator or matrix. The operators are selected according to the rule

$$[Q, R] = i\hbar\{q, r\} \tag{3.15}$$

where Q and R are the operators selected for the dynamical variables q and r, $[Q, R]$ is the commutator of Q with R, and $\{q, r\}$ is the Poisson bracket of q and r.

Some of the important classical observables and the corresponding operators are given in Table 3.1.

Table 3.1 Important Observables and Their Operators

Observable	*Classical form*	*Operator*
Coordinates	x, y, z	x, y, z
Momentum	$\boldsymbol{p}$	$-i\hbar\nabla$
Energy	E	$i\hbar\frac{\partial}{\partial t}$
Time	t	t
Kintetic energy	$\frac{p^2}{2m}$	$-\frac{\hbar^2}{2m}\nabla^2$
Hamiltonian	H	$-\frac{\hbar^2}{2m}\nabla^2 + V(r)$

3.5.3 Postulate 3—Expectation Value

When a system is in a state described by the wave function Ψ, the expectation value of any observable a whose operator is A is given by

$$\langle a \rangle = \int_{-\infty}^{\infty} \Psi^* A Y \, d\tau \tag{3.16}$$

3.5.4 Postulate 4—Eigenvalues

The possible values which a measurement of an observable whose operator is A can give are the eigenvalues a_i of the equation

$$A\Psi_i = a_i\Psi_i, \quad i = 1, 2, \ldots, n \tag{3.17}$$

The eigenfunctions Ψ_i form a complete set of n independent functions.

3.5.5 Postulate 5—Time Development of a Quantum System

The time development of a quantum system can be described by the evolution of state function in time by the time dependent Schrödinger equation

$$i\hbar \frac{\partial \Psi}{\partial t}(\boldsymbol{r}, t) = H\Psi(\boldsymbol{r}, t) \tag{3.18}$$

where H is the Hamiltonian operator of the system which is independent of time.

3.6 General Uncertainty Relation

The uncertainty (ΔA) in a dynamical variable A is defined as the root mean square deviation from the mean. Here, mean implies expectation value. So,

$$(\Delta A)^2 = \langle (A - \langle A \rangle)^2 \rangle = \langle A^2 \rangle - \langle A \rangle^2 \tag{3.19}$$

Now, consider two Hermitian operators, A and B. Let their commutator be

$$[A, B] = iC \tag{3.20}$$

The general uncertainty relation is given by

$$(\Delta A)(\Delta B) \geq \frac{\langle C \rangle}{2} \tag{3.21}$$

In the case of the variables x and p_x, $[x, p_x] = i\hbar$ and, therefore,

$$(\Delta x)(\Delta P_x) \geq \frac{\hbar}{2} \tag{3.22}$$

3.7 Dirac's Notation

To denote a state vector, Dirac introducted the symbol $|\ \rangle$, called the **ket vector** or, simply, **ket**. Different states such as $\psi_a(\mathbf{r})$, $\psi_b(\mathbf{r})$, ... are denoted by the kets $|a\rangle$, $|b\rangle$, ... Corresponding to every vector, $|a\rangle$, is defined as a conjugate vector $|a\rangle^*$, for which Dirac used the notation $\langle a|$, called a **bra vector** or simply **bra**. In this notation, the functions ψ_a and ψ_b are orthogonal if

$$\langle a|b\rangle = 0 \tag{3.23}$$

3.8 Equations of Motion

The equation of motion allows the determination of a system at a time from the known state at a particular time.

3.8.1 Schrödinger Picture

In this representation, the state vector changes with time but the operator remains constant. The state vector $|\psi_s(t)\rangle$ changes with time as follows:

$$i\hbar\frac{d}{dt}|\psi_s(t)\rangle = H|\psi_s(t)\rangle \tag{3.24}$$

Integration of this equation gives

$$|\psi_s(t)\rangle = e^{-iHt/\hbar}|\psi_s(0)\rangle \tag{3.25}$$

The time derivative of the expectation value of the operator is given by

$$\frac{d}{dt}\langle A_s\rangle = \frac{1}{i\hbar}\langle[A_s, H]\rangle + \frac{\partial A_s}{\partial t} \tag{3.26}$$

3.8.2 Heisenberg Picture

The operator changes with time while the state vector remains constant in this picture. The state vector $|\psi_H\rangle$ and operator A_H are defined by

$$|\psi_H\rangle = e^{iHt/\hbar}|\psi_s(t)\rangle \tag{3.27}$$

$$A_H(t) = e^{iHt/\hbar}A_s e^{iHt/\hbar} \tag{3.28}$$

From Eqs. (3.27) and (3.25), it is obvious that

$$|\psi_H\rangle = |\psi_s(0)\rangle \tag{3.29}$$

The time derivative of the operator A_H is

$$\frac{d}{dt}A_H = \frac{1}{i\hbar}[A_H, H] + \frac{\partial A_H}{\partial t} \tag{3.30}$$

3.8.3 Momentum Representation

In the momentum representation, the state function of a system $\Phi(\boldsymbol{p}, t)$ is taken as a function of the momentum and time. The momentum $\boldsymbol{p}$ is represented by the operator $\boldsymbol{p}$ itself and the posistion coordinate is represented by the operator $i\hbar\nabla_p$, where ∇_p is the gradient in the p-space. The equation of motion in the momentum representation is

$$i\hbar\frac{\partial}{\partial t}\Phi(\boldsymbol{p},t) = \left[\frac{p^2}{2m} + V(r)\right]\Phi(\boldsymbol{p},t) \tag{3.31}$$

For a one-dimensional system, the Fourier representation $\Psi(x, t)$ is given by

$$\Psi(x,\ t) = \frac{1}{\sqrt{2x}}\int_{-\infty}^{\infty}\Phi(k,t)\exp(ikx)\,dk \tag{3.32}$$

$$\Phi(k,\ t) = \frac{1}{\sqrt{2\pi}}\int_{-\infty}^{\infty}\Psi(x,t)\exp(-ikx)\,dk \tag{3.33}$$

Changing the variable from k to p, we get

$$\Psi(x,\ t) = \frac{1}{\sqrt{2\pi\hbar}}\int_{-\infty}^{\infty}\Phi(p,t)\exp\left(\frac{ipx}{\hbar}\right)dp \tag{3.34}$$

$$\Phi(p,\ t) = \frac{1}{\sqrt{2\pi\hbar}}\int_{-\infty}^{\infty}\Psi(x,t)\exp\left(-\frac{ipx}{\hbar}\right)dx \tag{3.35}$$

The probability density in the momentum representation is $|\Phi(\boldsymbol{p},\ t)|^2$.

PROBLEMS

3.1 A and B are two operators defined by $A\psi(x) = \psi(x) + x$ and $B\psi(x) = (d\psi/dx) + 2\psi(x)$. Check for their linearity.

Solution. An operator O is said to be linear if

$$O\,[c_1 f_1(x) + c_2 f_2(x)] = c_1 O f_1(x) + c_2 O f_2(x)$$

For the operator A,

$$A\,[c_1 f_1(x) + c_2 f_2(x)] = [c_1 f_1(x) + c_2 f_2(x)] + x$$

$$\text{LHS} = c_1 A f_1(x) + c_2 A f_2(x) = c_1 f_1(x) + c_2 f_2(x) + c_1 x + c_2 x$$

which is not equal to the RHS. Hence, the operator A is not linear.

$$B\,[c_1 f_1(x) + c_2 f_2(x)] = \frac{d}{dx}[c_1 f_1(x) + c_2 f_2(x)] + 2[c_1 f_1(x) + c_2 f_2(x)]$$

$$= c_1 \frac{d}{dx} f_1(x) + c_2 \frac{d}{dx} f_2(x) + 2c_1 f_1(x) + 2c_2 f_2(x)$$

$$= \frac{d}{dx} c_1 f_1(x) + 2c_1 f_1(x) + \frac{d}{dx} c_2 f_2(x) + 2c_2 f_2(x)$$

$$= c_1 B f_1(x) + c_2 B f_2(x)$$

Thus, the operator B is linear.

3.2 Prove that the operators $i(d/dx)$ and d^2/dx^2 are Hermitian.

Solution. Consider the integral $\int_{-\infty}^{\infty} \psi_m^* \left(i\frac{d}{dx} \right) \psi_n \, dx$. Integrating it by parts and remembering that ψ_m and ψ_n are zero at the end points, we get

$$\int_{-\infty}^{\infty} \psi_m^* \left(i\frac{d}{dx} \right) \psi_n \, dx = i[\psi_m^* \psi_n]_{-\infty}^{\infty} - i\int_{-\infty}^{\infty} \psi_n \frac{d}{dx} \psi_m^* \, dx$$

$$= \int_{-\infty}^{\infty} \left(i\frac{d}{dx} \psi_m \right)^* \psi_n \, dx$$

which is the condition for $i(d/dx)$ to be Hermitian. Therefore, id/dx is Hermitian.

$$\int_{-\infty}^{\infty} \psi_m^* \frac{d^2\psi_n}{dx^2} dx = \left[\psi_m^* \frac{d\psi_n}{dx} \right]_{-\infty}^{\infty} - \int_{-\infty}^{\infty} \frac{d\psi_n}{dx} \frac{d\psi_m^*}{dx} dx$$

$$= \left[\frac{d\psi_m^*}{dx} \psi_n \right]_{-\infty}^{\infty} + \int_{-\infty}^{\infty} \psi_n \frac{d^2\psi_m^*}{dx^2} dx = \int_{-\infty}^{\infty} \frac{d^2\psi_m^*}{dx^2} dx$$

Thus, d^2/dx^2 is Hermitian. The integrated terms in the above equations are zero since ψ_m and ψ_n are zero at the end points.

3.3 If A and B are Hermitian operators, show that (i) $(AB + BA)$ is Hermitian, and (ii) $(AB - BA)$ is non-Hermitian.

Solution.

(i) Since A and B are Hermitian, we have

$$\int \psi_m^* A\psi_n \, dx = \int A^* \psi_m^* \psi_n \, dx; \qquad \int \psi_m^* B\psi_n dx = \int B^* \psi_m^* \psi_n \, dx$$

$$\begin{aligned}\int \psi_m^*(AB + BA)\, \psi_n \, dx &= \int \psi_m^* \, AB\psi_n dx + \int \psi_m^* \, BA\psi_n dx \\ &= \int B^* A^* \psi_m^* \psi_n \, dx + \int A^* B^* \psi_m^* \psi_n \, dx \\ &= \int (AB + BA)^* \psi_m^* \psi_n \, dx\end{aligned}$$

Hence, $AB + BA$ is Hermitian.

(ii) $$\begin{aligned}\int \psi_m^*(AB - BA)\, \psi_n \, dx &= \int (B^*A^* - A^*B^*)\psi_m^* \psi_n \, dx \\ &= -\int (\mathrm{AB} - BA)^* \, \psi_m^* \psi_n \, dx\end{aligned}$$

Thus, $AB - BA$ is non-Hermitian.

3.4 If operators A and B are Hermitian, show that $i\,[A, B]$ is Hermitian. What relation must exist between operators A and B in order that AB is Hermitian?

Solution.

$$\begin{aligned}\int \psi_i^* i\,[A, B]\, \psi_n \, dx &= i \int \psi_m^* AB\psi_n \, dx - i \int \psi_m^* BA\psi_n \, dx \\ &= i \int B^* A^* \psi_m^* \psi_n \, dx - i \int A^* B^* \psi_m^* \psi_n \, dx \\ &= \int (i[A, B]\, \psi_m)^* \, \psi_n \, dx\end{aligned}$$

Hence, $i\,[A, B]$ is Hermitian.

For the product AB to be Hermitian, it is necessary that

$$\int \psi_m^* \, AB\psi_n dx = \int A^* B^* \psi_m^* \psi_n \, dx$$

Since A and B are Hermitian, this equation reduces to

$$\int B^* A^* \psi_m^* \psi_n \, dx = \int A^* B^* \psi_m^* \psi_n \, dx$$

which is possible only if $B^* A^* \psi_m^* = A^* B^* \psi_m^*$. Hence,

$$AB = BA$$

That is, for AB to be Hermitian, A must commute with B.

3.5 Prove the following commutation relations:

(i) $[[A, B], C] + [[B, C], A] + [[C, A], B] = 0.$

(ii) $\left[\dfrac{\partial}{\partial x}, \dfrac{\partial^2}{\partial x^2}\right]$

(iii) $\left[\dfrac{\partial}{\partial x}, F(x)\right]$

Solution.

(i) $[[A, B], C] + [[B, C], A] + [[C, A], B] = [A, B]\,C - C\,[A, B] + [B, C]\,A - A\,[B, C] + [C, A]\,B - B\,[C, A]$

$= ABC - BAC - CAB + CBA + BCA - CBA - ABC + ACB + CAB - ACB - BCA + BAC = 0$

(ii) $$\left[\frac{\partial}{\partial x}, \frac{\partial^2}{\partial x^2}\right]\psi = \left(\frac{\partial}{\partial x}\frac{\partial^2}{\partial x^2} - \frac{\partial^2}{\partial x^2}\frac{\partial}{\partial x}\right)\psi$$

$$= \left(\frac{\partial^3}{\partial x^3} - \frac{\partial^3}{\partial x^3}\right)\psi = 0$$

(iii) $$\left[\frac{\partial}{\partial x}, F(x)\right]\psi = \frac{\partial}{\partial x}(F\psi) - F\frac{\partial}{\partial x}\psi$$

$$= \frac{\partial F}{\partial x}\psi + F\frac{\partial \psi}{\partial x} - F\frac{\partial \psi}{\partial x} = \frac{\partial F}{\partial x}\psi$$

Thus, $$\left[\frac{\partial}{\partial x}, F(x)\right] = \frac{\partial F}{\partial x}$$

3.6 Show that the cartesian linear momentum components (p_1, p_2, p_3) and the cartesian components of angular momentum (L_1, L_2, L_3) obey the commutation relations (i) $[L_k, p_l] = i\hbar p_m$; (ii) $[L_k, p_k] = 0$, where k, l, m are the cyclic permutations of 1, 2, 3.

Solution.

(i) Angular momentum $L = \begin{vmatrix} \hat{k} & \hat{l} & \hat{m} \\ r_k & r_l & r_m \\ p_k & p_l & p_m \end{vmatrix}$

$$L_k = r_l p_m - r_m p_l = -i\hbar\left(r_l\frac{\partial}{\partial r_m} - r_m\frac{\partial}{\partial r_l}\right)$$

$$[L_k, p_l]\,\psi = -\hbar^2\left(r_l\frac{\partial}{\partial r_m} - r_m\frac{\partial}{\partial r_l}\right)\frac{\partial}{\partial r_l}\psi + \hbar^2\frac{\partial}{\partial r_l}\left(r_l\frac{\partial}{\partial r_m} - r_m\frac{\partial}{\partial r_l}\right)\psi$$

$$= -\hbar^2\left(r_l\frac{\partial}{\partial r_m}\frac{\partial \psi}{\partial r_l} - r_m\frac{\partial^2 \psi}{\partial r_l^2} - \frac{\partial \psi}{\partial r_m} - r_l\frac{\partial}{\partial r_l}\frac{\partial \psi}{\partial r_m} + r_m\frac{\partial^2 \psi}{\partial r_l^2}\right)$$

$$= \hbar^2\frac{\partial \psi}{\partial r_m} = i\hbar\left(-i\hbar\frac{\partial \psi}{\partial r_m}\right) = i\hbar p_m\psi$$

Hence, $[L_k, p_l] = i\hbar p_m$.

(ii) $$[L_k, p_k]\psi = -\hbar^2\left(r_l\frac{\partial}{\partial r_m} - r_m\frac{\partial}{\partial r_l}\right)\frac{\partial}{\partial r_k}\psi + \hbar^2\frac{\partial}{\partial r_k}\left(r_l\frac{\partial}{\partial r_m} - r_m\frac{\partial}{\partial r_l}\right)\psi = 0$$

$$= -\hbar^2\left(r_l\frac{\partial}{\partial r_m}\frac{\partial \psi}{\partial r_k} - r_m\frac{\partial}{\partial r_l}\frac{\partial \psi}{\partial r_k} - r_l\frac{\partial}{\partial r_k}\frac{\partial \psi}{\partial r_m} + r_m\frac{\partial}{\partial r_k}\frac{\partial \psi}{\partial r_l}\right) = 0$$

3.7 Show that (i) Operators having common set of eigenfunctions commute; (ii) commuting operators have common set of eigenfunctions.

Solution.

(i) Consider the operators A and B with the common set of eigenfunctions ψ_i, $i = 1, 2, 3, \ldots$ as

$$A\psi_i = a_i\psi_i, \qquad B\psi_i = b_i\psi_i$$

Then,

$$AB\psi_i = Ab_i\psi_i = a_ib_i\psi_i$$

$$BA\psi_i = Ba\psi_i = a_ib_i\psi_i$$

Since $AB\psi_i = BA\psi_i$, A commutes with B.

(ii) The eigenvalue equation for A is

$$A\psi_i = a_i\psi_i, \qquad i = 1, 2, 3, \ldots$$

Operating both sides from left by B, we get

$$BA\psi_i = a_iB\psi_i$$

Since B commutes with A,

$$AB\psi_i = a_iB\psi_i$$

i.e., $B\psi_i$ is an eigenfunction of A with the same eigenvalue a_i. If A has only nondegenerate eigenvalues, $B\psi_i$ can differ from ψ_i only by a multiplicative constant, say, b. Then,

$$B\psi_i = b_i\psi_i$$

i.e., ψ_i is a simultaneous eigenfunction of both A and B.

3.8 State the relation connecting the Poisson bracket of two dynamical variables and the value of the commutator of the corresponding operators. Obtain the value of the commutator $[x, p_x]$ and the Heisenberg's equation of motion of a dynamical variable which has no explicit dependence on time.

Solution. Consider the dynamical variables q and r. Let their operators in quantum mechanics be Q and R. Let $\{q, r\}$ be the Poisson bracket of the dynamical variables q and r. The relation connecting the Poisson bracket and the commutator of the corresponding operators is

$$[Q, R] = i\hbar\{q, r\} \tag{i}$$

The Poisson bracket $\{x, p_x\} = 1$. Hence,

$$[x, p_x] = i\hbar \tag{ii}$$

The equation of motion of a dynamical variable q in the Poisson bracket is

$$\frac{dq}{dt} = \{q, H\} \tag{iii}$$

Using Eq. (i), in terms of the operator Q, Eq. (iii) becomes

$$i\hbar\frac{dQ}{dt} = \{Q, H\} \tag{iv}$$

which is Heisenberg's equation of motion for the operator Q in quantum mechanics.

3.9 Prove the following commutation relations (i) $[L_k, r^2] = 0$, (ii) $[L_k, p^2] = 0$, where r is the radius vector, p is the linear momentum, and k, l, m are the cyclic permutations of 1, 2, 3.

Solution.

(i)
$$
\begin{aligned}
[L_k, r^2] &= [L_k, r_k^2 + r_l^2 + r_m^2] \\
&= [L_k, r_k^2] + [L_k, r_l^2] + [L_k, r_m^2] \\
&= r_k[L_k, r_k] + [L_k, r_k]r_k + r_l[L_k, r_l] + [L_k, r_l]r_l + r_m[L_k, r_m] + [L_k, r_m]r_m \\
&= 0 + 0 + r_l i\hbar r_m + i\hbar r_m r_l - r_m i\hbar r_l - i\hbar r_l r_m = 0
\end{aligned}
$$

(ii)
$$
\begin{aligned}
[L_k, p^2] &= [L_k, p_k^2] + [L_k, p_l^2] + [L_k, p_m^2] \\
&= p_k[L_k, p_k] + [L_k, p_k]p_k + p_l[L_k, p_l] + [L_k, p_l]p_l + p_m[L_k, p_m] + [L_k, p_m]p_m \\
&= 0 + 0 + i\hbar p_l p_m + i\hbar p_m p_l - i\hbar p_m p_l - i\hbar p_l p_m = 0
\end{aligned}
$$

3.10 Prove the following commutation relations:

(i) $[x, p_x] = [y, p_y] = [z, p_z] = i\hbar$

(ii) $[x, y] = [y, z] = [z, x] = 0$

(iii) $[p_x, p_y] = [p_y, p_z] = [p_z, p_x] = 0$

Solution.

(i) Consider the commutator $[x, p_x]$. Replacing x and p_x by the corresponding operators and allowing the commutator to operate on the function $\psi(x)$, we obtain

$$
\begin{aligned}
\left[x, -i\hbar\frac{d}{dx}\right]\psi(x) &= -i\hbar x\frac{d\psi}{dx} + i\hbar\frac{d(x\psi)}{dx} \\
&= -i\hbar x\frac{d\psi}{dx} + i\hbar\psi + i\hbar x\frac{d\psi}{dx} \\
&= i\hbar\psi
\end{aligned}
$$

Hence,

$$\left[x, -i\hbar\frac{d}{dx}\right] = [x, p_x] = i\hbar$$

Similarly,

$$[y, p_y] = [z, p_z] = i\hbar$$

(ii) Since the operators representing coordinates are the coordinates themselves,

$$[x, y] = [y, z] = [z, x] = 0$$

(iii)
$$
\begin{aligned}
[p_x, p_y]\,\psi(x, y) &= \left[-i\hbar\frac{\partial}{\partial x}, -i\hbar\frac{\partial}{\partial y}\right]\psi(x, y) \\
&= -\hbar^2\left[\frac{\partial^2}{\partial x\,\partial y} - \frac{\partial^2}{\partial y\,\partial x}\right]\psi(x, y)
\end{aligned}
$$

The right-hand side is zero as the order of differentiation can be changed. Hence the required result.

3.11 Prove the following:

(i) If ψ_1 and ψ_2 are the eigenfunctions of the operator A with the same eigenvalue, $c_1\psi_1 + c_2\psi_2$ is also an eigenfunction of A with the same eigenvalue, where c_1 and c_2 are constants.

(ii) If ψ_1 and ψ_2 are the eigenfunctions of the operator A with distinct eigenvalues, then $c_1\psi_1 + c_2\psi_2$ is not an eigenfunction of the operator A, c_1 and c_2 being constants.

Solution.

(i) We have

$$A\psi_1 = a_1\psi_1, \qquad A\psi_2 = a_1\psi_2$$
$$A(c_1\psi_1 + c_2\psi_2) = Ac_1\psi_1 + Ac_2\psi_2$$
$$= a_1(c_1\psi_1 + c_2\psi_2)$$

Hence, the required result.

(ii)
$$A\psi_1 = a_1\psi_1, \quad \text{and} \quad A\psi_2 = a_2\psi_2$$
$$A(c_1\psi_1 + c_2\psi_2) = Ac_1\psi_1 \mathrel{+}\text{m}\; Ac_2\psi_2$$
$$= a_1c_1\psi_1 + a_2c_2\psi_2$$

Thus, $c_1\psi_1 + c_2\psi_2$ is not an eigenfunction of the operator A.

3.12 For the angular momentum components L_x and L_y, check whether $L_xL_y + L_yL_x$ is Hermitian.

Solution. Since $i(d/dx)$ is Hermitian (Problem 3.2), $i(d/dy)$ and $i(d/dz)$ are Hermitian. Hence L_x and L_y are Hermitian. Since L_x and L_y are Hermitian,

$$\int \psi_m^*(L_xL_y + L_yL_x)\psi_n\,dx = \int (L_y^*L_x^* + L_x^*L_y^*)\psi_m^*\psi_n\,dx$$
$$= \int (L_xL_y + L_yL_x)^*\psi_m^*\psi_n\,dx$$

Thus, $L_xL_y + L_yL_x$ is Hermitian.

3.13 Check whether the operator $-i\hbar x(d/dx)$ is Hermitian.

Solution.

$$\int \psi_m^*\left(i\hbar x\frac{d}{dx}\right)\psi_n\,dx = \int \psi_m^* x\left(-i\hbar\frac{d}{dx}\right)\psi_n\,dx$$
$$= \int \left(-i\hbar\frac{d}{dx}\right)^* x^*\psi_m^*\psi_n\,dx$$
$$\neq \int \left(-i\hbar x\frac{d}{dx}\right)^* \psi_m^*\psi_n\,dx$$

Hence the given operator is not Hermitian.

3.14 If x and p_x are the coordinate and momentum operators, prove that $[x, p_x^n] = ni\hbar p_x^{n-1}$.

Solution.

$$[x, p_x^n] = [x, p_x^{n-1}p_x] = [x, p_x]\,p_x^{n-1} + p_x[x, p_x^{n-1}]$$
$$= i\hbar p_x^{n-1} + p_x([x, p_x]\,p_x^{n-2} + p_x[x, p_x^{n-2}])$$
$$= 2i\hbar p_x^{n-1} + p_x^2([x, p_x]\,p_x^{n-3} + p_x[x, p_x^{n-3}])$$
$$= 3i\hbar p_x^{n-1} + p_x^3[x, p_x^{n-3}]$$

Continuing, we have $[x, p_x^n] = ni\hbar p_x^{n-1}$

3.15 Show that the cartesian coordinates (r_1, r_2, r_3) and the cartesian components of angular momentum (L_1, L_2, L_3) obey the commutation relations.

(i) $[L_k, r_l] = i\hbar r_m$

(ii) $[L_k, r_k] = 0$, where k, l, m are cyclic permutations of 1, 2, 3.

Solution.

$$\text{(i)}\ [L_k, r_l]\psi = (L_k r_l - r_l L_k)\psi = -i\hbar\left[\left(r_l\frac{\partial}{\partial r_m} - r_m\frac{\partial}{\partial r_l}\right)r_l\psi - r_l\left(r_l\frac{\partial}{\partial r_m} - r_m\frac{\partial}{\partial r_l}\right)\psi\right]$$

$$= -i\hbar\left[r_l^2\frac{\partial\psi}{\partial r_m} - r_m\psi - r_m r_l\frac{\partial\psi}{\partial r_l} - r_l^2\frac{\partial\psi}{\partial r_m} + r_l r_m\frac{\partial\psi}{\partial r_l}\right]$$

$$= i\hbar r_m\psi$$

Hence, $[L_k, r_l] = i\hbar r_m$.

$$\text{(ii)}\ [L_k, r_k]\psi = -i\hbar\left[\left(r_l\frac{\partial}{\partial r_m} - r_m\frac{\partial}{\partial r_l}\right)r_k\psi - r_k\left(r_l\frac{\partial}{\partial r_m} - r_m\frac{\partial}{\partial r_l}\right)\psi\right] = 0$$

Thus, $[L_k, r_k] = 0$.

3.16 Show that the commutator $[x, [x, H]] = -\hbar^2/m$, where H is the Hamiltonian operator.

Solution.

$$\text{Hamiltonian } H = \frac{(p_x^2 + p_y^2 + p_z^2)}{2m}$$

Since

$$[x, p_y] = [x, p_z] = 0, \qquad [x, p_x] = i\hbar$$

we have

$$[x, H] = \frac{1}{2m}[x, p_x^2] = \frac{1}{2m}(p_x[x, p_x] + [x, p_x]p_x)$$

$$[x, H] = \frac{1}{2m}2i\hbar p_x = \frac{i\hbar}{m}p_x$$

$$[x, [x, H]] = \left[x, \frac{i\hbar p_x}{m}\right] = \frac{i\hbar}{m}[x, p_x] = -\frac{\hbar^2}{m}$$

3.17 Prove the following commutation relations in the momentum representation:

(i) $[x, p_x] = [y, p_y] = [z, p_z] = i\hbar$

(ii) $[x, y] = [y, z] = [z, x] = 0$

Solution.

$$\text{(i)}\ [x, p_x]f(p_x) = \left[i\hbar\frac{\partial}{\partial p_x}, p_x\right]f(p_x)$$

$$= i\hbar\frac{\partial}{\partial p_x}(p_x f) - i\hbar p_x\frac{\partial}{\partial p_x}f = i\hbar f$$

$[x, p_x] = i\hbar$

Similarly, $[y, p_y] = [z, p_z] = i\hbar$

$$\text{(ii)}\ [x, y]\, f(p_x, p_y) = (i\hbar)^2 \left[\frac{\partial}{\partial p_x}, \frac{\partial}{\partial p_y}\right] f(p_x, p_y)$$

$$= -\hbar^2 \left[\frac{\partial}{\partial p_x}\frac{\partial}{\partial p_y} - \frac{\partial}{\partial p_y}\frac{\partial}{\partial p_x}\right] f(p_x, p_y) = 0$$

since the order of differentiation can be changed. Hence, $[x, y] = 0$. Similarly, $[y, z] = [z, x] = 0$.

3.18 Evaluate the commutator (i) $[x, p_x^2]$, and (ii) $[xyz, p_x^2]$.

Solution.

(i) $[x, p_x^2] = [x, p_x]\, p_x + p_x\, [x, p_x]$

$$= i\hbar p_x + i\hbar p_x = 2i\hbar p_x$$

$$= 2i\hbar\left(-i\hbar \frac{d}{dx}\right) = 2\hbar^2 \frac{d}{dx}$$

(ii) $[xyz, p_x^2] = [xyz, p_x]\, p_x + p_x\, [xyz, px]$

$$= xy\,[z, p_x]\, p_x + [xy, p_x]\, zp_x + p_x x_y\, [z, p_x] + p_x\, [xy, p_x]\, z$$

Since $[z, p_x]$, the first and third terms on the right-hand side are zero. So,

$$[xyz, p_x^2] = x[y, px]\, zp_x + [x, px]\, yzp_x + p_x\, x[y, px]z + p_x\, [x, p_x]\, yz$$

The first and third terms on the right-hand side are zero since $[y, p_x] = 0$. Hence,

$$[xyz, p_x^2] = i\hbar yzp_x + i\hbar p_x yz = 2i\hbar yzp_x$$

where we have used the result

$$\frac{d}{dx}\,[yz\, f(x)] = yz\frac{\partial}{\partial x} f(x)$$

Substituting the operator for p_x, we get

$$[xyz, p_x^2] = 2\hbar^2 yz\frac{\partial}{\partial x}$$

3.19 Find the value of the operator products

(i) $\left(\dfrac{d}{dx} + x\right)\left(\dfrac{d}{dx} + x\right)$

(ii) $\left(\dfrac{d}{dx} + x\right)\left(\dfrac{d}{dx} - x\right)$

Solution.

(i) Allowing the product to operate on $f(x)$, we have

$$\left(\frac{d}{dx} + x\right)\left(\frac{d}{dx} + x\right) f(x) = \left(\frac{d}{dx} + x\right)\left(\frac{df}{dx} + xf\right)$$

$$= \frac{d^2 f}{dx^2} + x\frac{df}{dx} + f + x\frac{df}{dx} + x^2 f$$

$$= \left(\frac{d^2}{dx^2} + 2x\frac{d}{dx} + x^2 + 1\right) f$$

Dropping the arbitrary function $f(x)$, we get

$$\left(\frac{d}{dx} + x\right)\left(\frac{d}{dx} + x\right) = \frac{d^2}{dx^2} + 2x\frac{d}{dx} + x^2 + 1$$

(ii)
$$\left(\frac{d}{dx} + x\right)\left(\frac{d}{dx} - x\right)f = \left(\frac{d}{dx} + x\right)\left(\frac{df}{dx} - xf\right)$$

$$= \frac{d^2f}{dx^2} - x\frac{df}{dx} - f + x\frac{df}{dx} - x^2 f$$

$$\left(\frac{d}{dx} + x\right)\left(\frac{d}{dx} - x\right) = \frac{d^2}{dx^2} - x^2 - 1$$

3.20 By what factors do the operators $(x^2p_x^2 + p_x^2x^2)$ and $1/2(xp_x + p_xx)^2$ differ?

Solution. Allowing the operators to operate on the function f, we obtain

$$(x^2p_x^2 + p_x^2x^2)f = -\hbar^2\left[x^2\frac{\partial^2 f}{\partial x^2} + \frac{\partial^2(x^2f)}{\partial x^2}\right]$$

$$= -\hbar^2x^2\frac{\partial^2 f}{\partial x^2} - \hbar^2\frac{\partial}{\partial x}\frac{\partial(x^2 f)}{\partial x}$$

$$= -\hbar^2x^2\frac{\partial^2 f}{\partial x^2} - \hbar^2\frac{\partial}{\partial x}\left(2xf + x^2\frac{\partial f}{\partial x}\right)$$

$$= -\hbar^2\left(x^2\frac{\partial^2 f}{\partial x^2} + 2f + 2x\frac{\partial f}{\partial x} + x^2\frac{\partial^2 f}{\partial x^2} + 2x\frac{\partial f}{\partial x}\right)$$

$$= -\hbar^2\left(2x^2\frac{\partial^2}{\partial x^2} + 4x\frac{\partial}{\partial x} + 2\right)f$$

$$\frac{1}{2}(xp_x + p_xx)^2 f = -\frac{i\hbar}{2}(xp_x + p_xx)\left[x\frac{\partial f}{\partial x} + \frac{\partial(xf)}{\partial x}\right]$$

$$= -\frac{i\hbar}{2}(xp_x + p_xx)\left(2x\frac{\partial f}{\partial x} + f\right)$$

$$= -\frac{\hbar^2}{2}\left[x\frac{\partial}{\partial x}\left(2x\frac{\partial f}{\partial x}\right) + x\frac{\partial f}{\partial x} + \frac{\partial}{\partial x}\left(2x^2\frac{\partial f}{\partial x}\right) + \frac{\partial(xf)}{\partial x}\right]$$

$$= -\frac{\hbar^2}{2}\left(2x^2\frac{\partial^2 f}{\partial x^2} + 2x\frac{\partial f}{\partial x} + x\frac{\partial f}{\partial x} + 2x^2\frac{\partial^2 f}{\partial x^2} + 4x\frac{\partial f}{\partial x} + x\frac{\partial f}{\partial x} + f\right)$$

$$= -\frac{\hbar^2}{2}\left(8x\frac{\partial f}{\partial x} + 2x^2\frac{\partial^2 f}{\partial x^2} + f\right)$$

$$= -\hbar^2\left(2x^2\frac{\partial^2}{\partial x^2} + 4x\frac{\partial}{\partial x} + \frac{1}{2}\right)f$$

The two operators differ by a term $-(3/2)\hbar^2$.

3.21 The Laplace transform operator L is defined by $Lf(x) = \int_0^\infty e^{-sx} f(x)\,dx$

(i) Is the operator L linear?
(ii) Evaluate Le^{ax} if $s > a$.

Solution.
(i) Consider the function $f(x) = c_1 f_1(x) + c_2 f_2(x)$, where c_1 and c_2 are constants. Then,

$$L[c_1 f_1(x) + c_2 f_2(x)] = \int_0^\infty e^{-sx}[c_1 f_1(x) + c_2 f_2(x)]\,dx$$

$$= c_1 \int_0^\infty e^{-sx} f_1(x)\,dx + c_2 \int_0^\infty e^{-sx} f_2(x)\,dx$$

$$= c_1 L f_1(x) + c_2 L f_2(x)$$

Thus, the Laplace transform operator L is linear.

(ii) $$Le^{ax} = \int_0^\infty e^{-sx} e^{ax} dx = \int_0^\infty e^{-(s-a)x}\,dx = \left.\frac{e^{-(s-a)x}}{-(s-a)}\right]_0^\infty = \frac{1}{s-a}$$

3.22 The operator e^A is defined by

$$e^A = 1 + A + \frac{A^2}{2!} + \frac{A^3}{3!} + \cdots$$

Show that $e^D = T_1$, where $D = (d/dx)$ and T_1 is defined by $T_1 f(x) = f(x + 1)$

Solution. In the expanded form,

$$e^D = 1 + \frac{d}{dx} + \frac{1}{2!}\frac{d^2}{dx^2} + \frac{1}{3!}\frac{d^3}{dx^3} + \cdots \qquad \text{(i)}$$

$$e^D f(x) = f(x) + f'(x) + \frac{1}{2!} f''(x) + \frac{1}{3!} f'''(x) + \cdots \qquad \text{(ii)}$$

where the primes indicate differentiation. We now have

$$T_1 f(x) = f(x + 1) \qquad \text{(iii)}$$

Expanding $f(x + 1)$ by Taylor series, we get

$$f(x + 1) = f(x) + f'(x) + \frac{1}{2!} f''(x) + \cdots \qquad \text{(iv)}$$

From Eqs. (i), (iii) and (iv), we can write

$$e^D f(x) = T_1 f(x) \quad \text{or} \quad e^D = T_1$$

3.23 If an operator A is Hermitian, show that the operator $B = iA$ is anti-Hermitian. How about the operator $B = -iA$?

Solution. When A is Hermitian,

$$\int \psi^* A\psi\, d\tau = \int (A\psi)^* \psi\, d\tau$$

For the operator $B = iA$, consider the integral

$$\int \psi^* B\psi \, d\tau = \int \psi^* iA\psi \, d\tau$$

$$= i\int \psi^* A\psi \, d\tau = i\int A^* \psi^* \psi \, d\tau$$

$$= -\int (iA\psi)^* \psi \, d\tau = -\int (B\psi)^* \psi \, d\tau$$

Hence, $B = iA$ is anti-Hermitian. When $B = -iA$,

$$\int \psi^* B\psi \, d\tau = -i\int A^* \psi^* \psi \, d\tau$$

$$= \int (iA)^* \psi^* \psi \, d\tau$$

Thus, $B = -iA$ is Hermitian.

3.24 Find the eigenvalues and eigenfunctions of the operator d/dx.

Solution. The eigenvalue-eigenfunction equation is

$$\frac{d}{dx}\psi(x) = k\psi(x)$$

where k is the eigenvalue and $\psi(x)$ is the eigenfunction. This equation can be rewritten as

$$\frac{d\psi}{\psi} = k\,dx$$

Integrating $\ln \psi = kx + \ln c$, we get

$$\ln\left(\frac{\psi}{c}\right) = kx, \qquad \psi = ce^{kx}$$

where c and k are constants. If k is a real positive quantity, ψ is not an acceptable function since it tends to ∞ or $-\infty$ as $x \to \infty$ or $-\infty$. When k is purely imaginary, say ia,

$$\psi = ce^{iax}$$

The function ψ will be finite for all real values of a. Hence, $y = ce^{kx}$ is the eigenfunction of the operator d/dx with eigenvalues $k = ia$, where a is real.

3.25 Find the Hamiltonian operator of a charged particle in an electromagnetic field described by the vector potential $\boldsymbol{A}$ and the scalar potential ϕ.

Solution. The classical Hamiltonian of a charged particle in an electromagnetic field is given by

$$H = \frac{1}{2m}\left(\boldsymbol{p} - \frac{e}{c}\boldsymbol{A}\right)^2 + e\phi$$

Replacing $\boldsymbol{p}$ by its operator $-i\hbar\nabla$ and allowing the resulting operator equation to operate on function $f(r)$, we obtain

$$Hf(r) = \frac{1}{2m}\left(-i\hbar\nabla - \frac{e}{c}\boldsymbol{A}\right)\left(-i\hbar\nabla - \frac{e}{c}\boldsymbol{A}\right) f(r) + e\phi f(r)$$

$$= \frac{1}{2m}\left(-i\hbar\nabla - \frac{e}{c}\boldsymbol{A}\right)\left(-i\hbar\nabla f - \frac{e}{c}\boldsymbol{A}f\right) + e\phi f$$

$$= \frac{1}{2m}\left[-\hbar^2\nabla^2 f + \frac{ie\hbar}{c}\nabla(\boldsymbol{A}f) + \frac{ie\hbar}{c}\boldsymbol{A}\nabla f + \frac{e^2}{c^2}A^2 f\right] + e\phi f$$

$$= \frac{1}{2m}\left[-\hbar^2\nabla^2 f + \frac{ie\hbar}{c}(\nabla\cdot\boldsymbol{A})f + \frac{ie\hbar}{c}\boldsymbol{A}\cdot\nabla f + \frac{ie\hbar}{c}\boldsymbol{A}\cdot\nabla f + \frac{e^2}{c^2}A^2 f\right] + e\phi f$$

$$= \left[-\frac{\hbar^2}{2m}\nabla^2 + \frac{ie\hbar}{2mc}\nabla\cdot\boldsymbol{A} + \frac{ie\hbar}{mc}\boldsymbol{A}\cdot\nabla + \frac{e^2}{2mc^2}\boldsymbol{A}^2 + e\phi\right] f$$

Hence, the operator representing the Hamiltonian is

$$H = -\frac{\hbar^2}{2m}\nabla^2 + \frac{ie\hbar}{2mc}\nabla\cdot\boldsymbol{A} + \frac{ie\hbar}{mc}\boldsymbol{A}\cdot\nabla + \frac{e^2}{2mc^2}\boldsymbol{A}^2 + e\phi$$

3.26 The wavefunction of a particle in a state is $\psi = N\exp(-x^2/2\alpha)$, where $N = (1/\pi\alpha)^{1/4}$. Evaluate $(\Delta x)\,(\Delta p)$.

Solution. For evaluating $(\Delta x)\,(\Delta p)$, we require the values of $\langle x\rangle$, $\langle x^2\rangle$, $\langle p\rangle$ and $\langle p^2\rangle$. Since ψ is symmetrical about $x = 0$, $\langle x\rangle = 0$. Now,

$$\langle x^2\rangle = N^2\int_{-\infty}^{\infty} x^2\exp\left(\frac{-x^2}{\alpha}\right)dx = \frac{\alpha}{2}$$

$$\langle p\rangle = -i\hbar N^2\int_{-\infty}^{\infty}\exp\left(\frac{-x^2}{2\alpha}\right)\frac{d}{dx}\exp\left(\frac{-x^2}{2\alpha}\right)dx$$

$$= \text{constant}\int_{-\infty}^{\infty} x\exp\left(\frac{-x^2}{\alpha}\right)dx$$

$= 0$ since the integral is odd.

$$\langle p^2\rangle = (-i\hbar)^2 N^2\int_{-\infty}^{\infty}\exp\left(\frac{-x^2}{2\alpha}\right)\frac{d^2}{dx^2}\exp\left(\frac{-x^2}{2\alpha}\right)dx$$

$$= \frac{\hbar^2N^2}{\alpha}\int_{-\infty}^{\infty}\exp\left(\frac{-x^2}{\alpha}\right)dx - \frac{\hbar^2N^2}{\alpha^2}\int_{-\infty}^{\infty}x^2\exp\left(\frac{-x^2}{\alpha}\right)dx$$

$$= \frac{\hbar^2}{\alpha} - \frac{\hbar^2}{2\alpha} = \frac{\hbar^2}{2\alpha}$$

Refer the Appendix. Also,

$$(\Delta x)^2 (\Delta p)^2 = \langle x^2 \rangle \langle p^2 \rangle = \frac{\alpha}{2} \frac{\hbar^2}{2\alpha} = \frac{\hbar^2}{4}$$

$$(\Delta x)(\Delta p) = \frac{\hbar}{2}$$

3.27 Show that the linear momentum is not quantized.

Solution. The operator for the x-component of linear momentum is $-i\hbar\,(d/dx)$. Let $\psi_k(x)$ be its eigenfunction corresponding to the eigenvalue a_k. The eigenvalue equation is

$$-i\hbar \frac{d}{dx} \psi_k(x) = a_k \psi_k(x)$$

$$\frac{d\psi_k(x)}{\psi_k(x)} = \frac{i}{\hbar} a_k\, dx$$

Integrating, we get

$$\psi_k(x) = C \exp\left(\frac{i}{\hbar} a_k x\right)$$

where C is a constant. The function $\psi_k(x)$ will be finite for all real values of a_k. Hence, all real values of a_k are proper eigenvalues and they form a continuous spectrum. In other words, the linear momentum is not quantized.

3.28 Can we measure the kinetic and potential energies of a particle simultaneously with arbitrary precision?

Solution. The operator for kinetic energy, $T = -(\hbar^2/2m)\,\nabla^2$. The Operator for potential energy, $V = V(r)$. Hence,

$$\left[-\frac{\hbar^2}{2m}\nabla^2, V\right]\psi = -\frac{\hbar^2}{2m}\nabla^2(V\psi) - V\left(-\frac{\hbar^2}{2m}\nabla^2\right)\psi$$

$$= -\frac{\hbar^2}{2m}(\nabla^2 V)\psi \neq 0$$

Since the operators of the two observables do not commute, simultaneous measurement of both is not possible. Simultaneous measurement is possible if V is constant or linear in coordinates.

3.29 If the wave function for a system is an eigenfunction of the operator associated with the observable A, show that $\langle A^n \rangle = \langle A \rangle^n$.

Solution. Let the eigenfunctions and eigenvalues of the operator A associated with the observable A be ψ and α, respectively. Then,

$$\langle A^n \rangle = \int \psi^* A^n \psi\, d\tau = \int \psi^* A^{n-1} A\psi\, d\tau$$

$$= \alpha \int \psi^* A^{n-1} \psi\, d\tau = \alpha^2 \int \psi^* A^{n-2} \psi\, d\tau$$

$$= \alpha^n \int \psi^* \psi\, d\tau = \alpha^n$$

$$\langle A^n \rangle = \left(\int \psi^* A\psi\, d\tau\right)^n = \left(\alpha \int \psi^* \psi\, d\tau\right)^n = \alpha_n$$

Thus, $\langle A^n \rangle = \langle A \rangle^n$.

3.30 The wave function ψ of a system is expressed as a linear combination of normalized eigenfunctions ϕ_i, $i = 1, 2, 3, \ldots$ of the operator α of the observable A as $\psi = \sum_i c_i \phi_i$. Show that

$$\langle A^n \rangle = \sum_i |c_i|^2\, a_i^n, \qquad \alpha\phi_i = a_i \phi_i, \qquad i = 1, 2, 3, \ldots$$

Solution.

$$\psi = \sum_i c_i \phi_i, \qquad c_i = \int_{-\infty}^{\infty} \phi_i^* \psi^*\, d\tau, \qquad i = 1, 2, 3, \ldots$$

$$\langle A^n \rangle = \int_{-\infty}^{\infty} \psi^* \alpha^n \psi\, d\tau = \sum_i \sum_j c_i^* c_j \int_{-\infty}^{\infty} \phi_i^* \alpha^n \phi_j\, d\tau$$

$$= \sum_i \sum_j c_i^* c_j a_j^n \int_{-\infty}^{\infty} \phi_i^* \phi_j\, d\tau = \sum_i |c_i|^2\, a_i^n$$

since the ϕ's are orthogonal.

3.31 The Hamiltonian operator of a system is $H = -(d^2/dx^2) + x^2$. Show that $Nx \exp(-x^2/2)$ is an eigenfunction of H and determine the eigenvalue. Also evaluate N by normalization of the function.

Solution.

$$\psi = Nx \exp(-x^2/2),\ N \textit{ being a constant}$$

$$H\psi = \left(-\frac{d^2}{dx^2} + x^2\right) Nx \exp\left(-\frac{x^2}{2}\right)$$

$$= Nx^3 \exp\left(-\frac{x^2}{2}\right) \frac{d}{dx}\left[\exp\left(-\frac{x^2}{2}\right) - x^2 \exp\left(-\frac{x^2}{2}\right)\right]$$

$$= 3Nx \exp\left(-\frac{x^2}{2}\right) = 3\psi$$

Hence, the eigenvalue of H is 3. The normalization condition gives

$$N^2 \int_{-\infty}^{\infty} x^2 e^{-x^2}\, dx = 1$$

$$N^2 \frac{\sqrt{\pi}}{2} = 1 \text{ (refer the Appendix)}$$

$$N = \left(\frac{2}{\sqrt{\pi}}\right)^{1/2}$$

The normalized function $\psi = \left(\dfrac{2}{\sqrt{\pi}}\right)^{1/2} x \exp\left(-\dfrac{x^2}{2}\right)$.

3.32 If A is a Hermitian operator and ψ is its eigenfunction, show that (i) $\langle A^2 \rangle = \int |A\psi|^2 \, d\tau$ and (ii) $\langle A^2 \rangle \geq 0$.

Solution.

(i) Let the eigenvalue equation for the operator be

$$A\psi = a\psi$$

Let us assume that ψ is normalized and a is real. Since the operator A is Hermitian,

$$\langle A^2 \rangle = \int \psi^* A^2 \psi \, d\tau = \int A^* \psi^* A\psi \, d\tau$$

$$= \int |A\psi|^2 \, d\tau$$

(ii) Replacing $A\psi$ by $a\psi$, we get

$$\langle A^2 \rangle = \int |a\psi|^2 \, d\tau = \int |a|^2 |\psi|^2 \, d\tau$$

$$= |a|^2 \int |\psi|^2 \, d\tau = |a|^2$$

$$\geq 0$$

3.33 Find the eigenfunctions and nature of eigenvalues of the operator

$$\frac{d^2}{dx^2} + \frac{2}{x}\frac{d}{dx}$$

Solution. Let ψ be the eigenfunction corresponding to the eigenvalue λ. Then the eigenvalue equation is given by

$$\left(\frac{d^2}{dx^2} + \frac{2}{x}\frac{d}{dx} \right) \psi = \lambda\psi$$

Consider the function $u = x\psi$. Differentiating with respect to x, we get

$$\frac{du}{dx} = \psi + x\frac{d\psi}{dx}$$

$$\frac{d^2u}{dx^2} = \frac{d\psi}{dx} + \frac{d\psi}{dx} + x\frac{d^2\psi}{dx^2} = 2\frac{d\psi}{dx} + x\frac{d^2\psi}{dx^2}$$

Dividing throughout by x, we obtain

$$\frac{1}{x}\frac{d^2u}{dx^2} = \left(\frac{2}{x}\frac{d}{dx} + \frac{d^2}{dx^2} \right) \psi$$

Combining this equation with the first of the above two equations, we have

$$\frac{1}{x}\frac{d^2u}{dx^2} = \lambda\psi \quad \text{or} \quad \frac{d^2u}{dx^2} = \lambda u$$

The solution of this equation is

$$u = c_1 e^{\sqrt{\lambda} x} + c_2 e^{-\sqrt{\lambda} x}$$

where c_1 and c_2 are constants.

For u to be a physically acceptable function, $\sqrt{\lambda}$ must be imaginary, say, $i\beta$. Also, at $x = 0$, $u = 0$. Hence, $c_1 + c_2 = 0$, $c_1 = -c_2$. Consequently,

$$u = c_1 (e^{i\beta x} - e^{-i\beta x}), \qquad \psi = \frac{1}{x} c_1 (e^{i\beta x} - e^{-i\beta x})$$

$$\psi = c \frac{\sin \beta x}{x}$$

3.34 (i) Prove that the function $\psi = \sin (k_1 x) \sin (k_2 y) \sin (k_3 z)$ is an eigenfunction of the Laplacian operator and determine the eigenvalue. (ii) Show that the function $\exp (i\boldsymbol{k} \cdot \boldsymbol{r})$ is simultaneously an eigenfunction of the operators $-i\hbar\nabla$ and $-\hbar^2\nabla^2$ and find the eigenvalues.

Solution.

(i) The eigenvalue equation is

$$\nabla^2 \psi = \left(\frac{\partial^2}{\partial x^2} + \frac{\partial^2}{\partial y^2} + \frac{\partial^2}{\partial z^2} \right) \sin k_1 x \sin k_2 y \sin k_3 z$$

$$= -(k_1^2 + k_2^2 + k_3^2) \sin k_1 x \sin k_2 y \sin k_3 z$$

Hence, ψ is an eigenfunction of the Laplacian operator with the eigenvalue $-(k_1^2 + k_2^2 + k_3^2)$.

(ii)

$$-i\hbar\nabla e^{i(\boldsymbol{k} \cdot \boldsymbol{r})} = \hbar \boldsymbol{k} e^{i\boldsymbol{k} \cdot \boldsymbol{r}}$$

$$-\hbar^2\nabla^2 e^{i(\boldsymbol{k} \cdot \boldsymbol{r})} = +\hbar^2 k^2 e^{i(\boldsymbol{k} \cdot \boldsymbol{r})}$$

That is, $\exp (i\boldsymbol{k} \cdot \boldsymbol{r})$ is a simultaneous eigenfunction of the operators $-i\hbar\nabla$ and $-\hbar^2\nabla^2$, with eigenvalues $\hbar k$ and $\hbar^2 k^2$, respectively.

3.35 Obtain the form of the wave function for which the uncertainty product $(\Delta x)(\Delta p) = \hbar/2$.

Solution. Consider the Hermitian operators A and B obeying the relation

$$[A, B] = iC \tag{i}$$

For an operator R, we have (refer Problem 3.30)

$$\int |R\psi|^2 \, d\tau \geq 0 \tag{ii}$$

Then, for the operator $A + imB$, m being an arbitrary real number,

$$\int (A - imB)^* \psi^* (A + imB)\psi \, d\tau \geq 0 \tag{iii}$$

Since A and B are Hermitian, Eq. (iii) becomes

$$\int \psi^*(A - imB)(A + imB)\psi \, d\tau \geq 0$$

$$\int \psi^*(A^2 - mC + m^2B^2)\psi \, d\tau \geq 0$$

$$\langle A^2 \rangle - m\langle C \rangle + m^2 \langle B^2 \rangle \geq 0 \tag{iv}$$

The value of m, for which the LHS of Eq. (iv) is minimum, is when the derivative on the LHS with respect to m is zero, i.e.,

$$0 = -\langle C \rangle + 2m \langle B^2 \rangle \quad \text{or} \quad m = \frac{\langle C \rangle}{2\langle B^2 \rangle} \tag{v}$$

When the LHS of (iv) is minimum,

$$(A + imB)\,\psi = 0 \tag{vi}$$

Since

$$[A - \langle A\rangle, B - \langle B\rangle] = [A, B] = iC$$

Eq. (vi) becomes

$$[(A - \langle A\rangle) + im\,(B - \langle B\rangle)]\,\psi = 0 \tag{vii}$$

Identifying x with A and p with B, we get

$$[(x - \langle x\rangle) + im\,(p - \langle p\rangle)]\;\psi = 0, \qquad m = \frac{\hbar}{2(\Delta p)^2}$$

Substituting the value of m and repalcing p by $-i\hbar(d/dx)$, we obtain

$$\frac{d\psi}{dx} + \left[\frac{2(\Delta p)^2}{\hbar^2}(x - \langle x\rangle) - \frac{i\langle p\rangle}{\hbar}\right]\psi = 0$$

$$\frac{d\psi}{\psi} = -\left[\frac{2(\Delta p)^2}{\hbar^2}(x - \langle x\rangle) - \frac{i\langle p\rangle}{\hbar}\right]dx$$

Integrating and replacing Δp by $\hbar/2(\Delta x)$, we have

$$\ln\,\psi = -\frac{2(\Delta p)^2}{\hbar^2}\,\frac{(x - \langle x\rangle)^2}{2} + \frac{i\langle p\rangle x}{\hbar} + \ln N$$

$$\psi = N\exp\left[-\frac{(x - \langle x\rangle)^2}{4(\Delta x)^2} + \frac{i\langle p\rangle x}{\hbar}\right]$$

Normalization of the wave function is straightforward, which gives

$$\psi = \left(\frac{1}{2\pi(\Delta x)^2}\right)^{1/4}\exp\left[-\frac{(x - \langle x\rangle)^2}{4(\Delta x)^2} + \frac{i\langle p\rangle x}{\hbar}\right]$$

3.36 (i) Consider the wave function

$$\psi(x) = A\exp\left(-\frac{x^2}{a^2}\right)\exp\,(ikx)$$

where A is a real constant: (i) Find the value of A; (ii) calculate $\langle p\rangle$ for this wave function.

Solution.

(i) The normalization condition gives

$$A^2\int_{-\infty}^{\infty}\exp\left(-\frac{2x^2}{a^2}\right)dx = 1$$

$$A^2\left(\frac{\pi}{2/a^2}\right)^{1/2} = 1 \quad\text{or}\quad A^2\left(\frac{\pi}{2}\right)^{1/2}a = 1$$

$$A = \sqrt{\frac{2}{\pi}}\,a$$

(ii) $\langle p \rangle = \int \psi^* \left(-i\hbar \frac{d}{dx} \right) \psi \, dx$

$$= (-i\hbar)A^2 \int_{-\infty}^{\infty} \exp\left(-\frac{x^2}{a^2} \right) e^{-ikx} \left(-\frac{2x}{a^2} + ik \right) \exp\left(-\frac{x^2}{a^2} \right) e^{-ikx} \, dx$$

$$= (-i\hbar)\left(-\frac{2}{a^2} \right) \int_{-\infty}^{\infty} \exp\left(\frac{-2x^2}{a^2} \right) x \, dx + (-i\hbar)(ik) \, A^2 \int_{-\infty}^{\infty} \exp\left(\frac{-2x^2}{a^2} \right) dx$$

In the first term, the integrand is odd and the integral is from $-\infty$ to ∞. Hence the integral vanishes.

$$\langle p \rangle = \hbar k \quad \text{(refer the appendix)}$$

since $A^2 \int_{-\infty}^{\infty} \exp\left(\frac{-2x^2}{a^2} \right) dx = 1.$

3.37 The normalized wave function of a particle is $\psi(x) = A \exp(iax - ibt)$, where A, a and b are constants. Evaluate the uncertainty in its momentum.

Solution.

$$\psi(x) = Ae^{i(ax - bt)}$$

$$(\Delta p)^2 = \langle p^2 \rangle - \langle p \rangle^2$$

$$\langle p \rangle = -i\hbar \int \psi^* \frac{d}{dx} \psi \, dx = \hbar a \int \psi^* \psi \, dx = \hbar a$$

$$\langle p^2 \rangle = -\hbar^2 \int \psi^* \frac{d^2}{dx^2} \psi \, dx$$

$$= -\hbar^2 A^2 \int e^{-i(ax-bt)} \frac{d^2}{dx^2} e^{i(ax-bt)} \, dx$$

$$= -\hbar^2 (ia)^2 \int \psi^* \psi \, dx = \hbar^2 a^2$$

$$(\Delta p)^2 = \langle p^2 \rangle - \langle p \rangle^2 = \hbar^2 a^2 - \hbar^2 a^2 = 0$$

$$(\Delta p) = 0$$

3.38 Two normalized degenerate eigenfunctions $\psi_1(x)$ and $\psi_2(x)$ of an observable satisfy the condition $\int_{-\infty}^{\infty} \psi_1^* \psi_2 \, dx = a$, where a is real. Find a normalized linear combination of ψ_1 and ψ_2, which is orthogonal to $\psi_1 - \psi_2$.

Solution. Let the linear combination of ψ_1 and ψ_2 be

$$\psi = c_1 \psi_1 + c_2 \psi_2 \; (c_1, \, c_2 \text{ are real constants})$$

$$\int_{-\infty}^{\infty} (c_1 \psi_1 + c_2 \psi_2)^* \, (c_1 \psi_1 + c_2 \psi_2) \, dx = 1$$

$$c_1^2 + c_2^2 + 2c_1 c_2 a = 1$$

As the combination ψ is orthogonal to $\psi_1 - \psi_2$,

$$\int (\psi_1 - \psi_2)^* (c_1\psi_1 + c_2\psi_2)\, dx = 0$$

$$c_1 - c_2 + c_2 a - c_1 a = 0$$

$$(c_1 - c_2)(1 - a) = 0 \quad \text{or} \quad c_1 = c_2$$

With this condition, the earlier condition on c_1 and c_2 takes the form

$$c_2^2 + c_2^2 + 2c_2^2 a = 1 \quad \text{or} \quad c_2 = \frac{1}{\sqrt{2 + 2a}}$$

Then, the required linear combination is

$$\psi = \frac{\psi_1 + \psi_2}{\sqrt{2 + 2a}}$$

3.39 The ground state wave function of a particle of mass m is given by $\psi(x) = \exp(-\alpha^2 x^4/4)$, with energy eigenvalue $\hbar^2\alpha^2/m$. What is the potential in which the particle moves?

Solution. The Schrödinger equation of the system is given by

$$\left(-\frac{\hbar^2}{2m}\frac{d^2}{dx^2} + V\right) e^{-\alpha^2 x^4/4} = \frac{\hbar^2\alpha^2}{m} e^{-\alpha^2 x^4/4}$$

$$-\frac{\hbar^2}{2m}(-3\alpha^2 x^2 + \alpha^4 x^6)\, e^{-\alpha^2 x^4/4} + V e^{-\alpha^2 x^4/4} = \frac{\hbar^2\alpha^2}{m} e^{-\alpha^2 x^4/4}$$

$$V = \frac{\hbar^2}{2m}\alpha^4 x^6 - \frac{3}{2}\frac{\hbar^2}{2m}\alpha^2 x^2 + \frac{\hbar^2\alpha^2}{m}$$

3.40 An operator A contains time as a parameter. Using time-dependent Schrödinger equation for the Hamiltonian H, show that

$$\frac{d\langle A\rangle}{dt} = \frac{i}{\hbar}\langle [H, A]\rangle + \left\langle \frac{\partial A}{\partial t} \right\rangle$$

Solution. The ket $|\psi_s(t)\rangle$ varies in accordance with the time-dependent Schrödinger equation

$$i\hbar \frac{\partial}{\partial t}|\psi_s(t)\rangle = H|\psi_s(t)\rangle \tag{i}$$

As the Hamiltonian H is independent of time, Eq. (3.24) can be integrated to give

$$|\psi_s(t)\rangle = \exp(-iHt/\hbar)|\psi_s(0)\rangle \tag{ii}$$

Here, the operator $\exp(-iHt/\hbar)$ is defined by

$$\exp\left(-\frac{iHt}{\hbar}\right) = \sum_{n=0}^{\infty}\left(-\frac{iHt}{\hbar}\right)^n \frac{1}{n!} \tag{iii}$$

Equation (ii) reveals that the operator $\exp(-iHt/\hbar)$ changes the ket $|\psi_s(0)\rangle$ into ket $|\psi_s(t)\rangle$. Since H is Hermitian and t is real, this operator is unitary and the norm of the ket remains unchanged. The Hermitian adjoint of Eq. (i) is

$$-i\hbar \frac{\partial}{\partial t} \langle \psi_s(t)| = \langle \psi_s(t)| H^\dagger = \langle \psi_s(t)| H \tag{iv}$$

whose solution is

$$\langle \psi_s(t)| = \langle \psi_s(0)| \exp\left(\frac{iHt}{\hbar}\right) \tag{v}$$

Next we consider the time derivative of expectation value of the operator A_s. The time derivative of $\langle A_s \rangle$ is given by

$$\frac{d}{dt}\langle A_s \rangle = \frac{d}{dt}\langle \psi_s(t)| A_s |\psi_s(t)\rangle \tag{vi}$$

where A_s is the operator representing the observable A. Replacing the factors $\frac{d}{dt}|\psi_s(t)\rangle$ and $\frac{d}{dt}\langle \psi_s(t)|$ and using Eqs. (i) and (iv), we get

$$\frac{d}{dt}\langle A_s \rangle = \frac{1}{i\hbar}\langle \psi_s(t)| A_s H - HA_s |\psi_s(t)\rangle + \langle \psi_s(t) \left| \frac{\partial A_s}{\partial t} \right| \psi_s(t)\rangle$$

$$\frac{d}{dt}\langle A_s \rangle = \frac{1}{i\hbar}[A_s H] + \left\langle \frac{\partial A_s}{\partial t} \right\rangle \tag{vii}$$

3.41 A particle is constrained in a potential $V(x) = 0$ for $0 \le x \le a$ and $V(x) = \infty$ otherwise. In the x-representation, the wave function of the particle is given by

$$\psi(x) = \sqrt{\frac{2}{a}} \sin \frac{2\pi x}{a}$$

Determine the momentum function $\Phi(p)$.

Solution. From Eq. (3.35),

$$\Phi(p) = \frac{1}{\sqrt{2\pi\hbar}} \int_{-\infty}^{\infty} \psi(x) \exp\left(-\frac{ipx}{\hbar}\right) dx$$

In the present case, this equation can be reduced to

$$\Phi(p) = \frac{1}{\sqrt{\pi\hbar a}} I$$

where

$$I = \int_0^a \sin \frac{2\pi x}{a} e^{(-ipx/\hbar)}\, dx$$

Integrating by parts, we obtain

$$I = \left[-\frac{\hbar}{ip} \sin \frac{2\pi x}{a} e^{(-ipx/\hbar)} \right]_0^a - \int_0^a \left(-\frac{\hbar}{ip}\right) e^{(-ipx/\hbar)} \frac{2\pi}{a} \cos \frac{2\pi x}{a}\, dx$$

Since the integrated term is zero,

$$I = \frac{2\pi\hbar}{ipa}\left[\cos\frac{2\pi x}{a}\left(-\frac{\hbar}{ip}\right)e^{(-ipx/\hbar)}\right]_0^a - \frac{2\pi\hbar}{ipa}\int_0^a\left(-\frac{\hbar}{ip}\right)e^{(-ipx/\hbar)}\left(-\frac{2\pi}{a}\right)\sin\frac{2\pi x}{a}\,dx$$

$$= \frac{2\pi\hbar}{ipa}\left(-\frac{\hbar}{ip}\right)[e^{(-ipx/\hbar)} - 1] + \frac{4\pi^2\hbar^2}{a^2p^2}I$$

$$I\left(1 - \frac{4\pi^2\hbar^2}{a^2p^2}\right) = \frac{2\pi\hbar^2}{ap^2}[e^{(-ipx/\hbar)} - 1]$$

$$I = \frac{2\pi a\hbar^2}{a^2p^2 - 4\pi^2\hbar^2}[e^{(-ipa/\hbar)} - 1]$$

With this value of I,

$$\Phi(p) = \frac{1}{\sqrt{\pi\hbar a}}\frac{2\pi a\hbar^2}{a^2p^2 - 4\pi^2\hbar^2}[e^{(-ipa/\hbar)} - 1]$$

$$= \frac{2\pi^{1/2}a^{1/2}\hbar^{3/2}}{a^2p^2 - 4\pi^2\hbar^2}[e^{(-ipa/\hbar)} - 1]$$

3.42 A particle is in a state $|\psi\rangle = (1/\pi)^{1/4}\exp(-x^2/2)$. Find Δx and Δp_x. Hence evaluate the uncertainty product $(\Delta x)(\Delta p_x)$.

Solution. For the wave function, we have

$$\langle x\rangle = \left(\frac{1}{\pi}\right)^{1/2}\int_{-\infty}^{\infty} xe^{-x^2}dx = 0$$

since the integrand is an odd function of x. Now,

$$\langle x^2\rangle = \left(\frac{1}{\pi}\right)^{1/2}\int_{-\infty}^{\infty} x^2e^{-x^2}dx = 2\left(\frac{1}{\pi}\right)^{1/2}\frac{\sqrt{\pi}}{4} = \frac{1}{2} \quad \text{(see Appendix)}$$

$$(\Delta x)^2 = \langle x^2\rangle - \langle x^2\rangle = \frac{1}{2}$$

$$\langle p_x\rangle = \left(\frac{1}{\pi}\right)^{1/2}\int_{-\infty}^{\infty}\exp\left(-\frac{x^2}{2}\right)\left(-i\hbar\frac{d}{dx}\right)\exp\left(-\frac{x^2}{2}\right)dx$$

$$= i\hbar\left(\frac{1}{\pi}\right)^{1/2}\int_{-\infty}^{\infty} xe^{-x^2}\,dx = 0$$

$$\langle p_x^2\rangle = \left(\frac{1}{\pi}\right)^{1/2} \int_{-\infty}^{\infty} \exp\left(-\frac{x^2}{2}\right) (-i\hbar)^2 \frac{d^2}{dx^2} \exp\left(-\frac{x^2}{2}\right) dx$$

$$= \left(\frac{1}{\pi}\right)^{1/2} \hbar^2 \int_{-\infty}^{\infty} e^{-x^2}\, dx - \left(\frac{1}{\pi}\right)^{1/2} \hbar^2 \int_{-\infty}^{\infty} x^2 e^{-x^2}\, dx$$

$$= \left(\frac{1}{\pi}\right)^{1/2} \hbar^2 \pi^{1/2} - \left(\frac{1}{\pi}\right)^{1/2} \hbar^2 \frac{\pi^{1/2}}{2} = \frac{\hbar^2}{2} \quad \text{(see Appendix)}$$

$$(\Delta p_x)^2 = \langle p_x^2\rangle - \langle p_x\rangle^2 = \frac{\hbar^2}{2}$$

The uncertainty product

$$(\Delta x)(\Delta p_x) = \frac{\hbar}{2}$$

3.43 For a one-dimensional bound particle, show that

(i) $\frac{d}{dt} \int_{-\infty}^{\infty} \Psi^*(x, t)\, \Psi(x, t)\, dx = 0$, Ψ need not be a stationary state.

(ii) If the particle is in a stationary state at a given time, then it will always remain in a stationary state.

Solution.

(i) Consider the Schrödinger equation and its complex conjugate form:

$$i\hbar \frac{\partial \Psi(x,t)}{\partial t} = \left[-\frac{\hbar^2}{2m} \frac{\partial^2}{\partial x^2} + V(x)\right] \Psi(x, t)$$

$$-i\hbar \frac{\partial \Psi^*(x,t)}{\partial t} = \left[-\frac{\hbar^2}{2m} \frac{\partial^2}{\partial x^2} + V(x)\right] \Psi^*(x, t)$$

Multiplying the first equation by Ψ^* and the second by Ψ from LHS and subtracting the second from the first, we have

$$i\hbar\left[\Psi^* \frac{\partial \Psi}{\partial t} + \Psi \frac{\partial \Psi^*}{\partial t}\right] = -\frac{\hbar^2}{2m}\left[\Psi^* \frac{\partial^2 \Psi}{\partial x^2} - \Psi \frac{\partial^2 \Psi^*}{\partial x^2}\right]$$

$$\frac{\partial}{\partial t}(\Psi^*\Psi) = \frac{i\hbar}{2m}\left[\frac{\partial}{\partial x}\left(\Psi^* \frac{\partial \Psi}{\partial x} - \Psi \frac{\partial \Psi^*}{\partial x}\right)\right]$$

Integrating over x, we get

$$\int_{-\infty}^{\infty} \frac{\partial}{\partial t}(\Psi^*\Psi)\, dx = \frac{i\hbar}{2m}\left[\Psi^* \frac{\partial \Psi}{\partial x} - \Psi \frac{\partial \Psi^*}{\partial x}\right]_{-\infty}^{\infty}$$

$$\frac{\partial}{\partial t} \int_{-\infty}^{\infty} (\Psi^*\Psi)\, dx = \frac{i\hbar}{2m}\left[\Psi^* \frac{\partial \Psi}{\partial x} - \Psi \frac{\partial \Psi^*}{\partial x}\right]_{-\infty}^{\infty}$$

Since the state is bound, $\Psi = 0$ as $x \to \pm\infty$. Hence, the RHS of the above equation is zero. The integrated quantity will be a function of time only. Therefore,

$$\frac{d}{dt}\int_{-\infty}^{\infty} \Psi^*(x, t)\, \Psi(x, t)\, dx = 0$$

(ii) Let the particle be in a stationary state at $t = 0$, H be its Hamiltonian which is time independent, and E be its energy eigenvalue. Then,

$$H\Psi(x, 0) = E\Psi(x, 0)$$

Using Eq. (3.25), we have

$$\Psi(x,t) = \exp\left(-\frac{iHt}{\hbar}\right) \Psi(x,0)$$

Operating from left by H and using the commutability of H with exp $(-iHt/\hbar)$, we have

$$H\Psi(x, t) = \exp\left(-\frac{iHt}{\hbar}\right) H\Psi(x,0)$$

$$= E\exp\left(-\frac{iHt}{\hbar}\right) \Psi(x,0) = E\Psi(x,t)$$

Thus, $\Psi(x, t)$ represents a stationary state at all times.

3.44 The solution of the Schrödinger equation for a free particle of mass m in one dimension is $\Psi(x, t)$. At $t = 0$,

$$\Psi(x, 0) = A \exp\left(\frac{-x^2}{a^2}\right)$$

Find the probability amplitude in momentum space at $t = 0$ and at time t.

Solution.

(i) From Eq. (3.35),

$$\Phi(p, 0) = \frac{1}{\sqrt{2\pi\hbar}} \int_{-\infty}^{\infty} \Psi(x,0) \exp\left(-\frac{ipx}{\hbar}\right) dx$$

$$= \frac{A}{\sqrt{2\pi\hbar}} \int_{-\infty}^{\infty} \exp\left(-\frac{x^2}{a^2} - \frac{ipx}{\hbar}\right) dx$$

$$= \frac{A}{\sqrt{2\pi\hbar}} \int_{-\infty}^{\infty} \exp\left(\frac{-x^2}{a^2}\right) \cos\left(\frac{px}{\hbar}\right) dx$$

Here, the other term having sin $(px/\hbar)$ reduces to zero since the integrand is odd. Using the standard integral, we get

$$\Phi(p, 0) = \frac{Aa}{\sqrt{2\hbar}} \exp\left(-\frac{p^2a^2}{4\hbar^2}\right)$$

The Schrödinger equation in the momentum space equation (3.31) is

$$i\hbar \frac{\partial}{\partial t} \Phi(p,t) = \frac{p^2}{2m} \Phi(p,t)$$

$$\frac{\partial}{\partial t} \Phi(p,t) = \frac{-ip^2}{2m\hbar} \Phi(p,t)$$

$$\frac{d\Phi}{\Phi} = \left(-\frac{ip^2}{2m\hbar}\right) dt$$

Integrating and taking the exponential, we obtain

$$\Phi(p,t) = B \exp\left(\frac{-ip^2 t}{2m\hbar}\right)$$

At $t = 0$, $\Phi(p, 0) = B$. Hence,

$$\Phi(p,t) = \frac{Aa}{\sqrt{2\hbar}} \exp\left(\frac{-p^2 a^2}{4\hbar^2} \frac{-ip^2 t}{2m\hbar}\right)$$

3.45 Write the time-dependent Schrödinger equation for a free particle in the momentum space and obtain the form of the wave function.

Solution. The Schrödinger equation in the momentum space is

$$i\hbar \frac{\partial \Phi(\mathbf{p},t)}{\partial t} = \frac{p^2}{2m} \Phi(\mathbf{p},t)$$

$$\frac{\partial \Phi}{\partial t} = \frac{-ip^2}{2\hbar m} \Phi(\mathbf{p},t)$$

$$\frac{d\Phi}{\Phi} = \frac{-ip^2}{2\hbar m} dt$$

Integrating, we get

$$\ln \Phi = \frac{-ip^2 t}{2\hbar m} + \text{constant}$$

$$\Phi(\mathbf{p},t) = A \exp\left(\frac{-ip^2 t}{2\hbar m}\right), \quad \text{with } A \text{ as constant}$$

When $t = 0$, $\Phi(\mathbf{p}, t)$. Hence,

$$\Phi(\mathbf{p}, t) = \Phi(\mathbf{p},0) \exp\left(\frac{-ip^2 t}{2\hbar m}\right)$$

which is a form of the wave function in the momentum space.

3.46 The normalized state function ϕ of a system is expanded in terms of its energy eigenfunctions as $\phi = \sum_i c_i \psi_i(\mathbf{r})$, c_i's being constants. Show that $|c_i|^2$ is the probability for the occurrence of the energy eigenvalue E_i in a measurement.

Solution. The expectation value of the Hamiltonian operator H is

$$\langle H\rangle = \langle \phi | H | \phi \rangle = \sum_i \sum_j c_i^* e_j \langle \psi_i | H | \psi_j \rangle$$

$$= \sum_i \sum_j c_i^* c_j \langle \psi_i | E_j | \psi_j \rangle$$

$$= \sum_i |c_i|^2 E_i$$

Let ω_i be the probability for the occurrence of the eignevalue E_i. Then,

$$\langle H\rangle = \sum_i \omega_i E_i$$

Since E_i's are constants from the above two equations for $\langle H\rangle$,

$$\omega_i = |c_i|^2$$

3.47 Show that, if the Hamiltonian H of a system does not depend explicitly on time, the ket $|\psi(t)\rangle$ varies with time according to

$$|\psi(t)\rangle = \exp\left(-\frac{iHt}{\hbar}\right)|\psi(0)\rangle$$

Solution. The time-dependent Schrödinger equation for the Hamiltonian operator H is

$$i\hbar \frac{d}{dt}\bigg| \psi(t)\rangle = H|\psi(t)\rangle$$

Rearranging, we get

$$\frac{d|\psi(t)\rangle}{|\psi(t)\rangle} = \frac{H}{i\hbar}\, dt$$

Integrating, we obtain

$$\ln |\psi(t)\rangle = \frac{Ht}{i\hbar} + C, \text{ with } C \text{ as constant,}$$

$$C = \ln |\psi(0)\rangle$$

Substituting the value of C, we have

$$\ln \frac{|\psi(t)\rangle}{|\psi(0)\rangle} = \frac{Ht}{i\hbar}$$

$$\frac{|\psi(t)\rangle}{|\psi(0)\rangle} = \exp\left(-\frac{iHt}{\hbar}\right)$$

$$|\psi(t)\rangle = \exp\left(-\frac{iHt}{\hbar}\right)|\psi(0)\rangle$$

3.48 Show that, if P, Q and R are the operators in the Schrödinger equation satisfying the relation $[P, Q] = R$, then the corresponding operators P_H, Q_H and R_H of the Heisenberg picture satisfy the relation $[P_H, Q_H] = R_H$.

Solution. The operator in the Heisenberg picture A_H corresponding to the operator A_S in the Schrödinger equation is given by

$$A_H(t) = e^{iHt/\hbar} A_S e^{-iHt/\hbar}$$

By the Schrödinger equation,

$$PQ - QP = R$$

Inserting $e^{-iHt/\hbar} e^{-iHt/\hbar} = 1$ between quantities, we obtain

$$Pe^{-iHt/\hbar} e^{iHt/\hbar} Q - Qe^{-iHt/\hbar} e^{iHt/\hbar} P = R$$

Pre-multiplying each term by $e^{iHt/\hbar}$ and post-multiplying by $e^{-iHt/\hbar}$, we get

$$e^{iHt/\hbar} Pe^{-iHt/\hbar} Qe^{-iHt/\hbar} - e^{iHt/\hbar} Qe^{-iHt/\hbar} e^{iHt/\hbar} Pe^{-iHt/\hbar} = e^{iHt/\hbar} Re^{-iHt/\hbar}$$

$$P_H Q_H - Q_H P_H = R_H$$

$$[P_H, Q_H] = R_H$$

3.49 Show that the expectation value of an observable, whose operator does not depend on time explicitly, is a constant with zero uncertainty.

Solution. Let the operator associated with the observable be A and its eigenvalue be a_n. The wave function of the system is

$$\Psi_n(\boldsymbol{r}, t) = \psi_n(\boldsymbol{r}) \exp\left(-\frac{iE_n t}{\hbar}\right)$$

The expectation value of the operator A is

$$\langle A \rangle = \int_{-\infty}^{\infty} \psi_n^*(\boldsymbol{r}) \exp\left(\frac{iE_n t}{\hbar}\right) A\psi_n(\boldsymbol{r}) \exp\left(\frac{iE_n t}{\hbar}\right) d\tau$$

$$= \int_{-\infty}^{\infty} \psi_n^*(\boldsymbol{r})\, A\psi_n(\boldsymbol{r})\, d\tau = a_n \int_{-\infty}^{\infty} \psi_n^*(\boldsymbol{r})\, \psi_n(\boldsymbol{r})\, d\tau$$

$$= a_n$$

That is, the expectation value of the operator A is constant. Similarly,

$$\langle A^2 \rangle = \int_{-\infty}^{\infty} \psi_n^*(\boldsymbol{r})\, A^2 \psi_n(\boldsymbol{r})\, d\tau = a_n^2$$

$$\text{Uncertainty } (\Delta A) = \langle A^2 \rangle - \langle A \rangle^2 = a_n^2 - a_n^2 = 0$$

3.50 For the one-dimensional motion of a particle of mass m in a potential $V(x)$, prove the following relations:

$$\frac{d\langle x \rangle}{dt} = \frac{\langle p_x \rangle}{m}, \qquad \frac{d\langle p_x \rangle}{dt} = -\left\langle \frac{dV}{dx} \right\rangle$$

Explain the physical significance of these results also.

Solution. If an operator A has no explicit dependence on time, from Eq. (3.26),

$$i\hbar \frac{d}{dt}\langle A \rangle = \langle [A, H] \rangle, \ H \text{ being the Hamiltonian operator}$$

Since $H = \frac{p_x^2}{2m} + V(x)$, we have

$$i\hbar \frac{d}{dt}\langle x\rangle = \left\langle \left[x, \frac{p_x^2}{2m} + V \right] \right\rangle$$

$$\left[x, \frac{p_x^2}{2m} + V \right] = \frac{1}{2m}[x, p_x^2] + [x, V(x)]$$

$$= \frac{1}{2m}[x, p_x]p_x + \frac{1}{2m}p_x[x, p_x]$$

$$= 2\frac{i\hbar}{2m}p_x = i\hbar\frac{p_x}{m}$$

Consequently,

$$\frac{d\langle x\rangle}{dt} = \frac{\langle p_x\rangle}{m}$$

For the second relation, we have

$$i\hbar\frac{d}{dt}\langle p_x\rangle = \langle [p_x, H]\rangle$$

$$[p_x, H] = \frac{1}{2m}[p_x, p_x^2] + [p_x, V] = [p_x, V(x)]$$

Allowing $[p_x, V(x)]$ to operate on $\psi(x)$, we get

$$\left[-i\hbar\frac{\partial}{\partial x}, V(x) \right]\psi = -i\hbar\frac{\partial}{\partial x}(V\psi) + i\hbar V\frac{\partial}{\partial x}\psi$$

$$= -i\hbar\frac{\partial V}{\partial x}\psi$$

Hence,

$$i\hbar\frac{d}{dt}\langle p_x\rangle = i\hbar\left\langle -\frac{dV}{dx}\right\rangle \quad \text{or} \quad \frac{d}{dt}\langle p_x\rangle = \left\langle -\frac{dV}{dx}\right\rangle$$

In the limit, the wave packet reduces to a point, and hence

$$\langle x\rangle = x, \qquad \langle p_x\rangle = p_x$$

Then the first result reduces to

$$m\frac{dx}{dt} = p_x$$

which is the classical equation for momentum. Since $-(\partial V/\partial x)$ is a force, when the wave packet reduces to a point, the second result reduces to Newton's Second Law of Motion.

3.51 Find the operator for the velocity of a charged particle of charge e in an electromagnetic field.

Solution. The classical Hamiltonian for a charged particle of charge e in an electromagnetic field is

$$H = \frac{1}{2m}\left(\boldsymbol{p} - \frac{e}{c}\boldsymbol{A}\right)^2 + e\phi$$

where $\boldsymbol{A}$ is the vector potential and ϕ is the scalar potential of the field. The operator representing the Hamiltonian (refer Problem 3.23)

$$H = -\frac{\hbar^2}{2m}\nabla^2 + \frac{ie\hbar}{2mc}\nabla \cdot \boldsymbol{A} + \frac{ie\hbar}{mc}\boldsymbol{A} \cdot \nabla + \frac{e^2 \boldsymbol{A}^2}{2mc^2} + e\phi$$

For our discussion, let us consider the x-component of velocity. In the Heisenberg picture, for an operator A not having explicit dependence on time, we have

$$\frac{dA}{dt} = \frac{1}{i\hbar}[A, H]$$

Applying this relation for the x coordinate of the charged particle, we obtain

$$\frac{dx}{dt} = \frac{1}{i\hbar}[x, H]$$

As x commutes with the second, fourth and fifth terms of the above Hamiltonian, we have

$$\begin{aligned} \frac{dx}{dt} &= \frac{1}{i\hbar}\left[x, \frac{-\hbar^2}{2m}\frac{d^2}{dx^2} + \frac{ie\hbar}{mc}A_x\frac{d}{dx}\right] \\ &= \frac{1}{i\hbar}\left[x, \frac{-\hbar^2}{2m}\frac{d^2}{dx^2}\right] + \frac{1}{i\hbar}\left[x, \frac{ie\hbar}{mc}A_x\frac{d}{dx}\right] \end{aligned}$$

$$\begin{aligned} \left[x, \frac{-\hbar^2}{2m}\frac{d^2}{dx^2}\right]\psi &= -\frac{\hbar^2}{2m}x\frac{d^2\psi}{dx^2} + \frac{\hbar^2}{2m}\frac{d}{dx}\frac{d(x\psi)}{dx} \\ &= -\frac{\hbar^2}{2m}x\frac{d^2\psi}{dx^2} + \frac{\hbar^2}{2m}\left(x\frac{d^2\psi}{dx^2} + 2\frac{d\psi}{dx}\right) \\ &= \frac{\hbar^2}{m}\frac{d\psi}{dx} \end{aligned}$$

$$\begin{aligned} \left[x, \frac{ie\hbar}{mc}A_x\frac{d}{dx}\right] &= \frac{ie\hbar}{mc}\left[xA_x\frac{d\psi}{dx} - A_x\frac{d(x\psi)}{dx}\right] \\ &= -\frac{ie\hbar}{mc}A_x\psi \end{aligned}$$

Substituting these results, we get

$$\frac{dx}{dt} = \frac{1}{i\hbar}\frac{\hbar^2}{m}\frac{d}{dx} - \frac{1}{i\hbar}\frac{ie\hbar}{mc}A_x = \frac{1}{m}\left[-i\hbar\frac{d}{dx} - \frac{e}{c}A_x\right] = \frac{1}{m}\left(p_x - \frac{e}{c}A_x\right)$$

Including the other two components, the operator for

$$\mathbf{v} = \frac{1}{m}\left(\boldsymbol{p} - \frac{e}{c}\boldsymbol{A}\right), \quad p = i\hbar\nabla$$

3.52 For the momentum and coordinate operators, prove the following: (i) $\langle p_x x\rangle - \langle xp_x\rangle = -i\hbar$, (ii) for a bound state, the expectation value of the momentum operator $\langle p\rangle$ is zero.

Solution.

(i)
$$\langle px\rangle = \int \psi^*\left(-i\hbar\frac{d}{dx}\right)(x\psi)\,dx$$

$$= -i\hbar\int\left(\psi^*\psi + \psi^* x\frac{d\psi}{dx}\right)dx$$

$$= -i\hbar\int\psi^*\psi\,dx - i\hbar\int\psi^* x\left(\frac{d}{dx}\right)\psi\,dx$$

$$= -i\hbar + \int\psi^* x\left(-i\hbar\frac{d}{dx}\right)\psi\,dx$$

$$= -i\hbar + \langle xp\rangle$$

$$\langle px\rangle - \langle xp\rangle = -i\hbar$$

(ii) The expectation value of $\boldsymbol{p}$ for a bound state defined by the wave function ψ_n is

$$\langle p\rangle = \int\psi_n^*(-i\hbar\nabla)\psi_n\,d\tau$$

If ψ_n is odd, $\nabla\psi_n$ is even and the integrand becomes odd. The value of the integral is then zero. If ψ_n is even, $\nabla\psi_n$ is odd and the integrand is again odd. Therefore, $\langle p\rangle = 0$.

3.53 Substantiate the statement: "Eigenfunctions of a Hermitian operator belonging to distinct eigenvalues are orthogonal" by taking the time-independent Schrödinger equation of a one-dimensional system.

Solution. The time-independent Schrödinger equation of a system in state n is

$$\frac{d^2\psi_n}{dx^2} + \frac{2m}{\hbar^2}[E_n - V(x)]\psi_n = 0 \tag{i}$$

The complex conjugate equation of state k is

$$\frac{d^2\psi_k^*}{dx^2} + \frac{2m}{\hbar^2}[E_k - V(x)]\psi_k^* = 0 \tag{ii}$$

Multiplying the first by ψ_k^* and the second by ψ_n from LHS and subtracting, we get

$$\psi_k^*\frac{d^2\psi_n}{dx^2} - \psi_n\frac{d^2\psi_k^*}{dx^2} + \frac{2m}{\hbar^2}(E_n - E_k)\psi_k^*\psi_n = 0 \tag{iii}$$

Integrating Eq. (iii) over all values of x, we obtain

$$\frac{2m}{\hbar^2}(E_k - E_n) \int_{-\infty}^{\infty} \psi_k^* \psi_n \, dx = \int_{-\infty}^{\infty} \left(\psi_k^* \frac{d^2\psi_n}{dx^2} - \psi_n \frac{d^2\psi_k^*}{dx^2} \right) dx$$

$$= \left[\psi_k^* \frac{d\psi_n}{dx} - \psi_n \frac{d\psi_k^*}{dx} \right]_{\infty}^{\infty}$$

Since $\psi \to 0$ as $x \to \infty$, the RHS is zero. Consequently,

$$\int_{-\infty}^{\infty} \psi_k^* \psi_n \, dx = 0$$

Hence the statement.

3.54 Find the physical dimensions of the wave function $\psi(\mathbf{r})$ of a particle moving in three dimensional space.

Solution. The wave function of a particle moving in a three-dimensional box of sides a, b and c is given by (refer Problem 5.1)

$$\psi(\mathbf{r}) = \sqrt{\frac{8}{abc}} \sin \frac{n_1 \pi x}{a} \sin \frac{n_2 \pi y}{b} \sin \frac{n_3 \pi z}{c}$$

As the sine of a quantity is dimensionless, $\psi(\mathbf{r})$ has the physical dimension of $(length)^{-3/2}$.

3.55 A and B are Hermitian operators and $AB - BA = iC$. Prove that C is a Hermitian operator.

Solution.

$$\text{Operator } C = \frac{1}{i}(AB - BA) = -i(AB - BA)$$

$$C^* = i\,(A^*B^* - B^*A^*)$$

Consider the integral

$$\int \psi_i^* C \psi_n \, d\tau = -i \int \psi_m^* (AB - BA)\, \psi_n \, d\tau$$

$$= -i \int (B^*A^* - A^*B^*)\, \psi_n^* \psi_n \, d\tau$$

$$= i \int (A^*B^* - B^*A^*)\, \psi_m^* \psi_n \, d\tau$$

$$= \int C^* \psi_m^* \psi_n \, d\tau$$

Thus the operator C is Hermitian.

3.56 Consider a particle of mass m moving in a spherically symmetric potential $V = kr$, where k is a positive constant. Estimate the ground state energy using the uncertainty principle.

Solution. The uncertainty principle states that

$$(\Delta p)(\Delta x) \geq \frac{\hbar}{2}$$

Since the potential is spherically symmetric, $\langle p \rangle = \langle r \rangle = 0$. Hence,

$$\langle \Delta r \rangle^2 = \langle r^2 \rangle, \qquad \langle \Delta p \rangle^2 = \langle p^2 \rangle$$

We can then assume that

$$\Delta r \cong r, \qquad \Delta p \cong p$$

$$(\Delta p)(\Delta r) = \frac{\hbar}{2} \quad \text{or} \quad \Delta p = \frac{\hbar}{2(\Delta r)}$$

$$\text{Energy } E = \frac{p^2}{2m} + kr = \frac{(\Delta p)^2}{2m} + k(\Delta r)$$

$$= \frac{\hbar^2}{8m(\Delta r)^2} + k(\Delta r)$$

For the energy to be minimum, $[\partial E/\partial(\Delta r)] = 0$, and hence

$$-\frac{\hbar^2}{4m(\Delta r)^3} + k = 0 \quad \text{or} \quad \Delta r = \left(\frac{\hbar^2}{4mk}\right)^{1/3}$$

Substituting this value of Δr in the energy equation, we get

$$E = \frac{3}{2}\left(\frac{k^2\hbar^2}{4m}\right)^{1/3}$$

3.57 If the Hamiltonian of a system H = $(p_x^2/2m) + V(x)$, obtain the value of the commutator $[x, H]$. Hence, find the uncertainty product (Δx) (ΔH).

Solution.

$$[x, H] = \left[x, \frac{p_x^2}{2m}\right] + [x, V(x)]$$

$$= \frac{1}{2m}[x, p_x]p_x + \frac{1}{2m}p_x[x, p_x]$$

$$= \frac{1}{2m}(i\hbar)p_x + \frac{1}{2m}p_x(i\hbar)$$

$$= i\frac{\hbar}{m}p_x \tag{i}$$

Consider the operators A and B. If

$$[A, B] = iC \tag{ii}$$

the general uncertainty relation states that

$$(\Delta A)(\Delta B) = \frac{\langle C \rangle}{2} \tag{iii}$$

Identifying A with x, B with H and C with p_x, we can write

$$(\Delta x)(\Delta H) \geq \frac{\hbar}{2m}\langle p_x \rangle$$

3.58 If L_z is the z-component of the angular momentum and ϕ is the polar angle, show that $[\phi, L_z] = i\hbar$ and obtain the value of $(\Delta\phi)(\Delta L_z)$.

Solution. The z-component of angular momentum in the spherical polar coordinates is given by

$$L_z = -i\hbar \frac{d}{d\phi}$$

$$[\phi, L_z] = \left[\phi, -i\hbar \frac{d}{d\phi}\right] = -i\hbar\left[\phi, \frac{d}{d\phi}\right]$$

Allowing the commutator to operate on a function $f(\phi)$, we get

$$\left[\phi, \frac{d}{d\phi}\right] f = \phi \frac{df}{d\phi} - \frac{d(\phi f)}{d\phi}$$

$$= \phi \frac{df}{d\phi} - \phi \frac{df}{d\phi} - f = -f$$

Hence,

$$\left[\phi, \frac{d}{d\phi}\right] = -1$$

With this value of $[\phi, (d/d\phi)]$, we have

$$[\phi, L_z] = i\hbar$$

Comparing this with the general uncertainty relation, we get

$$[A, B] = iC, \qquad (\Delta A)(\Delta B) \geq \frac{\langle C \rangle}{2}$$

$$(\Delta\phi)(\Delta L_z) \geq \frac{\hbar}{2}$$

3.59 Find the probability current density $\mathbf{j}(\mathbf{r}, t)$ associated with the charged particle of charge e and mass m in a magnetic field of vector potential $\mathbf{A}$ which is real.

Solution. The Hamiltonian operator of the system is (refer Problem 3.23)

$$H = \frac{1}{2m}\left(\mathbf{p} - \frac{e}{c}\mathbf{A}\right)^2 = -\frac{\hbar^2}{2m}\nabla^2 + \frac{ie\hbar}{2mc}(\nabla \cdot A) + \frac{ie\hbar}{mc}(A \cdot \nabla) + \frac{e^2A^2}{2mc^2}$$

The time-dependent Schrödinger equation is

$$i\hbar\frac{\partial\Psi}{\partial t} = -\frac{\hbar^2}{2m}\nabla^2\Psi + \frac{ie\hbar}{2mc}(\nabla \cdot A)\Psi + \frac{ie\hbar}{mc}A \cdot \nabla\Psi + \frac{e^2A^2}{2mc^2}\Psi$$

Its complex conjugate equation is

$$-i\hbar\frac{\partial\Psi^*}{\partial t} = -\frac{\hbar^2}{2m}\nabla^2\Psi^* - \frac{ie\hbar}{2mc}(\nabla \cdot A)\Psi^* - \frac{ie\hbar}{mc}A \cdot \nabla\Psi^* + \frac{e^2A^2}{2mc^2}\Psi^*$$

Multiplying the first equation by Ψ^* from left and the complex conjugate equation by Ψ and subtracting, we get

$$i\hbar\left(\Psi^*\frac{\partial\Psi}{\partial t}+\Psi\frac{\partial\Psi^*}{\partial t}\right) = -\frac{\hbar^2}{2m}[\Psi^*\nabla^2\Psi - \Psi\nabla^2\Psi^*] + \frac{ie\hbar}{2mc}[\Psi^*(\nabla\cdot\boldsymbol{A})\Psi + \Psi(\nabla\cdot\boldsymbol{A})\Psi^*]$$

$$+\frac{ie\hbar}{2mc}[\Psi^*(\nabla\Psi)\cdot\boldsymbol{A} + \Psi(\nabla\Psi^*)\cdot\boldsymbol{A}]$$

$$\frac{\partial}{\partial t}(\Psi^*\Psi) = \frac{i\hbar}{2m}[\nabla\cdot(\Psi^*\nabla\Psi - \Psi\nabla\Psi^*)] + \frac{e}{mc}\Psi^*\Psi(\nabla\boldsymbol{A}) + \frac{e}{mc}[\Psi^*\boldsymbol{A}\cdot\nabla\Psi + \Psi\boldsymbol{A}\cdot\nabla\Psi^*]$$

$$\frac{\partial}{\partial t}(\Psi^*\Psi) = \nabla\cdot\left[\frac{i\hbar}{2m}(\Psi^*\nabla\Psi - \Psi\nabla\Psi^*) + \frac{e}{mc}(\Psi^*\Psi\boldsymbol{A})\right]$$

Defining the probability current density vector $\boldsymbol{j}(\boldsymbol{r}, t)$ by

$$\boldsymbol{j}(\boldsymbol{r}, t) = \frac{i\hbar}{2m}(\Psi\nabla\Psi^* - \Psi^*\nabla\Psi) - \frac{e}{mc}(\Psi^*\Psi\boldsymbol{A})$$

the above equation reduces to

$$\frac{\partial}{\partial t}P(\boldsymbol{r}, t) + \nabla\cdot\boldsymbol{j}(\boldsymbol{r}, t) = 0$$

which is the familiar equation of continuity for probability.

3.60 The number operator N_k is defined by $N_k = a_k^\dagger a_k$, where $a_k^\dagger$ and a_k obey the commutation relations

$$[a_k, a_l^\dagger] = \delta_{kl}, \qquad [a_k, a_l] = [a_k^\dagger, a_l^\dagger] = 0$$

Show that (i) the commutator $[N_k, N_l] = 0$, and (ii) all positive integers including zero are the eigenvalues of N_k.

Soultion. The number operator N_k is defined by

$$N_k = a_k^\dagger a_k$$

(i)
$$\begin{aligned}[N_k, N_l] &= [a_k^\dagger a_k, a_l^\dagger a_l] = [a_k^\dagger a_k, a_l^\dagger]\, a_l + a_l^\dagger [a_k^\dagger a_k, a_l] \\ &= a_k^\dagger [a_k, a_l^\dagger]\, a_l + [a_k^\dagger, a_l^\dagger]\, a_k a_l + a_l^\dagger a_k^\dagger [a_k a_l] + a_l^\dagger [a_k^\dagger, a_l]\, a_k \\ &= a_k^\dagger \delta_{kl} a_l + 0 + 0 + a_l^\dagger(-\delta_{kl})\, a_k \\ &= a_k^\dagger a_k - a_k^\dagger a_k = 0\end{aligned}$$

(ii) Let the eigenvalue equation of N_k be

$$N_k\psi(n_k) = n_k\psi(n_k)$$

where n_k is the eigenvalue. Multiplying from left by $\psi^*(n_k)$ and integrating over the entire space, we get

$$\begin{aligned}n_k &= \int \psi^*(n_k)\, N_k\psi(n_k)\, d\tau \\ &= \int \psi^*(n_k)\, a_k^\dagger a_k\, \psi(n_k)\, d\tau \\ &= \int |a_k\psi(n_k)|^2\, d\tau \ge 0\end{aligned}$$

Thus, the eigenvalues of N_k are all positive integers, including zero.

3.61 For a system of fermions, the creation ($a_k^\dagger$) and annihilation (a) operators obey the anticommutation relations

$$[a_k, a_k^\dagger]_+ = \delta_{kl}, \qquad [a_k, a_l]_+ = [a_k^\dagger, a_l^\dagger]_+ = 0$$

Show that the eigenvalues of the number operator N_k defined by $N_k = a_k^\dagger a_k$ are 0 and 1.

Solution. Since $[a_k, a_k^\dagger]_+ = \delta_{kl}$, we have

$$[a_k, a_k^\dagger]_+ = a_k a_k^\dagger + a_k^\dagger a_k = 1$$

$$a_k a_k^\dagger = 1 - a_k^\dagger a_k \tag{i}$$

Also,

$$[a_k, a_k]_+ = [a_k^\dagger, a_k^\dagger]_+ = 0$$

$$a_k a_k = a_k^\dagger a_k^\dagger = 0 \tag{ii}$$

$$N_k^2 = a_k^\dagger a_k a_k^\dagger a_k = a_k^\dagger (a_k a_k^\dagger) a_k$$

$$= a_k^\dagger (1 - a_k^\dagger a_k) a_k = a_k^\dagger a_k - a_k^\dagger a_k^\dagger a_k a_k$$

$$= N_k \tag{iii}$$

since the second term is zero. If n_k is the eigenvalue of N_k, Eq (iii) is equivalent to

$$n_k^2 = n_k \quad \text{or} \quad n_k^2 - n_k = 0$$

$$n_k(n_k - 1) = 0 \tag{iv}$$

which gives

$$n_k = 0, 1$$

Thus, the eigenvalues of N_k are 0 and 1.

CHAPTER 4

One-Dimensional Systems

In this chapter, we shall apply the basic ideas developed so far to some simple one-dimensional systems. In each case, we solve the time-independent Schrödinger equation

$$-\frac{\hbar^2}{2m}\frac{d^2\psi(x)}{dx^2} + V(x)\psi(x) = E\psi(x)$$

to obtain the energy eigenvalues E and the energy eigenfunctions.

4.1 Infinite Square Well Potential

(a) Potential $V(x) = \begin{cases} 0, & -a \le x \le a \\ \infty, & \text{otherwise} \end{cases}$ (4.1)

This potential is illustrated in Fig. 4.1(a). Now, the energy eigenvalues are given by

$$E_n = \frac{\pi^2\hbar^2 n^2}{8ma^2}, \qquad n = 1, 2, 3, \ldots \tag{4.2}$$

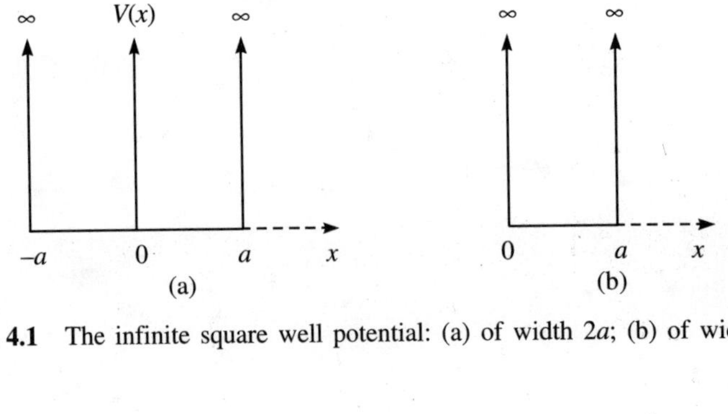

Fig. 4.1 The infinite square well potential: (a) of width $2a$; (b) of width a.

and the energy eigenfunctions by

$$\psi_n(x) = \begin{cases} \dfrac{1}{\sqrt{a}} \cos \dfrac{n\pi x}{2a}, & n = 1, 3, 5 \ldots \\ \dfrac{1}{\sqrt{a}} \sin \dfrac{n\pi x}{2a}, & n = 2, 4, 6 \ldots \end{cases} \tag{4.3}$$

A general solution is a linear combination of these two solutions.

(b) Potential $V(x) = \begin{cases} 0, & 0 \le x \le a \\ \infty, & \text{otherwise} \end{cases}$

which is illustrated in Fig. 4.1(b). Again, the energy eigenvalues

$$E_n = \frac{\pi^2 \hbar^2 n^2}{2ma^2}, \qquad n = 1, 2, 3, \ldots \tag{4.4}$$

and the energy eigenfunction

$$\psi_n = \sqrt{\frac{2}{a}} \sin \frac{n\pi x}{a}, \qquad n = 1, 2, 3, \ldots \tag{4.5}$$

4.2 Square Well Potential with Finite Walls

$$\text{Potential } V(x) = \begin{cases} V_0, & x < -a \\ 0, & -a < x < a \\ V_0, & x > a \end{cases} \tag{4.6}$$

Case (i): $E < V_0$. The wave function inside the well can either be symmetric or anti-symmetric about the origin. The continuity of the wave function and derivative give

Symmetric case: $ka \tan ka = \alpha a$ (4.7)

Antisymmetric case: $ka \cot ka = -\alpha a$ (4.8)

where

$$k^2 = \frac{2mE}{\hbar^2}, \qquad \alpha^2 = \frac{2m(V_0 - E)}{\hbar^2} \tag{4.9}$$

The energy eigenvalues are obtained by solving Eqs. (4.7) and (4.8) graphically. The solutions give the following results regarding the number of bound states in the well:

$$\begin{aligned} &\text{One (symmetric)} \quad \text{if } 0 < V_0 a^2 < \frac{\pi^2\hbar^2}{8m} \\ &\text{Two (1-symmetric, 1-antisymmetric)} \quad \text{if } \frac{\pi^2\hbar^2}{8m} < V_0 a^2 < \frac{4\pi^2\hbar^2}{8m} \\ &\text{Three (two-symmetric, one anti-symmetric)} \quad \text{if } \frac{4\pi^2\hbar^2}{8m} < V_0 a^2 < \frac{9\pi^2\hbar^2}{8m} \end{aligned} \tag{4.10}$$

Case (ii): $E > V_0$. In this case, the particle is not bound and the wave function is sinusoidal in all the regions.

4.3 Square Potential Barrier

The potential is defined by

$$V(x) = V_0 \quad \text{for } 0 < x < a \qquad (4.11)$$
$$V(x) = 0, \text{ otherwise}$$

Consider a stream of particles of mass m, the energy $E < V_0$ approaching the square barrier from the left. A portion of the particles is reflected back and the rest is transmitted. For a broad high barrier, the transmission coefficient T is given by

$$T = \frac{16k^2\alpha^2 e^{-2\alpha a}}{(\alpha^2 + k^2)^2} = \frac{16E(V_0 - E)e^{-2\alpha a}}{V_0^2} \qquad (4.12)$$

where k and α have the same definitions as in Eq. (4.9).

4.4 Linear Harmonic Oscillator

4.4.1 The Schrödinger Method

The solution of the Schrödinger equation for the linear harmonic oscillator potential $V = (1/2)kx^2$, where $k = m\omega^2$, gives the energy eigenvalues

$$E_n = \left(n + \frac{1}{2}\right)h\nu = \left(n + \frac{1}{2}\right)\hbar\omega, \qquad n = 0, 1, 2, \ldots \qquad (4.13)$$

The normalized eigenfunctions are

$$\psi_n(y) = \left(\frac{\alpha}{2^n n!\sqrt{\pi}}\right)^{1/2} H_n(y)e^{-y^2/2} \qquad (4.14)$$

where

$$y = \alpha x \quad \text{and} \quad \alpha = \left(\frac{m\omega}{\hbar}\right)^{1/2} \qquad (4.15)$$

$$\psi_0(x) = \left(\frac{\alpha}{\sqrt{\pi}}\right)^{1/2} \exp\left(-\frac{\alpha^2 x^2}{2}\right) \qquad (4.16)$$

$$\psi_1(x) = \left(\frac{\alpha}{2\sqrt{\pi}}\right)^{1/2} (2\alpha x)\exp\left(-\frac{\alpha^2 x^2}{2}\right) \qquad (4.17)$$

4.4.2 The Operator Method

The operator method is based on the basic commutation relation $[x, p] = i\hbar$, where x and p are the coordinate and momentum operators. The creation ($a^\dagger$) and annihilation (a) operators are defined by

$$a^\dagger = \left(\frac{m\omega}{2\hbar}\right)^{1/2} x - i\left(\frac{1}{2m\hbar\omega}\right)^{1/2} p \tag{4.18}$$

$$a = \left(\frac{m\omega}{2\hbar}\right)^{1/2} x + i\left(\frac{1}{2m\hbar\omega}\right)^{1/2} p \tag{4.19}$$

In terms of $a^\dagger$ and a, the Hamiltonian of a linear harmonic oscillator

$$H = \frac{\hbar\omega}{2}(aa^\dagger + a^\dagger a) \tag{4.20}$$

Also, we have

$$a|n\rangle = \sqrt{n}\,|n-1\rangle, \qquad a^\dagger|n\rangle = \sqrt{n+1}\,|n+1\rangle \tag{4.21}$$

With these concepts, one can easily get the energy eigenvalues of a linear harmonic oscillator.

4.5 The Free Particle

The free-particle Schrödinger equation

$$\frac{d^2\psi}{dx^2} = -k^2\psi, \qquad k^2 = \frac{2mE}{\hbar^2} \tag{4.22}$$

has the solutions

$$\psi(x) = Ae^{ikx} \quad \text{and} \quad \psi(x) = Ae^{-ikx} \tag{4.23}$$

As the normalization in the usual sense is not possible, one has to do either box normalization or delta function normalization, which are, respectively,

$$\psi(x) = \frac{1}{\sqrt{L}}\,e^{ikx} \quad \text{and} \quad \psi(x) = \frac{1}{\sqrt{2\pi}}\,e^{-ikx} \tag{4.24}$$

where L is the size of the box.

PROBLEMS

4.1 Obtain the energy eigenvalues and eigenfunctions of a particle trapped in the potential $V(x) = 0$ for $0 \le x \le a$ and $V(x) = \infty$ otherwise. Show that the wave functions for the different energy levels of the particle trapped in the square well are orthogonal.

Solution. The Schrödinger equation is

$$-\frac{\hbar^2}{2m}\frac{d^2\psi(x)}{dx^2} + V\psi(x) = E\psi(x), \qquad 0 \le x \le a$$

$$\frac{d^2\psi(x)}{dx^2} = -k^2\psi(x), \qquad k^2 = \frac{2mE}{\hbar^2}$$

$$\psi(x) = A \sin kx + B \cos kx, \qquad 0 \le x \le a$$

$$\psi(0) = 0 \text{ gives } B = 0 \quad \text{or} \quad \psi(x) = A \sin kx$$

$$\psi(a) = 0 \text{ gives } A \sin ka = 0 \quad \text{or} \quad \sin ka = 0$$

$$ka = n\pi \quad \text{or} \quad E_n = \frac{n^2\pi^2\hbar^2}{2ma^2}, \qquad n = 1, 2, \ldots$$

$$\psi(x) = \sqrt{2/a}\, \sin \frac{n\pi x}{a}$$

$$\int_0^a \psi_i^* \psi_n \, dx = \frac{2}{a}\int_0^a \sin\frac{m\pi x}{a} \sin\frac{n\pi x}{a}\, dx$$

$$= \frac{2}{\pi}\int_0^\pi \sin ny \sin my \, dy, \qquad y = \frac{\pi x}{a}$$

$$= \frac{1}{\pi}\int_0^\pi [\cos (n-m)y - \cos (n+m)y]\, dy = 0$$

4.2 Consider a particle of mass m moving in a one dimensional potential specified by

$$V(x) = \begin{cases} 0, & -2a < x < 2a \\ \infty, & \text{otherwise} \end{cases}$$

Find the energy eigenvalues and eigenfunctions.

Solution. The time-independent Schrödinger equation for the region $-2a < x < 2a$ (Fig. 4.2) is

$$\frac{d^2\psi}{dx^2} + k^2\psi = 0, \qquad k^2 = \frac{2mE}{\hbar^2}$$

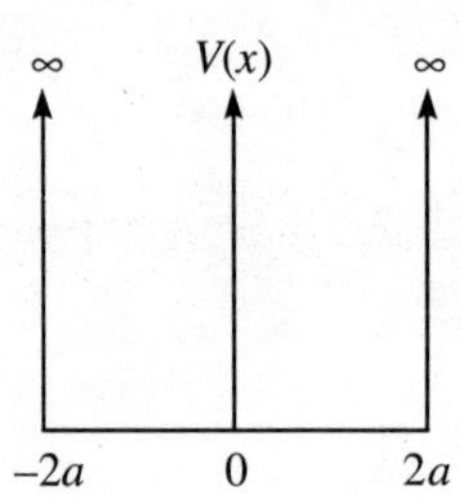

Fig. 4.2 Infinite square well of bottom.

Its solution is

$$\psi(x) = A \sin kx + B \cos kx$$

At $x = \pm 2a$, $V(x) = \infty$. Hence, $\psi(\pm 2a) = 0$.
Application of this boundary condition gives

$$A \sin (2ka) + B \cos (2ka) = 0$$

$$-A \sin (2ka) + B \cos (2ka) = 0$$

From the above two relations,

$$A \sin (2ka) = 0, \qquad B \cos (2ka) = 0$$

Now, two possibilities arise: $A = 0$, $B \neq 0$ and $A \neq 0$, $B = 0$.
The first condition gives

$$\cos (2ka) = 0; \quad 2ka = \frac{n\pi}{2}, \qquad n = 1, 3, 5, \ldots$$

$$k^2 = \frac{n^2\pi^2}{16a^2} = \frac{2mE_n}{\hbar^2}$$

$$E_n = \frac{n^2\pi^2\hbar^2}{32ma^2}, \qquad n = 1, 3, 5, \ldots$$

$$\psi_n = B \cos \frac{n\pi x}{4a}, \qquad n = 1, 3, 5, \ldots$$

Normalization yields

$$\psi_n = \frac{1}{\sqrt{2a}} \cos \frac{n\pi x}{4a}, \qquad n = 1, 3, 5, \ldots$$

The condition $A \neq 0$, $B = 0$ leads to

$$E_n = \frac{n^2\pi^2\hbar^2}{32ma^2}, \qquad n = 2, 4, 6, \ldots$$

$$\psi_n = \frac{1}{\sqrt{2a}} \sin \frac{n\pi x}{4a}, \qquad n = 2, 4, 6, \ldots$$

4.3 For an electron in a one-dimensional infinite potential well of width 1 Å, calculate (i) the separation between the two lowest energy levels; (ii) the frequency and wavelength of the photon corresponding to a transition between these two levels; and (iii) in what region of the electromagnetic spectrum is this frequency/wavelength?

Solution.

(i) From Eq. (4.2),

$$E_n = \frac{\pi^2\hbar^2 n^2}{8ma^2}, \qquad 2a = 1 \text{ Å} = 10^{-10} \text{ m}$$

$$E_2 - E_1 = \frac{3\pi^2\hbar^2}{8ma^2} = \frac{3 \times \pi^2 \times (1.055 \times 10^{-34} \text{ J s})^2 \times 4}{8\,(9.1 \times 10^{-31} \text{ kg})\, 10^{-20} \text{ m}^2}$$

$$= 1.812 \times 10^{-17} \text{ J} = 113.27 \text{ eV}$$

(ii)

$$h\nu = 1.812 \times 10^{-17}\ \text{J}$$

$$\nu = 2.7 \times 10^{16}$$

$$\lambda = \frac{c}{\nu} = \frac{3 \times 10^8\ \text{ms}^{-1}}{2.7 \times 10^{16}\ \text{s}^{-1}} = 1.1 \times 10^{-8}\ \text{m}$$

(iii) This frequency falls in the vacuum ultraviolet region.

4.4 Show that the energy and the wave function of a particle in a square well of finite depth V_0 reduces to the energy and the wave function of a square well with rigid walls in the limit $V_0 \to \infty$.

Solution. For a well of finite depth V_0, Eq. (4.7) gives

$$\tan ka = \frac{\alpha}{k}, \qquad k^2 = \frac{2mE}{\hbar^2}, \qquad \alpha^2 = \frac{2m}{\hbar^2}(V_0 - E)$$

$$\tan ka = \sqrt{\frac{V_0 - E}{E}} \quad \text{or} \quad \underset{V_0 \to \infty}{\text{Lt}} \tan ka \to \infty$$

$$ka = \frac{n\pi}{2} \quad \text{or} \quad k^2a^2 = \frac{n^2\pi^2}{4}$$

$$E_n = \frac{\pi^2\hbar^2 n^2}{8ma^2} \quad \text{[which is the same as Eq. (4.2).]}$$

The wave functions in the different regions will be

$$\psi(x) = \begin{cases} Ae^{\alpha x}, & x < -a \\ B \sin kx + C \cos kx, & -a < x < a \\ De^{-\alpha x}, & x > a \end{cases}$$

When $V_0 \to \infty$, $\alpha \to \infty$, and the wave function reduces to

$$\psi(x) = \begin{cases} 0, & x < -a \\ A \sin kx + B \cos kx, & -a < x < a \\ 0, & x > a \end{cases}$$

which is the wave function of a particle in a square well with rigid walls.

4.5 Calculate the expectation values of position $\langle x \rangle$ and of the momentum $\langle p_x \rangle$ of the particle trapped in the one-dimensional box of Problem 4.1.

Solution.

$$\langle x \rangle = \frac{2}{a}\int_0^a \sin\frac{n\pi x}{a}\, x \sin\frac{n\pi x}{a}\, dx$$

$$= \frac{2}{a}\int_0^a x \sin^2\frac{n\pi x}{a}\, dx = \frac{1}{a}\int_0^a x\left(1 - \cos\frac{2n\pi x}{a}\right) dx$$

$$= \frac{1}{a}\int_0^a x\, dx - \frac{1}{a}\int_0^a x \cos\frac{2n\pi x}{a}\, dx$$

As the second term vanishes when integrated by parts,

$$\langle x \rangle = \frac{a}{2}$$

$$\langle p_x \rangle = \frac{2}{a}\int_0^a \sin\frac{n\pi x}{a}\left(-i\hbar\frac{d}{dx}\right)\sin\frac{n\pi x}{a}\,dx$$

$$= -i\hbar\frac{2n\pi}{a^2}\int_0^a \sin\frac{n\pi x}{a}\cos\frac{n\pi x}{a}\,dx$$

$$= -i\hbar\frac{n\pi}{a^2}\int_0^a \sin\frac{2\pi nx}{a}\,dx = 0$$

4.6 An electron in a one-dimensional infinite potential well, defined by $V(x) = 0$ for $-a \le x \le a$ and $V(x) = \infty$ otherwise, goes from the $n = 4$ to the $n = 2$ level. The frequency of the emitted photon is 3.43×10^{14} Hz. Find the width of the box.

Solution.

$$E_n = \frac{\pi^2\hbar^2 n^2}{8ma^2}, \qquad m = 9.1 \times 10^{-31}\text{ kg}$$

$$E_4 - E_2 = \frac{12\pi^2\hbar^2}{8ma^2} = h\nu$$

$$a^2 = \frac{3h}{8m\nu} = \frac{3(6,626 \times 10^{-34}\text{ J s})}{8(9.1 \times 10^{-31}\text{ kg})(3.43 \times 10^{14}\text{ s}^{-1})}$$

$$= 79.6 \times 10^{-20}\text{ m}^2$$

$$a = 8.92 \times 10^{-10}\text{ m} \quad \text{or} \quad 2a = 17.84 \times 10^{-10}\text{ m}$$

4.7 A particle of mass m trapped in the potential $V(x) = 0$ for $-a \le x \le a$ and $V(x) = \infty$ otherwise. Evaluate the probability of finding the trapped particle between $x = 0$ and $x = a/n$ when it is in the nth state.

Solution. Wave function $\psi(x) = \sqrt{\frac{2}{a}}\sin\frac{n\pi x}{a}$ (refer Problem 1)

Probability density $P(x) = \frac{2}{a}\sin^2\frac{n\pi x}{a}$

Required probability $P = \int_0^{a/n} P(x)\,dx = \frac{2}{a}\int_0^{a/n}\sin^2\frac{n\pi x}{a}\,dx$

$$P = \frac{1}{a}\int_0^{a/n}\left(1 - \cos\frac{2n\pi x}{a}\right)dx = \frac{1}{n}$$

4.8 An alpha particle is trapped in a nucleus of radius 1.4×10^{-15} m. What is the probability that it will escape from the nucleus if its energy is 2 MeV? The potential barrier at the surface of the nucleus is 4 MeV and the mass of the α-particle = 6.64×10^{-27} kg.

Solution. Transmission coefficient $T = 16\frac{E}{V_0}\left(1 - \frac{e}{V_0}\right)\exp\left[-\frac{2a}{\hbar}\sqrt{2m(V_0 - E)}\right]$

Mass of alpha particle = 6.64×10^{-27} kg

$$\sqrt{2m(V_0 - E)} = [2(6.64 \times 10^{-27}\ \text{kg})(2 \times 10^6\ \text{eV})(1.6 \times 10^{-19}\ \text{J/eV})]^{1/2}$$

$$= 6.52 \times 10^{-20}\ \text{kg m s}^{-1}$$

$$\frac{2a}{\hbar}\sqrt{2m(V_0 - E)} = \frac{2(2.8 \times 10^{-15}\ \text{m})}{1.05 \times 10^{-34}\ \text{J s}} \times 6.52 \times 10^{-20}\ \text{kg m s}^{-1} = 3.477$$

$$T = 16 \times \frac{1}{2} \times \frac{1}{2} \times \exp(-3.477) = 0.124$$

4.9 The wave function of a particle confined in a box of length a is

$$\psi(x) = \sqrt{\frac{2}{a}}\sin\frac{\pi x}{a}, \qquad 0 \le x \le a$$

Calculate the probability of finding the particle in the region $0 < x < a/2$.

Solution. The required probability $P = \frac{2}{a}\int_0^{a/2} \sin^2\frac{\pi x}{a}\,dx$

$$= \frac{1}{a}\int_0^{a/2}\left(1 - \cos\frac{2\pi x}{a}\right)dx$$

$$= \frac{1}{a}\int_0^{a/2} dx - \frac{1}{a}\int_0^{a/2}\cos\frac{2\pi x}{a}\,dx = \frac{1}{2}$$

4.10 Find $\langle x\rangle$ and $\langle p\rangle$ for the nth state of the linear harmonic oscillator.

Solution. For the harmonic oscillator, $\psi_n(x) = AH_n(x)\exp(-m\omega x^2/2\hbar)$

$$\langle x\rangle = A^2\int_{-\infty}^{\infty} H_n^2(x)\, x\exp\left(-\frac{m\omega x^2}{\hbar}\right)dx = 0$$

since the integrand is an odd function of x.

$$\langle p\rangle = -i\hbar A^2\int_{-\infty}^{\infty} H_n(x)\exp\left(-\frac{m\omega x^2}{2\hbar}\right)\frac{d}{dx}\left[H_n\exp\left(-\frac{m\omega x^2}{2\hbar}\right)\right]dx$$

$$= -i\hbar A^2\int_{-\infty}^{\infty}\left[H_n H_n \exp\left(-\frac{m\omega x^2}{\hbar}\right) - \frac{m\omega x}{\hbar}H_n^2\exp\left(-\frac{m\omega x^2}{\hbar}\right)\right]dx$$

$$= 0$$

since both the integrand terms are odd functions of x. Here, $H'_n = dH_n/dx$.

4.11 For the nth state of the linear harmonic oscillator, evaluate the uncertainty product (Δx) (Δp).

Solution. According to the Virial theorem, the average values of the kinetic and potential energies of a classical harmonic oscillator are equal. Assuming that this holds for the expectation values of the quantum oscillator, we have

$$\frac{1}{2m}\langle p_x^2\rangle = \frac{1}{2}k\langle x^2\rangle = \frac{\hbar\omega}{2}\left(n+\frac{1}{2}\right) \qquad k = m\omega^2$$

Hence,

$$\langle p_x^2\rangle = m\hbar\omega\left(n+\frac{1}{2}\right), \qquad \langle x^2\rangle = \frac{\hbar}{m\omega}\left(n+\frac{1}{2}\right)$$

$$(\Delta x)^2 = \langle x^2\rangle - \langle x\rangle^2 = \langle x^2\rangle \quad \text{[refer Problem 4.10]}$$

$$(\Delta p_x)^2 = \langle p_x^2\rangle$$

$$(\Delta x)^2(\Delta p_x)^2 = \left(n+\frac{1}{2}\right)^2\hbar^2, \qquad (\Delta x)(\Delta p_x) = \left(n+\frac{1}{2}\right)\hbar$$

4.12 A harmonic oscillator is in the ground state. (i) Where is the probability density maximum? (ii) What is the value of maximum probability density?

Solution.

(i) The ground state wave function

$$\psi_0(x) = \left(\frac{m\omega}{\hbar\pi}\right)^{1/4}\exp\left(\frac{-m\omega x^2}{2\hbar}\right)$$

The probability density

$$P(x) = \psi_0^*\psi_0 = \left(\frac{m\omega}{\hbar\pi}\right)^{1/2}\exp\left(-\frac{m\omega^2 x^2}{\hbar}\right)$$

$P(x)$ will be maximum at the point where

$$\frac{dP}{dx} = 0 = \left(\frac{m\omega}{\hbar\pi}\right)^{1/2}\left(-\frac{m\omega}{\hbar}\right)2x\exp\left(-\frac{m\omega^2 x^2}{\hbar}\right)$$

$$x = 0$$

Thus, the probability density is maximum at $x = 0$.

(ii) $P(0) = \left(\dfrac{m\omega}{\hbar\pi}\right)^{1/2}$

4.13 A 1 eV electron got trapped inside the surface of a metal. If the potential barrier is 4.0 eV and the width of the barrier is 2 Å, calculate the probability of its transmission.

Solution. If L is the width of the barrier, the transmission coefficient

$$T = 16\frac{E}{V}\left(1-\frac{E}{V}\right)\exp\left[-\frac{2L}{\hbar}\sqrt{2m(V-E)}\right]$$

$$= 16\times\frac{1}{4}\times\frac{3}{4}\times\exp\left(-\frac{2\times2\times10^{-10}\,\text{m}}{1.05\times10^{-34}\,\text{J s}}\sqrt{2(9.1\times10^{-31}\,\text{kg})(3\times1.6\times10^{-19}\,\text{J})}\right)$$

$$= 0.085$$

4.14 An electron is in the ground state of a one-dimensional infinite square well with $a = 10^{-10}$ m. Compute the force that the electron exerts on the wall during an impact on either wall.

Solution. The force on the wall

$$F = -\frac{dE_n}{da}$$

The energy of the ground state

$$E_1 = \frac{\pi^2\hbar^2}{2ma^2}$$

and hence the force on the wall

$$F = -\frac{dE_1}{da}\bigg|_{a=10^{-10}} = \frac{\pi^2\hbar^2}{ma^3}\bigg|_{a=10^{-10}}$$

$$= \frac{\pi^2(1.054 \times 10^{-34}\ \text{J s})^2}{(9.1 \times 10^{-31}\ \text{kg})(10^{-10}\ \text{m})^3}$$

$$= 1.21 \times 10^{-7}\ \text{N}$$

4.15 Show that the probability density of the linear harmonic oscillator in an arbitrary superposition state is periodic with the period equal to the period of the oscillator.

Solution. The time-dependent wave function of the linear harmonic oscillator in a superposition state is

$$\Psi(x,t) = \sum_n C_n\psi_n(x)\exp(-iE_nt/\hbar)$$

where $\psi_n(x)$ is the time-independent wave function of the harmonic oscillator in the nth state. The probability density

$$P(x,t) = |\Psi(x,t)|^2 = \sum_m\sum_n C_m^*C_n\psi_m^*\psi_n \exp[i(E_m - E_n)t/\hbar)]$$

It is obvious that $P(x, t)$ is dependent on time. Let us investigate what happens to $P(x, t)$ if t is replaced by $t + 2\pi/\omega$. It follows that

$$\exp\left[\frac{i(E_m - E_n)}{\hbar}\left(t + \frac{2\pi}{\omega}\right)\right] = \exp\left[\frac{i(E_m - E_n)t}{\hbar}\right]\exp\left[\frac{i(E_m - E_n)}{\hbar}\frac{2\pi}{\omega}\right]$$

$$= \exp\left[\frac{i(E_m - E_n)t}{\hbar}\right]$$

since $(E_m - E_n)$ is an integral multiple of $\hbar\omega$, i.e., $P(x, t)$ is periodic with period $2\pi/\omega$, the period of the linear harmonic oscillator.

4.16 For harmonic oscillator wave functions, find the value of $(\psi_k, x\psi_n)$.

Solution. For Hermite polynomials,

$$H_{n+1}(y) - 2yH_n(y) + 2nH_{n-1}(y) = 0$$

Substituting the values of H_{n+1}, H_n and H_{n-1} in terms of the oscillator wave functions, [(Eq. 4.14)], and dropping $e^{y^2/2}(\hbar\pi/m\omega)^{1/4}$ from all terms, we get

$$[2^{n+1}(n+1)!]^{1/2}\psi_{n+1} - 2y(2^n n!)^{1/2}\psi_n + 2n[2^{n-1}(n-1)!]^{1/2}\psi_{n-1} = 0$$

$$(n+1)]^{1/2}\psi_{n+1} - \sqrt{2}\, y\psi_n + n^{1/2}\psi_{n-1} = 0$$

Since $y = (m\omega/\hbar)^{1/2}\, x$, the inner product of this equation with ψ_k gives

$$(n+1)^{1/2}(\psi_k, \psi_{n+1}) - (2m\omega/\hbar)^{1/2}(\psi_k, x\psi_n) + n^{1/2}(\psi_k, \psi_{n-1}) = 0$$

$$(\psi_k, \psi_n) = \left[\frac{(n+1)\hbar}{2m\omega}\right]^{1/2}(\psi_k, \psi_{n+1}) + \left(\frac{n\hbar}{2m\omega}\right)^{1/2}(\psi_k, \psi_{n-1})$$

$$(\psi_k, x\psi_n) = \begin{cases} \sqrt{\hbar(n+1)/2m\omega} & \text{if } k = n+1 \\ \sqrt{\hbar n/2m\omega} & \text{if } k = n-1 \\ 0 & \text{if } k \neq n \pm 1 \end{cases}$$

4.17 Evaluate $\langle x^2\rangle$, $\langle p^2\rangle$, $\langle V\rangle$ and $\langle T\rangle$ for the states of a harmonic oscillator.

Solution. From Problem 4.16,

$$(n+1)^{1/2}\psi_{n+1} - \left(\frac{2m\omega}{\hbar}\right)^{1/2} x\psi_n + n^{1/2}\psi_{n-1} = 0$$

Multiplying from left by x and then taking the inner product of the resulting equation with ψ_n, we get

$$(n+1)^{1/2}(\psi_n, x\psi_{n+1}) - \left(\frac{2m\omega}{\hbar}\right)^{1/2}(\psi_n, x^2\psi_n) + n^{1/2}(\psi_n, x\psi_{n-1}) = 0$$

Using the results of Problem 4.16, we obtain

$$\sqrt{n+1}\sqrt{\frac{\hbar(n+1)}{2m\omega}} - \sqrt{\frac{2m\omega}{\hbar}}(\psi_n, x^2\psi_n) + \sqrt{n}\sqrt{\frac{\hbar n}{2m\omega}} = 0$$

$$\sqrt{\frac{\hbar n}{2m\omega}}(2n+1) = \sqrt{\frac{2m\omega}{\hbar}}(\psi_n, x^2\psi_n)$$

$$\langle x^2\rangle = (\psi_n, x^2\psi_n) = \frac{\hbar}{2m\omega}(2n+1)$$

$$\langle p^2\rangle = -\hbar^2\left(\psi_n, \frac{d^2\psi_n}{dx^2}\right)$$

The Schrödinger equation for harmonic oscillator is

$$\frac{d^2\psi_n}{dx^2} = -\frac{2mE_n}{\hbar^2}\psi_n + \frac{m^2\omega^2x^2}{\hbar^2}\psi_n$$

Substituting this value of $d^2\psi_n/dx^2$ and using the result for $\langle x^2\rangle$, we get

$$\langle p^2\rangle = 2mE_n\,(\psi_n, \psi_n) - m^2\omega^2\,(\psi_n, x^2\psi_n)$$

$$\langle p^2\rangle = 2mE_n - m^2\omega^2 \frac{\hbar}{2m\omega}(2n+1)$$

$$= (2n+1)\,m\hbar\omega - \frac{(2n+1)}{2}\,m\hbar\omega$$

$$= \frac{(2n+1)}{2}\,m\hbar\omega = m\left(n + \frac{1}{2}\right)\hbar\omega$$

Expectation value of potential energy $= \frac{1}{2}k\langle x^2\rangle$

$$\langle V\rangle = \frac{1}{2}\left(n + \frac{1}{2}\right)\hbar\omega = \frac{E_n}{2}$$

The expectation value of kinetic energy

$$\langle T\rangle = \frac{1}{2m}\langle p^2\rangle = \frac{1}{2}\left(n + \frac{1}{2}\right)\hbar\omega = \frac{E_n}{2}$$

4.18 Show that the zero point energy of (1/2) $\hbar\omega$ of a linear harmonic oscillator is a manifestation of the uncertainty principle.

Solution. The average position and momentum of a classical harmonic oscillator bound to the origin is zero. According to Ehrenfest's theorem, this rule must be true for the quantum mechanical case also. Hence,

$$(\Delta x)^2 = \langle x^2\rangle - \langle x\rangle^2 = \langle x^2\rangle$$

$$(\Delta p)^2 = \langle p^2\rangle - \langle p\rangle^2 = \langle p^2\rangle$$

For the total energy E,

$$\langle E\rangle = \frac{1}{2m}\langle p^2\rangle + \frac{1}{2}k\,\langle x^2\rangle, \qquad k = m\omega^2$$

$$= \frac{1}{2m}\langle \Delta p^2\rangle + \frac{1}{2}k\,\langle \Delta x\rangle^2$$

Replacing $\langle \Delta p\rangle^2$ with the help of the relation

$$\langle \Delta p\rangle^2\,\langle \Delta x\rangle^2 \geq \frac{\hbar^2}{4}$$

$$\langle E\rangle \geq \frac{\hbar^2}{8m\,(\Delta x)^2} + \frac{1}{2}k\,\langle \Delta x\rangle^2$$

For the RHS to be minimum, the differential of $\langle E\rangle$ with respect to $\langle \Delta x\rangle^2$ must be zero, i.e.,

$$\frac{\hbar^2}{8m\,(\Delta x)^4_{\min}} + \frac{1}{2}k = 0 \quad \text{or} \quad (\Delta x)^2_{\min} = \frac{\hbar^2}{2m\omega}$$

$$\langle E\rangle_{\min} = \frac{\hbar^2}{8m}\frac{2m\omega}{\hbar} + \frac{1}{2}m\omega^2\frac{\hbar}{2m\omega} = \frac{1}{2}\hbar\omega$$

4.19 A stream of particles of mass m and energy E move towards the potential step $V(x) = 0$ for $x < 0$ and $V(x) = V_0$ for $x > 0$. If the energy of the particles $E > V_0$, show that the sum of fluxes of the transmitted and reflected particles is equal to the flux of incident paricles.

Solution. The Schrödinger equation for regions 1 and 2 (see Fig. 4.3) are

$$\frac{d^2\psi_1}{dx^2} + k_0^2\psi = 0, \qquad k_0^2 = \frac{2mE}{\hbar^2}, \qquad x < 0$$

$$\frac{d^2\psi_2}{dx^2} + k^2\psi = 0, \qquad k^2 = \frac{2m(E - V_0)}{\hbar^2}, \qquad x > 0$$

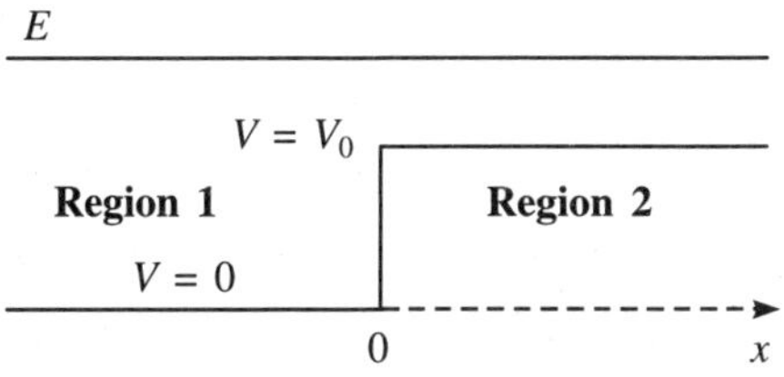

Fig. 4.3 Potential step.

The solutions of the two equations are

$$\psi_1 = e^{ik_0x} + Ae^{-ik_0x}, \qquad x < 0$$

$$\psi_2 = Be^{ikx}, \qquad x > 0$$

For convenience, the amplitude of the incident wave is taken as 1. The second term in ψ_1, a wave travelling from right to left, is the reflected wave whereas ψ_2 is the transmitted wave. It may be noted that in region 2 we will not have a wave travelling from right to left. The continuity conditions on ψ and its derivative at $x = 0$ give

$$1 + A = B, \qquad k_0(1 - A) = kB$$

Simplifying, we get

$$A = \frac{k_0 - k}{k_0 + k}, \qquad B = \frac{2k_0}{k_0 + k}$$

Flux of particles for the incident wave (see Problem 2.22) $= \dfrac{k_0\hbar}{m}$

Magnitude of flux of particles for the reflected wave $= \dfrac{k_0\hbar}{m}|A|^2$

Flux of particles for the transmitted wave $= \dfrac{k\hbar}{m}|B|^2$

The sum of reflected and transmitted flux is given by

$$\frac{\hbar}{m}[k_0|A|^2 + k|B|^2] = \frac{\hbar k_0}{m}\left[\frac{(k_0 - k)^2}{(k_0 + k)^2} + \frac{4kk_0}{(k_0 + k)^2}\right] = \frac{\hbar k_0}{m}$$

which is the incident flux.

4.20 A stream of particles of mass m and energy E move towards the potential step of Problem 4.19. If the energy of particles $E < V_0$, show that there is a finite probability of finding the particles in the region $x > 0$. Also, determine the flux of (i) incident particles, (ii) reflected particles, and (iii) the particles in region 2. Comment on the results.

Solution. The Schrödinger equation and its solution for the two regions (see Fig. 4.3) are

$$\frac{d^2\psi_1}{dx^2} + k_0^2\psi_1 = 0, \qquad k_0^2 = \frac{2mE}{\hbar^2}, \qquad x < 0$$

$$\frac{d^2\psi_2}{dx^2} - \gamma^2\psi_2 = 0, \qquad \gamma^2 = \frac{2m(V_0 - E)}{\hbar^2}, \qquad x > 0$$

$$\psi_1 = e^{ik_0x} + Be^{-ik_0x}, \qquad x < 0$$

$$\psi_2 = Ce^{-\gamma x}, \qquad x > 0$$

The solution $e^{\gamma x}$ in region 2 is left out as it diverges and the region is an extended one. The continuity condition at $x = 0$ gives

$$1 + B = C, \qquad ik_0(1 - B) = -\gamma C$$

Solving, we get

$$B = \frac{ik_0 + \gamma}{ik_0 - \gamma}, \qquad C = \frac{2ik_0}{ik_0 - \gamma}$$

The reflection coefficient

$$R = |B|^2 = \left(\frac{ik_0 + \gamma}{ik_0 - \gamma}\right)\left(\frac{-ik_0 + \gamma}{-ik_0 - \gamma}\right) = 1$$

$$\text{Reflected flux} = -\frac{\hbar k_0}{m}|B|^2 = -\frac{\hbar k_0}{m}$$

The negative sign indicates that it is from right to left. Since ψ_2 is real, the transmitted flux = 0 and, therefore, the transmission coefficient $T = 0$. However, the wave function in the region $x > 0$ is given by

$$\psi_2 = \frac{2ik_0}{ik_0 - \gamma}\, e^{-\gamma x}$$

Therefore, the probability that the particle is found in the region $x > 0$ is finite. Due to the uncertainty in energy, the total energy may even be above V_0.

4.21 A beam of 12 eV electrons is incident on a potential barrier of height 30 eV and width 0.05 nm. Calculate the transmission coefficient.

Solution. The transmission coefficient T is given by

$$T = \frac{16E(V_0 - E)}{V_0^2} \exp\left[-\frac{2a}{\hbar}\sqrt{2m(V_0 - E)}\right]$$

$$\frac{16E(V_0 - E)}{V_0^2} = \frac{16 \times 12 \times 18}{30 \times 30} = 3.84$$

$$\frac{2a}{\hbar}\sqrt{2m(V_0 - E)} = \frac{2(0.05 \times 10^{-9}\,\text{m})}{(1.054 \times 10^{-34}\,\text{J s})} \times 2 \times (9.1 \times 10^{-31}\,\text{kg})(18 \times 1.6 \times 10^{-19}\,\text{J})^{1/2}$$

$$= 2.172$$

$$T = \frac{3.84}{\exp(2.172)} = \frac{3.84}{8.776} = 0.44$$

4.22 For the nth state of the linear harmonic oscillator, what range of x values is allowed classically? In its ground state, show that the probability of finding the particle outside the classical limits is about 16 per cent.

Solution. At the classical turning points, the oscillator has only potential energy. Hence, at the turning points,

$$\frac{1}{2}m\omega^2 x^2 = \left(n + \frac{1}{2}\right)\hbar\omega$$

$$x = \pm\left[\frac{(2n+1)\hbar}{m\omega}\right]^{1/2}$$

The allowed range of x values are

$$-\left[\frac{(2n+1)\hbar}{m\omega}\right]^{1/2} < x < \left[\frac{(2n+1)\hbar}{m\omega}\right]^{1/2}$$

When the oscillator is in the ground state, the turning points are $-\left(\dfrac{\hbar}{m\omega}\right)^{1/2}$ and $\left(\dfrac{\hbar}{m\omega}\right)^{1/2}$

The ground state wave function is

$$\psi_0(x) = \left(\frac{m\omega}{\pi\hbar}\right)^{1/4} \exp\left(-\frac{m\omega x^2}{2\hbar}\right)$$

The probability for the particle to be outside, the classical limits are

$$P = 2\int_{(\hbar/m\omega)^{1/2}}^{\infty} |\psi_0|^2\, dx = 2\left(\frac{m\omega}{\pi\hbar}\right)^{1/2} \int_{(\hbar/m\omega)^{1/2}}^{\infty} \exp\left(-\frac{m\omega x^2}{\hbar}\right) dx$$

$$= \frac{2}{\pi^{1/2}}\int_1^{\infty} e^{-y^2}\, dy = \frac{2}{\pi^{1/2}} \times 0.1418 = 0.1599 = 16\%$$

4.23 An electron moves in a one-dimensional potential of width 8 Å and depth 12 eV. Find the number of bound states present.

Solution. If follows from Eq. (4.10) that, if the width is $2a$, Then

(a) One bound state exists if $0 < V_0 a^2 < \pi^2\hbar^2/8m$.

(b) Two bound states exist if $\pi^2\hbar^2/8m < V_0 a^2 < 4\pi^2\hbar^2/8m$.

(c) Three bound states exist if $4\pi^2\hbar^2/8m < V_0 a^2 < 9\pi^2\hbar^2/8m$.

(d) Four bound states exist if $9\pi^2\hbar^2/8m < V_0 a^2 < 16\pi^2\hbar^2/8m$, ...

In the given case, the width is 8Å, and hence a = 4Å = 4 × 10^{-10} m. Therefore,

$$V_0 a^2 = (12 \times 1.6 \times 10^{-19}\ \text{J})\ (16 \times 10^{-20}\ \text{m}^2) = 307.2 \times 10^{-39}\ \text{kg m}^4\ \text{s}^{-2}$$

$$\frac{\pi^2\hbar^2}{8\text{m}} = \frac{\pi^2(1.05\times10^{-34}\ \text{J s})^2}{8(9.1\times10^{-31}\ \text{kg})} = 14.96\times10^{-39}\ \text{kg m}^4\text{s}^{-2}$$

$V_0 a^2$ = 307.2 × 10^{-39} kg m^4 s^{-2} lies between $\dfrac{16\,\pi^2\hbar^2}{8m}$ and $\dfrac{25\,\pi^2\hbar^2}{8m}$

Thus, the number of bound states present is 5.

4.24 A linear harmonic oscillator is in the first excited state. (i) At what point is its probability density maximum? (ii) What is the value of maximum probability density?

Solution. The harmonic oscillator wave function in the n = 1 state is

$$\psi_1(x) = \left(\frac{\alpha}{2\sqrt{\pi}}\right)^{1/2} 2\alpha x \exp\left(\frac{-a^2x^2}{2}\right) \qquad \alpha = \left(\frac{m\omega}{\hbar}\right)^{1/2}$$

(i) Probability density $P(x) = \psi\psi^* = \dfrac{2\alpha^3}{\sqrt{\pi}}\, x^2 \exp(-\alpha^2 x^2)$

$P(x)$ is maximum when dP/dx = 0, and hence

$$0 = \frac{2\alpha^3}{\sqrt{\pi}}(2x - 2\alpha^2 x^3) \quad \text{or} \quad x = \pm\frac{1}{\alpha}$$

(ii) Maximum value of $P(x) = \dfrac{2\alpha}{\sqrt{\pi}}\dfrac{1}{e} = \dfrac{2\alpha}{\sqrt{\pi}}\dfrac{1}{2.718} = 0.415\alpha$

4.25 Sketch the probability density $|\psi|^2$ of the linear harmonic oscillator as a function of x for n = 10. Compare the result with that of the classical oscillator of the same total energy and discuss the limit $n \to \infty$.

Solution. Figure 4.4 illustrates the probability $|\psi_{10}|^2$ (n = 10: solid curve). For n = 0, the probability is maximum at x = 0. As the quantum number increases, the maximum probability moves towards the extreme positions. This can be seen from the figure. For a classical oscillator, the probability of finding the oscillator at a given point is inversely proportional to it s velocity at that point. The total energy

$$E = \frac{1}{2}m\text{v}^2 + \frac{1}{2}kx^2 \quad \text{or} \quad \text{v} = \sqrt{\frac{2E - kx^2}{m}}$$

Therefore, the classical probability

$$P_c \propto \sqrt{\frac{m}{2E - kx^2}}$$

This is minimum at x = 0 and maximum at the extreme positions. Figure 4.4 also shows the classical probability distribution (dotted line) for the same energy. Though the two distributions become more and more similar for high quantum numbers, the rapid oscillations of $|\psi_{10}|^2$ is still a discrepancy.

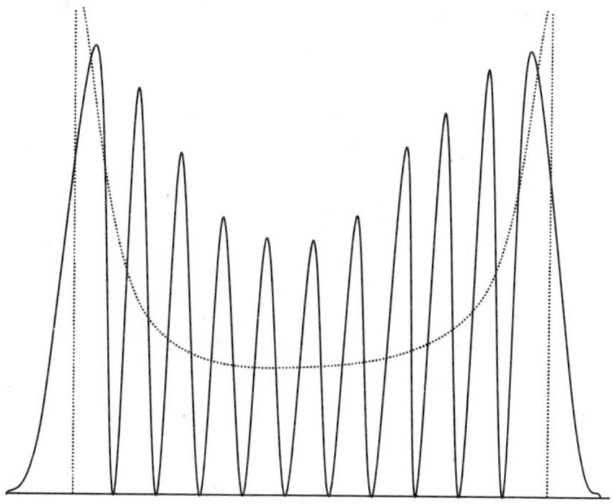

Fig. 4.4 The probability density $|\psi|^2$ for the state $n = 10$ (solid curve) and for a classical oscillator of the same total energy (broken curve).

4.26 Calculate the energy levels and wave functions of a particle of mass m moving in the one-dimensional potential well defined by

$$V(x) = \begin{cases} \infty & \text{for } x < 0 \\ \dfrac{1}{2} m\omega^2 x^2 & \text{for } x > 0 \end{cases}$$

Solution. The harmonic oscillator wave function is given by Eq. (4.14). As $H_1(x)$, $H_3(x)$, $H_5(x)$... are zero at $x = 0$, $\psi(0) = 0$ for odd quantum numbers. However, for $n = 0, 2, 4, \ldots$, $\psi(0) \neq 0$, but finite. The given potential is the same as the simple harmonic oscillator for $x > 0$ and $V(x) = \infty$ for $x < 0$. Hence, $\psi(0)$ has to be zero. Therefore, the even quantum number solutions are not physically acceptable. Consequently, the energy eigenvalues and eigenfunctions are the same as the simple harmonic oscillators with $n = 1, 3, 5, \ldots$

4.27 The strongest *IR* absorption band of $^{12}C^{16}O$ molecule occurs at 6.43×10^{13} Hz. If the reduced mass of $^{12}C^{16}O$ is 1.385×10^{-26} kg, calculate (i) the approximate zero point energy, and (ii) the force constant of the CO bond.

Solution. Zero point energy $\varepsilon_0 = (1/2)h\nu_0$, and hence

$$\varepsilon_0 = \frac{1}{2}(6.626 \times 10^{-34}\text{ J s})(6.43 \times 10^{13}\text{ s}^{-1})$$

$$= 21.30 \times 10^{-21}\text{ J} = 0.133\text{ eV}$$

The force constant $k = 4\pi^2 \nu_0^2 \mu$, and therefore,

$$k = 4\pi^2 \times (6.43 \times 10^{13}\text{ s}^{-1})^2 (1.1385 \times 10^{-26}\text{ kg})$$

$$= 1860\text{ N m}^{-1}$$

4.28 A particle of mass m confined to move in a potential $V(x) = 0$ for $0 \le x \le a$ and $V(x) = \infty$ otherwise. The wave function of the particle at time $t = 0$ is given by

$$\psi(x,0) = A \sin\frac{5\pi x}{a} \cos\frac{2\pi x}{a}$$

(i) Normalize $\psi(x, 0)$, (ii) Find $\psi(x, t)$, (iii) Is $\psi(x, t)$ a stationary state?

Solution. Given

$$\psi(x,0) = A \sin\frac{5\pi x}{a} \cos\frac{2\pi x}{a} = \frac{A}{2}\left(\sin\frac{7\pi x}{a} + \sin\frac{3\pi x}{a}\right)$$

(i) The normalization condition gives

$$\frac{A^2}{4}\int_0^a \left(\sin\frac{7\pi x}{a} + \sin\frac{3\pi x}{a}\right)^2 dx = 1$$

$$\frac{A^2}{4}\int_0^a \left(\sin^2\frac{7\pi x}{a} + \sin^2\frac{3\pi x}{a} + 2\sin\frac{7\pi x}{a}\sin\frac{3\pi x}{a}\right) dx = 1$$

$$\frac{A^2}{4}\left(\frac{a}{2} + \frac{a}{2}\right) = 1 \quad \text{or} \quad A = \frac{2}{\sqrt{a}}$$

Normalized $\psi(x, 0)$ is

$$\psi(x, 0) = \frac{1}{\sqrt{a}}\left(\sin\frac{7\pi x}{a} + \sin\frac{3\pi x}{a}\right)$$

For a particle in an infinite square well, the eigenvalues and eigenfunctions are

$$E_n = \frac{n^2\pi^2\hbar^2}{2ma^2}, \qquad \phi_n(x) = \left(\frac{2}{a}\right)^{1/2} \sin\frac{n\pi x}{a}, \qquad n = 1, 2, 3, \ldots$$

Hence,

$$\psi(x, 0) = \frac{1}{\sqrt{2}}(\phi_7 + \phi_3) = \frac{1}{\sqrt{a}}\left(\sin\frac{7\pi x}{a} + \sin\frac{3\pi x}{a}\right)$$

(ii) The time dependence of a state is given by

$$\psi(x, t) = \psi(x, 0)\, e\,(-iEt/\hbar)$$

Hence, $\psi(x, t)$ in this case is

$$\psi(x, t) = \frac{1}{\sqrt{2}}[\phi_7 \exp(-iE_7 t/\hbar) + \phi_3 \exp(-iE_3 t/\hbar)]$$

(iii) It is not a stationary state since $\psi(x, t)$ is a superposition state.

4.29 Consider a particle of mass m in the one-dimensional short range potential

$$V(x) = -V_0\delta(x), \qquad V_0 > 0$$

where $\delta(x)$ is the Dirac delta function. Find the energy of the system.

Solution. The Schrödinger equation for such a potential is

$$-\frac{\hbar^2}{2m}\frac{d^2\psi(x)}{dx^2} - V_0\delta(x)\psi(x) = E\psi(x)$$

$$\frac{d^2\psi}{dx^2} + \frac{2mE\psi}{\hbar^2} = -\frac{2mV_0}{\hbar^2}\delta(x)\psi$$

Since the potential is attractive, when $E < 0$, the equation to be solved is

$$\frac{d^2\psi}{dx^2} - k^2\psi = -\frac{2mV_0}{\hbar^2}\delta(x)\psi, \qquad k^2 = \frac{2m|E|}{\hbar^2}$$

The solution everywhere except at $x = 0$ must satisfy the equation

$$\frac{d^2\psi}{dx^2} - k^2\psi = 0$$

and for the solution to vanish at $x \to \pm\infty$, we must have

$$\psi(x) = \begin{cases} e^{-kx}, & x > 0 \\ e^{kx}, & x > 0 \end{cases} \tag{i}$$

The normalization factor is assumed to be unity. Integrating the original equation from $-\lambda$ to $+\lambda$, λ being an arbitrarily small positive number, we get

$$\left(\frac{d\psi}{dx}\right)_{-\lambda}^{\lambda} - k^2\int_{-\lambda}^{\lambda}\psi\,dx = -\frac{2mV_0}{\hbar^2}\int_{-\lambda}^{\lambda}\delta(x)\,\psi(x)\,dx$$

The integral on the RHS becomes $-(2mV_0/\hbar_2)\ \psi(0)$ (refer the Appendix). Hence, in the limit $\lambda \to 0$, the above equation becomes

$$\left(\frac{d\psi}{dx}\right)_{x=0+} - \left(\frac{d\psi}{dx}\right)_{x=0-} = -\frac{2mV_0}{\hbar^2}\psi(0)$$

Substituting the values of the LHS from Eq. (i), we get

$$-k\psi(0) - k\psi(0) = -\frac{2mV_0}{\hbar^2}\psi(0)$$

$$k = \frac{mV_0}{\hbar^2} \quad \text{or} \quad \frac{2m\,|E|}{\hbar^2} = \frac{m^2V_0^2}{\hbar^4}$$

$$|E| = \frac{mV_0^2}{2\hbar^2} \quad \text{or} \quad E = -\frac{mV_0^2}{2\hbar^2}$$

4.30 Consider the one-dimensional problem of a particle of mass m in a potential $V = \infty$ for $x < 0$; $V = 0$ for $0 \le x \le a$, and $V = V_0$ for $x > a$ (see Fig. 4.5). Obtain the wave functions and show that the bound state energies ($E < V_0$) are given by

$$\tan\frac{\sqrt{2mE}}{\hbar}a = -\sqrt{\frac{E}{V_0 - E}}$$

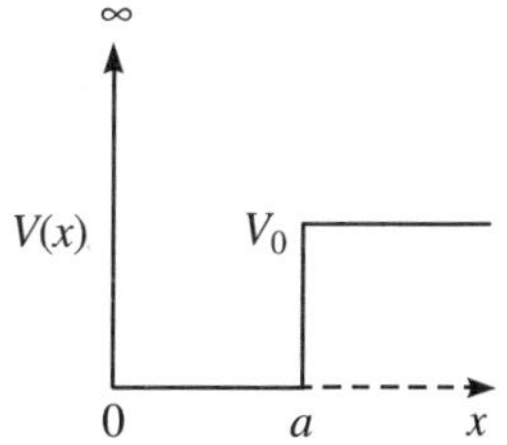

Fig. 4.5 Potential defined in Problem 4.30.

Solution. The Schrödinger equation for the different regions are

$$\frac{d^2\psi}{dx^2} + k^2\psi = 0, \qquad k^2 = \frac{2mE}{\hbar^2}, \qquad 0 \le x \le a$$

$$\frac{d^2\psi}{dx^2} - k_1^2\psi = 0, \qquad k_1^2 = \frac{2m}{\hbar^2}(V_0 - E), \qquad x > a$$

The solution of these equations are

$$\psi = A \sin kx + B \cos kx, \qquad 0 \le x \le a$$

$$\psi = Ce^{-k_1x} + De^{k_1x}, \qquad x > a$$

where A, B, C and D are constants. Applying the boundary conditions $\psi = 0$ at $x = 0$ and $\psi \to 0$ as $x \to \infty$, we get

$$\psi = A \sin kx, \qquad 0 \le x \le a$$

$$\psi = Ce^{-k_1x}, \qquad x > 0$$

The requirement that ψ and $d\psi/dx$ are continuous at $x = a$ gives

$$A \sin ka = Ce^{-k_1a}$$

$$Ak \cos ka = Ck_1e^{-k_1a}$$

Dividing one by the other, we obtain

$$\tan ka = -\frac{k}{k_1}$$

$$\tan\left(\frac{\sqrt{2mE}a}{\hbar}\right) = -\left(\frac{E}{V_0 - E}\right)^{1/2}$$

4.31 Consider a stream of particles of mass m, each moving in the positive x-direction with kinetic energy E towards the potential barrier. Then,

$$V(x) = 0 \qquad \text{for } x \le 0$$

$$V(x) = \frac{3E}{4} \qquad \text{for } x > 0$$

Find the fraction of the particles reflected at $x = 0$.

Solution. The Schrödinger equations for the different regions are

$$\frac{d^2\psi}{dx^2} + k^2\psi = 0, \qquad k^2 = \frac{2mE}{\hbar^2}, \qquad x \le 0 \tag{i}$$

$$\frac{d^2\psi}{dx^2} - \frac{2m}{\hbar^2}\left(\frac{3E}{4} - E\right)\psi = 0, \qquad x > 0$$

$$\frac{d^2\psi}{dx^2} + \left(\frac{k}{2}\right)^2\psi = 0, \qquad x > 0 \tag{ii}$$

The solution of equation (i) is

$$y = e^{ikx} + re^{-ikx}, \qquad x \leq 0$$

where r is the amplitude of the reflected wave since e^{-ikx} represents a wave travelling in the negative x-direction. The solution of equation (ii) is

$$\psi = te^{ikx/2}, \qquad x > 0$$

where t is the amplitude of the transmitted wave. It is also oscillatory since the height of the barrier is less than the kinetic energy of the particle. As the wave function is continuous at $x = 0$,

$$1 + r = t$$

Since the derivative $d\psi/dx$ is continuous at $x = 0$,

$$(1 - r) = \frac{t}{2}$$

Solving the two equations, $r = 1/3$ and hence one-ninth of the particle is reflected at $x = 0$.

4.32 An electron of mass m is contained in a cube of side a, which is fairly large. If it is in an electromagnetic field characterized by the vector potential $\boldsymbol{A} = B_0 x\hat{y}$, $\hat{y}$ being the unit vector along the y-axis, determine the energy levels and eigenfunctions.

Solution. The Hamiltonian operator of the electron having charge $-e$ is

$$H = \frac{1}{2m}\left[p_x^2 + \left(p_y + \frac{B_0 ex}{c}\right)^2 + p_z^2\right]$$

where p_x, p_y, p_z are operators. We can easily prove the following commutation relations:

$$[H, p_y] = [H, p_z] = 0, \qquad [H, p_x] \neq 0$$

Hence, p_y and p_z are constants. The Schrödinger equation is

$$\frac{1}{2m}\left(-\hbar^2 \frac{d^2}{dx^2} + \frac{B_0^2 e^2 x^2}{c^2} + \frac{2B_0 e p_y x}{c} + p_y^2 + p_z^2\right)\psi = E\psi$$

$$\frac{-\hbar^2}{2m}\frac{d^2\psi}{dx^2} + \left(\frac{B_0^2 e^2 x^2}{2mc^2} + \frac{B_0 e p_y x}{mc} + \frac{p_y^2}{2m}\right)\psi = \left(E - \frac{1}{2m}p_z^2\right)\psi$$

we now introduce a new variable x_1 defined by

$$x_1 = x + \frac{cp_y}{B_0 e}$$

$$x_1^2 = x^2 + \frac{2cp_y x}{B_0 e} + \frac{c^2 p_y^2}{B_0^2 e^2}$$

Multiplying by $B_0^2 e^2/(2mc^2)$, we get

$$\frac{B_0^2 e^2 x_1^2}{2mc^2} = \frac{B_0^2 e^2 x^2}{2mc^2} + \frac{B_0 e p_y x}{mc} + \frac{p_y^2}{2m}$$

In terms of the new variable, the Schrödinger equation takes the form

$$-\frac{\hbar^2}{2m}\frac{d^2\psi}{dx_1^2}+\frac{1}{2}\frac{B_0^2e^2x_1^2\psi}{mc^2}=\left(E-\frac{1}{2m}p_z^2\right)\psi$$

The form of this equation is similar to that of the Schrödinger equation for a simple harmonic oscillator. Hence, the energy eigenvalues are

$$E-\frac{1}{2m}p_z^2=\left(n+\frac{1}{2}\right)\hbar\omega,\qquad n=0,\,1,\,2,\,\ldots$$

$$E=\left(n+\frac{1}{2}\right)\hbar\omega+\frac{1}{2m}p_z^2,\qquad n=0,\,1,\,2,\,\ldots$$

where

$$m\omega^2=\frac{B_0^2e^2}{mc^2}\quad\text{or}\quad\omega=\frac{B_0e}{mc}$$

The eigenfunctions are given by

$$\psi_n(x_1)=\left[\left(\frac{\alpha}{\pi}\right)^{1/2}\frac{1}{2^n n!}\right]^{1/2}H_n(\sqrt{\alpha}x_1)\exp(-\alpha x_1^2/2)$$

where

$$\alpha=\frac{m\omega}{\hbar}=\frac{B_0e}{c\hbar}$$

4.33 An electron is confined in the ground state of a one-dimensional harmonic oscillator such that $\Delta x=10^{-10}$ m. Assuming that $\langle T\rangle=\langle V\rangle$, find the energy in electron volts required to excite it to its first excited state.

Solution. Given $\langle T\rangle=\langle V\rangle$. Hence,

$$E_0=\langle T\rangle+\langle V\rangle=2\langle V\rangle=m\omega^2\langle x^2\rangle$$

$$\frac{\hbar\omega}{2}=m\omega^2\langle x^2\rangle\quad\text{or}\quad\omega=\frac{\hbar}{2m\langle x^2\rangle}$$

For harmonic oscillator, $\langle x\rangle=0$ and, therefore,

$$\Delta x=\sqrt{\langle(x-\langle x\rangle)^2\rangle}=\sqrt{\langle x^2\rangle}=10^{-10}\text{ m}$$

The energy required to excite the electron to its first excited state is

$$\Delta E=\hbar\omega=\frac{\hbar^2}{2m\langle x^2\rangle}$$

$$=\frac{(1.05\times10^{-34}\text{ J s})^2}{2(9.1\times10^{-31}\text{ kg})10^{-20}\text{ m}^2}=6.05769\times10^{-19}\text{ J}$$

$$=\frac{6.05769\times10^{-19}\text{ J}}{1.6\times10^{-19}\text{ J/eV}}=3.79\text{ eV}$$

4.34 An electron having energy $E = 1$ eV is incident upon a rectangular barrier of potential energy $V_0 = 2$ eV. How wide must the barrier be so that the transmission probability is 10^{-3}?

Solution. The transmission probability

$$T \cong \frac{16E(V_0 - E)}{V_0^2} e^{-2\alpha a}, \qquad \alpha \cong \frac{1}{\hbar}\sqrt{2m(V_0 - E)}$$

$$T = 4e^{-2\alpha a} \quad \text{or} \quad \ln\frac{T}{4} = -2\alpha a$$

$$-8.294 = -2\alpha a$$

$$\alpha = \frac{\sqrt{2(9.1 \times 10^{-31}\text{ kg})\,1\text{ eV}(1.6 \times 10^{-19}\text{ J/eV})}}{1.05 \times 10^{-34}\text{ J s}}$$

$$= 5.1393 \times 10^9\text{ m}^{-1}$$

$$a = \frac{8.294}{2 \times 5.1395 \times 10^9\text{ m}^{-1}} = 0.8069 \times 10^{-9}\text{ m}$$

$$= 8.1 \times 10^{-8}\text{ cm}$$

4.35 A particle of mass m confined to move in a potential $V(x) = 0$ for $0 \le x \le a$ and $V(x) = \infty$ otherwise. The wave function of the particle at time $t = 0$ is

$$\psi(x, 0) = A\left(2\sin\frac{\pi x}{a} + \sin\frac{3\pi x}{a}\right)$$

(i) Normalize $\psi(x, 0)$; (ii) find $\psi(x, t)$.

Solution. For a particle, in the potential given, the energy eigenvalues and eigenfunctions are given by

$$E_n = \frac{n^2\pi^2\hbar^2}{2ma^2}, \qquad \phi_n(x) = \left(\frac{2}{a}\right)^{1/2}\sin\frac{n\pi x}{a}, \qquad n = 1, 2, 3, \ldots$$

(i)

$$1 = A^2\int_0^2 \left(2\sin\frac{\pi x}{a} + \sin\frac{3\pi x}{a}\right)^2 dx$$

$$1 = A^2\left(4\frac{a}{2} + \frac{a}{2}\right) \quad \text{or} \quad \frac{5a}{2}A^2 = 1$$

$$A = \sqrt{\frac{2}{5a}}$$

$$\Psi(x, 0) = \frac{1}{\sqrt{5}}\left(2\sqrt{\frac{2}{a}}\sin\frac{\pi x}{a} + \sqrt{\frac{2}{a}}\sin\frac{3\pi x}{a}\right)$$

$$= \frac{1}{\sqrt{5}}(2\phi_1 + \phi_3)$$

(ii)

$$\Psi(x, t) = \frac{1}{\sqrt{5}}\left(2\phi_1 e^{-iE_1 t/\hbar} + \phi_3 e^{-iE_3 t/\hbar}\right)$$

4.36 The force constant of HCl molecule is 480 Nm^{-1} and its reduced mass is 1.63×10^{-27} kg. At 300 K, what is the probability that the molecule is in its first excited vibrational state?

Solution. The vibrational energy of the molecule is given by

$$E_v = \left(v + \frac{1}{2}\right)\hbar\omega, \qquad v = 0, 1, 2, \ldots$$

$$\omega = \sqrt{\frac{k}{\mu}} = \sqrt{\frac{480\ \text{Nm}^{-1}}{1.63 \times 10^{-27}\ \text{kg}}} = 5.427 \times 10^{14}\ \text{s}^{-1}$$

The number of molecules in a state is proportional to

$$\exp\left(-\frac{v\hbar\omega}{kT}\right) = \exp(-vx)$$

where $x = \hbar\omega/kT$, where k is the Boltzmann constant. Now,

$$x = \frac{\hbar\omega}{kT} = \frac{(1.054 \times 10^{-34}\ \text{J s})(5.427 \times 10^{14}\ s^{-1})}{(1.38 \times 10^{-23}\ \text{J/k})\ 300\ \text{K}} = 13.8$$

The probability that the molecule is in the first excited state is

$$P_1 = \frac{e^{-x}}{\sum_v e^{-vx}} = \frac{e^{-x}}{1 + e^{-x} + e^{-2x} + \cdots}$$

$$= \frac{e^{-x}}{(1 - e^{-x})^{-1}} = e^{-x}(1 - e^{-x})$$

$$\cong e^{-x} = e^{-13.8} = 1.02 \times 10^{-6}$$

4.37 For a one-dimensional harmonic oscillator, using creation and annihilation operators, show that

$$(\Delta x)(\Delta p) = \left(n + \frac{1}{2}\right)\hbar$$

Solution. From Eqs. (4.18) and (4.19),

$$x = \sqrt{\frac{\hbar}{2m\omega}}(a + a^\dagger), \qquad p = i\sqrt{\frac{m\hbar\omega}{2m\omega}}(a^\dagger - a)$$

where a and $a^\dagger$ are annihilation and creation operators satisfying the conditions

$$a|n\rangle = \sqrt{n}\,|n-1\rangle \quad \text{and} \quad a^\dagger|n\rangle = \sqrt{n+1}\,|n+1\rangle$$

We have the relations

$$(\Delta x)^2 = \langle x^2\rangle - \langle x\rangle^2$$

$$\langle x\rangle = \langle n|x|n\rangle = \sqrt{\frac{\hbar}{2m\omega}}\,[\langle n|a|n\rangle + \langle n|a^\dagger|n\rangle]$$

$$= \sqrt{\frac{\hbar}{2m\omega}}\,[\sqrt{n}\,\langle n|n-1\rangle + \sqrt{n+1}\,\langle n|n+1\rangle] = 0$$

$$\langle x^2\rangle = \langle n|x^2|n\rangle = \frac{\hbar}{2m\omega}\langle n|(a + a^\dagger)(a + a^\dagger)|n\rangle$$

$$= \frac{\hbar}{2m\omega}[\langle n|a\,a|n\rangle + \langle n|a\,a^\dagger|n\rangle + \langle n|a^\dagger a|n\rangle + \langle n|a^\dagger a^\dagger|n\rangle]$$

$$= \frac{\hbar}{2m\omega}[0 + \sqrt{n+1}\sqrt{n+1} + \sqrt{n}\sqrt{n} + 0]$$

$$= \frac{\hbar}{2m\omega}(2n+1)$$

Similarly,

$$\langle n|p|n\rangle = 0, \qquad \langle n|p^2|n\rangle = \frac{m\hbar\omega}{2}(2n+1)$$

$$(\Delta p)^2 = \langle p^2\rangle = \frac{n\hbar\omega}{2}(2n+1)$$

$$(\Delta x)^2(\Delta p)^2 = \frac{\hbar(2n+1)}{2m\omega}\cdot\frac{m\hbar\omega(2n+1)}{2} = \left(n + \frac{1}{2}\right)^2\hbar^2$$

$$(\Delta x)(\Delta p) = \left(n + \frac{1}{2}\right)\hbar$$

4.38 A harmonic oscillator moves in a potential $V(x) = (1/2)kx^2 + cx$, where c is a constant. Find the energy eigenvalues.

Solution. The Hamiltonian of the system is given by

$$H = -\frac{\hbar^2}{2m}\frac{d^2}{dx^2} + \frac{1}{2}kx^2 + cx$$

$$= -\frac{\hbar^2}{2m}\frac{d^2}{dx^2} + \frac{1}{2}k\left(x + \frac{c}{k}\right)^2 - \frac{c^2}{2k}$$

Defining a new variable x_1 by

$$x_1 = x + \frac{c}{k}$$

we get

$$H = -\frac{\hbar^2}{2m}\frac{d^2}{dx_1^2} + \frac{1}{2}kx_1^2 - \frac{c^2}{2k}$$

The Schrödinger equation is

$$-\frac{\hbar^2}{2m}\frac{d^2\psi}{dx_1^2} + \frac{1}{2}kx_1^2\psi - \frac{c^2}{2k}\psi = E\psi$$

which can be modified as

$$-\frac{\hbar^2}{2m}\frac{d^2\psi}{dx_1^2} + \frac{1}{2}k_1^2\psi = \left(E + \frac{c^2}{2k}\right)\psi$$

The form of this equation is the same as the Schrödinger equation for a simple Harmonic oscillator. The energy eigenvalues are

$$E_n' = \left(n + \frac{1}{2}\right)\hbar\omega$$

$$E_n = \left(n + \frac{1}{2}\right)\hbar\omega - \frac{c^2}{2k}$$

4.39 An electron confined to the potential well $V(x) = (1/2)\,kx^2$, where k is a constant, is subjected to an electric field ε along the x-axis. Find the shift of the energy levels of the system.

Solution. The potential energy due to the electric field is $= -\mu \cdot \varepsilon = -(-ex) = e \in x$.

$$\text{Total Hamiltonian } H = -\frac{\hbar^2}{2m}\frac{d^2}{dx^2} + \frac{1}{2}kx^2 + e \in x$$

$$= -\frac{\hbar^2}{2m}\frac{d^2}{dx^2} + \frac{1}{2}k\left(x + \frac{e\varepsilon}{k}\right)^2 - \frac{e^2\varepsilon^2}{2k}$$

Proceeding as in Problem 4.38, the energy eigenvalues are

$$E_n = \left(n + \frac{1}{2}\right)\hbar\omega - \frac{e^2\varepsilon^2}{2k}$$

Hence, the energy shift due to the electric field is $e^2\varepsilon^2/2k$.

4.40 A particle of mass m is confined to a one-dimensional infinite square well of side $0 \le x < a$. At $t = 0$, the wave function of the system is

$$\Psi(x, 0) = c_1 \sin\frac{\pi x}{a} + c_2 \sin\frac{2\pi x}{a},$$

where c_1 and c_2 are the normalization constants for the respective states.

(i) What is the wave function at time t?

(ii) What is the average energy of the system at time t?

Solution. In an infinite square well $0 < x < a$, the energy eigenvalues and eigenfunctions are

$$E_n = \frac{n^2\pi^2\hbar^2}{2ma^2}, \qquad \psi_n = \sqrt{\frac{2}{a}}\sin\frac{n\pi x}{a}, \qquad n = 1, 2, 3, \ldots$$

(i) The wave function at time t is

$$\Psi(x, t) = \Psi(x, 0)\exp\left(\frac{-iE_n t}{\hbar}\right)$$

$$= c_1 \sin\frac{\pi x}{a}\exp\left(\frac{-iE_1 t}{\hbar}\right) + c_2 \sin\frac{2\pi x}{a}\exp\left(\frac{-iE_2 t}{\hbar}\right)$$

$$= c_1 \sin\frac{\pi x}{a}\exp\left(\frac{-i\pi^2\hbar t}{2ma^2}\right) + c_2 \sin\frac{2\pi x}{a}\exp\left(\frac{-i2\pi^2\hbar t}{ma^2}\right)$$

(ii) The average energy of the system at t is

$$\langle E\rangle = \left\langle \Psi(x,t)\left| i\hbar\frac{\partial}{\partial t}\right|\Psi(x,t)\right\rangle = i\hbar\left\langle \Psi\left|\frac{\partial}{\partial t}\right|\Psi\right\rangle$$

$$i\hbar\frac{\partial}{\partial t}\Psi = i\hbar\left(-\frac{i\pi^2\hbar}{2ma^2}\right)c_1 \sin\frac{\pi x}{a}\exp\left(\frac{-i\pi^2\hbar t}{2ma^2}\right) + i\hbar\left(-\frac{i2\pi^2\hbar}{ma^2}\right)c_2\sin\frac{2\pi x}{a}\exp\left(\frac{-i\pi^2\hbar t}{ma^2}\right)$$

Writing $\Psi = c_1\phi_1 + c_2\phi_2$, we get

$$\begin{aligned}\langle E\rangle &= \langle (c_1\phi_1 + c_2\phi_2)\,|\,(E_1c_1\phi_1 + E_2c_2\phi_2)\rangle\\ &= E_1\langle c_1\phi_1|c_1\phi_1\rangle + E_2\langle c_2\phi_2|c_2\phi_2\rangle\\ &= E_1 + E_2\end{aligned}$$

4.41 A particle in a box is in a superposition state and is described by the wave function

$$\Psi(x,t) = \frac{1}{\sqrt{a}}\left[\exp\left(\frac{-iE_1t}{\hbar}\right)\cos\frac{\pi x}{2a} + \exp\left(\frac{-iE_2t}{\hbar}\right)\sin\frac{2\pi x}{2a}\right], \qquad -a < x < a$$

where E_1 and E_2 are the energy eigenvalues of the first two states. Evaluate the expectation value of x.

Solution.

$$\langle x\rangle = \int_{-\infty}^{\infty} \Psi^*(x,t)\, x\Psi(x,t)\, dx$$

Substituting the values of Ψ and Ψ^*, we get

$$\langle x\rangle = \frac{1}{a}\int_{-a}^{a} x\cos^2\frac{\pi x}{2a}\,dx + \frac{1}{a}\int_{-a}^{a} x\sin^2\frac{2\pi x}{2a}\,dx$$

$$+\frac{1}{a}\{\exp[i(E_1 - E_2)t/\hbar)] + \exp[i(E_2 - E_1)t/\hbar)]\}\int_{-a}^{a} x\cos\frac{\pi x}{2a}\sin\frac{2\pi x}{2a}\,dx$$

The integrands in the first two terms are odd and hence will not contribute.

$$\int_{-a}^{a} x\cos\frac{\pi x}{2a}\sin\frac{2\pi x}{2a}\,dx = \int_{-a}^{a}\frac{x}{2}\left(\sin\frac{3\pi x}{2a} + \sin\frac{\pi x}{2a}\right)dx$$

Integrating each term by parts, we get

$$\begin{aligned}\int_{-a}^{a} x\sin\frac{3\pi x}{2a}\,dx &= -\frac{2a}{3\pi}\left(x\cos\frac{3\pi x}{2a}\right)_{-a}^{a} + \frac{2a}{3\pi}\left(\frac{2a}{3\pi}\sin\frac{3\pi x}{2a}\right)_{-a}^{a}\\ &= 0 + \frac{4a^2}{9\pi^2}(-1-1) = -\frac{8a^2}{9\pi^2}\end{aligned}$$

Similarly,

$$\int_{-a}^{a} x\sin\frac{\pi x}{2a}\,dx = \frac{8a^2}{\pi^2}$$

Substituting the values of the integral, we obtain

$$\int_{-a}^{a} x \cos \frac{\pi x}{2a} \sin \frac{2\pi x}{2a}\, dx = \frac{1}{2}\left(-\frac{8a^2}{9\pi^2} + \frac{8a^2}{\pi^2}\right) = \frac{32a^2}{9\pi^2}$$

Replacing the exponential by the cosine function, we get

$$\langle x \rangle = \frac{2}{a} \cos \frac{(E_2 - E_1)t}{\hbar}\left(\frac{32a^2}{9\pi^2}\right)$$

$$= \frac{64a}{9\pi^2} \cos \frac{(E_2 - E_1)t}{\hbar}$$

4.42 For a particle trapped in the potential well, $V(x) = 0$ for $-a/2 \le x \le a/2$ and $V(x) = \infty$ otherwise, the ground state energy and eigenfunction are

$$E_1 = \frac{\pi^2 \hbar^2}{2ma^2}, \qquad \psi_1 = \sqrt{\frac{2}{a}} \cos \frac{\pi x}{a}$$

Evaluate $\langle x \rangle$, $\langle x^2 \rangle$, $\langle p \rangle$, $\langle p^2 \rangle$ and the uncertainty product.

Solution.

$$\langle x \rangle = \frac{2}{a} \int_{-a/2}^{a/2} x \cos^2 \frac{\pi x}{a}\, dx = 0$$

since the integrand is an odd function.

$$\langle x^2 \rangle = \frac{2}{a} \int_{-a/2}^{a/2} x^2 \cos^2 \frac{\pi x}{a}\, dx = \frac{2}{a} \int_{-a/2}^{a/2} \frac{x^2}{2}\left(1 + \cos \frac{2\pi x}{a}\right) dx$$

$$= \frac{1}{a} \int_{-a/2}^{a/2} x^2\, dx + \frac{1}{a} \int_{-a/2}^{a/2} x^2 \cos \frac{2\pi x}{a}\, dx$$

When integrated by parts, the integrated quantity in the second term vanishes.

$$\langle x^2 \rangle = \frac{a^2}{12} - \frac{2}{a}\frac{a}{2\pi} \int_{-a/2}^{a/2} x \sin \frac{2\pi x}{a}\, dx$$

$$= \frac{a^2}{12} + \frac{2}{a}\frac{a}{2\pi}\frac{a}{2\pi}\left[x \cos \frac{2\pi x}{a}\right]_{-a/2}^{a/2} + \int_{-a/2}^{a/2} \cos \frac{2\pi x}{a}\, dx$$

The integral in the third term vanishes, and hence

$$\langle x^2 \rangle = \frac{a^2}{12} - \frac{a^2}{2\pi^2}$$

$$\langle p \rangle = \frac{2}{a} \int_{-a/2}^{a/2} \cos \frac{\pi x}{a}\left(-i\hbar \frac{d}{dx}\right) \cos \frac{\pi x}{a}\, dx$$

$$= \frac{2i\hbar}{\pi} \int_{-a/2}^{a/2} \cos \frac{\pi x}{a} \sin \frac{\pi x}{a}\, dx = 0$$

since the integrand is an odd function. Now,

$$\langle p^2\rangle = \frac{2}{a}\int_{-a/2}^{a/2} \cos\frac{\pi x}{a}\left(-i\hbar\frac{d}{dx}\right)\left(-i\hbar\frac{d}{dx}\right)\cos\frac{\pi x}{a}\,dx$$

Using the Schrödinger equation, we get

$$-\frac{\hbar^2}{2m}\frac{d^2}{dx^2}\psi_1(x) = E_1\psi_1(x)$$

$$\langle p^2\rangle = 2mE_1\int_{-a/2}^{a/2}\psi_i^*\psi_1\,dx = 2mE_1$$

$$= 2m\frac{\pi^2\hbar^2}{2ma^2} = \frac{\pi^2\hbar^2}{a^2}$$

$$(\Delta x)^2 = \langle x^2\rangle - \langle x\rangle^2 = \frac{a^2}{12x^2}(\pi^2-6)$$

$$(\Delta p)^2 = \langle p^2\rangle - \langle p\rangle^2 = \frac{\pi^2\hbar^2}{a^2}$$

The uncertainty product

$$(\Delta x)(\Delta p) = \sqrt{\frac{a^2(\pi^2-6)}{12\pi^2}\times\frac{\pi^2\hbar^2}{a^2}} = \sqrt{\frac{\pi^2-6}{12}}\,\hbar$$

4.43 In the simple harmonic oscillator problem, the creation ($a^\dagger$) and annihilation (a) operators are defined as

$$a^\dagger = \left(\frac{m\omega}{2\hbar}\right)^{1/2} x - i\left(\frac{1}{2m\hbar\omega}\right)^{1/2} p, \qquad a = \left(\frac{m\omega}{2\hbar}\right)^{1/2} x + i\left(\frac{1}{2m\hbar\omega}\right)^{1/2} p$$

Show that (i) $[a, a^\dagger] = 1$; (ii) $[a, H] = \hbar\omega a$, where H is the Hamiltonian operator of the oscillator; and (iii) $\langle n|a^\dagger a|n\rangle \geq 0$, where $|n\rangle$ are the energy eigenkets of the oscillator.

Solution.

$$\text{(i)}\ a a^\dagger = \left[\left(\frac{m\omega}{2\hbar}\right)^{1/2} x + i\left(\frac{1}{2m\hbar\omega}\right)^{1/2} p\right]\left[\left(\frac{m\omega}{2\hbar}\right)^{1/2} x - i\left(\frac{1}{2m\hbar\omega}\right) p\right]$$

$$= \frac{m\omega}{2\hbar}x^2 + \frac{1}{2m\hbar\omega}p^2 - \frac{i}{2\hbar}(xp - px)$$

$$= \frac{1}{\hbar\omega}\left(\frac{p^2}{2m} + \frac{1}{2}m\omega^2x^2\right) - \frac{i}{2\hbar}(i\hbar)$$

$$= \frac{H}{\hbar\omega} + \frac{1}{2} \qquad \text{(i)}$$

where H is the Hamiltonian operator of the simple harmonic oscillator. Simlarly,

$$a^{\dagger}a = \frac{H}{\hbar\omega} - \frac{1}{2} \qquad \text{(ii)}$$

$$[a, a^{\dagger}] = aa^{\dagger} - a^{\dagger}a = \frac{H}{\hbar\omega} + \frac{1}{2} - \frac{H}{\hbar\omega} + \frac{1}{2} = 1 \qquad \text{(iii)}$$

(ii) From Eqs. (i) and (ii),

$$aa^{\dagger} + a^{\dagger}a = \frac{2H}{\hbar\omega}$$

$$H = \frac{\hbar\omega}{2}(aa^{\dagger} + a^{\dagger}a) \qquad \text{(iv)}$$

$$\begin{aligned}[a, H] &= aH - Ha \\ &= \frac{\hbar\omega}{2}(aaa^{\dagger} + aa^{\dagger}a) - \frac{\hbar\omega}{2}(aa^{\dagger}a + a^{\dagger}aa) \\ &= \frac{\hbar\omega}{2}(a^2a^{\dagger} - a^{\dagger}a^2) = \frac{\hbar\omega}{2}\{a[a, a^{\dagger}] + [a, a^{\dagger}]a\}\end{aligned}$$

Substituting the value of $[a, a^{\dagger}] = 1$, we get

$$[a, H] = \hbar\omega a \qquad \text{(v)}$$

Similarly,

$$[a^{\dagger}, H] = -\hbar\omega a^{\dagger} \qquad \text{(vi)}$$

(iii) $\langle n|a^{\dagger}a|n\rangle = \langle n|a^{\dagger}|m\rangle\langle m|a|n\rangle$

$$\begin{aligned}&= \langle m|a|n\rangle^{\dagger}\langle m|a|n\rangle \\ &= |\langle m|a|n\rangle|^2 \geq 0\end{aligned} \qquad \text{(vii)}$$

4.44 Particles of mass m and charge e approach a square barrier defined by $V(x) = V_0$ for $0 < x < a$ and $V(x) = 0$ otherwise. The wave function in the region $0 < x < a$ is

$$\psi = Be^{\alpha x} + Ce^{-\alpha x}, \qquad \alpha = \frac{\sqrt{2m(V_0 - E)}}{\hbar}, \qquad E < V_0$$

(i) Explain why the exponentially increasing function $Be^{\alpha x}$ is retained in the wave function.
(ii) Show that the current density in this region is $(2\hbar\alpha e/m)\,[I_m(BC^*)]$.

Solution.

(i) It is true that $e^{ax} \to \infty$ as $x \to \infty$. However, it is also an acceptable solution since the barrier is of finite extent.

(ii) The probability current density

$$\begin{aligned}j_x &= \frac{i\hbar}{2m}\left(\psi\frac{d\psi^*}{dx} - \psi^*\frac{d\psi}{dx}\right) \\ &= \frac{i\hbar}{2m}\alpha\,[(Be^{\alpha x} + Ce^{-\alpha x})(B^*e^{\alpha x} - C^*e^{-\alpha x}) - (B^*e^{\alpha x} + C^*e^{-\alpha x})(Be^{\alpha x} - Ce^{-\alpha x})] \\ &= \frac{i\hbar\alpha}{2m}[-BC^* + CB^* + B^*C - C^*B] \\ &= \frac{i\hbar\alpha}{m}[B^*C - BC^*]\end{aligned}$$

Let $B = (B_r + iB_i)$ and $C = (C_r + iC_i)$. Then,

$$B^*C - BC^* = (B_r - iB_i)(C_r + iC_i) - (B_r + iB_i)(C_r - iC_i)$$

$$= 2i\,(B_rC_i - B_iC_r)$$

Hence,

$$j_x = \frac{i\hbar\alpha}{m}\,2i\,(B_rC_i - B_iC_r) = \frac{2\hbar\alpha}{m}\,(B_iC_r - B_rC_i)$$

$$= \frac{2\hbar\alpha}{m}\,(I_m(BC^*))$$

since

$$(B_r + iB_i)(C_r + iC_i) = (B_rC_r + B_iC_i) + i\,(B_iC_r - B_rC_i)$$

$$\text{Current density } J = \frac{2\hbar\alpha e}{m}\,(I_m(BC^*))$$

4.45 Consider particles of mass m and charge e approaching from left a square barrier defined by $V(x) = V_0$ for $0 < x < a$ and $V(x) = 0$ otherwise. The energy of the particle $E < V_0$. If the wave function

$$\psi(x) = e^{ikx} + Be^{-ikx}, \qquad x < 0, \qquad k^2 = \frac{2mE}{\hbar^2}$$

Show that the current density

$$J_x = \frac{e\hbar k}{m}\,(1 - |B|^2)$$

Solution. The probability current density

$$j_x = \frac{i\hbar}{2m}\left(\psi\frac{d\psi^*}{dx} - \psi^*\frac{d\psi}{dx}\right)$$

For the region $x < 0$, the Schrödinger equation is

$$-\frac{\hbar^2}{2m}\frac{d^2\psi}{dx^2} = E\psi \quad \text{or} \quad \frac{d^2\psi}{dx^2} = -k^2\psi$$

Here, the parameter k is real.

$$\psi\frac{d\psi^*}{dx} = ik\,(e^{ikx} + Be^{-ikx})(-e^{-ikx} + B^*e^{ikx})$$

$$= ik\,(-1 + |B|^2 + B^*e^{2ikx} - Be^{-2ikx})$$

$$\psi^*\frac{d\psi}{dx} = ik\,(e^{-ikx} + B^*e^{ikx})(e^{ikx} - Be^{-ikx})$$

$$= ik\,(1 - |B|^2 - Be^{-2ikx} + B^*e^{2ikx})$$

Hence,

$$j_x = \frac{i\hbar}{2m}\,ik(-2 + 2\,|B|^2) = \frac{\hbar k}{m}\,(1 - |B|^2)$$

Current density $$J_x = \frac{e\hbar k}{m}(1 - |B|^2)$$

4.46 Define the creation ($a^\dagger$) and annihilation (a) operators for a harmonic oscillator and show that

(i) $Ha|n\rangle = (E_n - \hbar\omega)a|n\rangle$ and $Ha^\dagger|n = (E_n + \hbar\omega)a^\dagger|n\rangle$.

(ii) $a|n\rangle = \sqrt{n}\,|n-1\rangle$ and $a^\dagger|n\rangle = \sqrt{n+1}\,|n+1\rangle$.

Solution.

(i) Creation and annihilation operators are defined in Problem 4.43, from which we have

$$[a, H] = \hbar\omega a, \qquad [a^\dagger, H] = -\hbar\omega a^\dagger$$

From the first relation,

$$Ha|n\rangle = aH|n\rangle - \hbar\omega a|n\rangle$$

$$= (E_n - \hbar\omega)\, a|n\rangle \tag{i}$$

Similarly, from the second relation,

$$H\, a^\dagger|n\rangle = (E_n + \hbar\omega)\, a^\dagger|n\rangle \tag{ii}$$

Since $E_n = [n + (1/2]\,\hbar\omega$, from Eq. (i),

$$Ha|n\rangle = [n - (1/2]\,\hbar\omega a|n\rangle \tag{iii}$$

For the $(n - 1)$ state, we have

$$H|n-1\rangle = E_{n-1}|n-1\rangle = \left(n - 1 + \frac{1}{2}\right)\hbar\omega|n-1\rangle$$

$$= \left(n - \frac{1}{2}\right)\hbar\omega|n-1\rangle \tag{iv}$$

Relations (iii) and (iv) are possible only if $a|n\rangle$ is a multiple of $|n - 1\rangle$, i.e.,

$$a|n\rangle = \alpha|n-1\rangle \tag{v}$$

$$\langle n|a^\dagger = \langle n-1|\alpha^*$$

Hence,

$$\langle n|a^\dagger a|n\rangle = \langle n-1||a|^2|n-1\rangle$$

Substituting the value of $a^\dagger a$, we get

$$|\alpha|^2 = \left\langle n\left|\frac{H}{\hbar\omega} - \frac{1}{2}\right|n\right\rangle = \left\langle n\left|n + \frac{1}{2} - \frac{1}{2}\right|n\right\rangle = n$$

$$\alpha = \sqrt{n}$$

Consequently,

$$a|n\rangle = \sqrt{n}\,|n-1\rangle \tag{vi}$$

Similarly,

$$a^\dagger|n\rangle = \sqrt{n+1}\,|n+1\rangle \tag{vii}$$

4.47 In the harmonic oscillator problem, the creation ($a^\dagger$) and annihilation (a) operators in dimensionless units ($\hbar = \omega = m = 1$) are defined by

$$a^\dagger = \frac{x - ip}{\sqrt{2}}, \qquad a = \frac{x + ip}{\sqrt{2}}$$

An unnormalized energy eigenfunction is $\psi_n = (2x^2 - 1) \exp(-x^2/2)$. What is its state? Find the eigenfunctions corresponding to the adjacent states.

Solution. We have

$$a|n\rangle = \sqrt{n}\,|n-1\rangle, \qquad a^\dagger|n\rangle = \sqrt{n+1}\,|n+1\rangle$$

$$aa^\dagger|n\rangle = a\sqrt{n+1}\,|n+1\rangle = (n+1)|n\rangle$$

Operators for $a^\dagger$ and a are

$$a^\dagger = \frac{1}{\sqrt{2}}\left(x - \frac{d}{dx}\right), \qquad a = \frac{1}{\sqrt{2}}\left(x + \frac{d}{dx}\right)$$

In the given case, substituting the values of a, $a^\dagger$ and $|n\rangle$,

$$aa^\dagger|\psi_n\rangle = \frac{1}{2}\left(x + \frac{d}{dx}\right)\left(x - \frac{d}{dx}\right)(2x^2 - 1)\exp(-x^2/2)$$

$$= \frac{1}{2}\left(x + \frac{d}{dx}\right)(4x^3 - 6x)\exp(-x^2/2)$$

$$= \frac{1}{2}(12x^2 - 6)\exp(-x^2/2) = 3\,(2x^2 - 1)\exp(-x^2/2)$$

$$= (2 + 1)|\psi_n\rangle$$

Hence, the quantum number corresponding to this state is 2. The adjacent states are the $n = 1$ and $n = 3$ states. Therefore,

$$\psi_1 = |1\rangle = \frac{1}{\sqrt{2}}\,a\,|2\rangle$$

$$= \frac{1}{\sqrt{2}}\frac{1}{\sqrt{2}}\left(x + \frac{d}{dx}\right)(2x^2 - 1)\exp(-x^2/2)$$

$$= \frac{1}{2}\,[2x^3 - x + 4x + (2x^2 - 1)(-x)]\exp(-x^2/2)$$

$$= 2x\exp(-x^2/2)$$

Substituting the values of a and $|2\rangle$, we get

$$\psi_3 = |3\rangle = \frac{1}{\sqrt{3}}\,a^\dagger\,|2\rangle = \frac{1}{\sqrt{3}}\frac{1}{\sqrt{2}}\left(x - \frac{d}{dx}\right)(2x^2 - 1)\exp -\frac{x^2}{2}$$

$$= \frac{1}{\sqrt{6}}\,[2x^3 - x - 4x - (2x^2 - 1)(-x)]\exp -\frac{x^2}{2}$$

$$= \frac{2}{\sqrt{6}}\,(2x^3 - 3x)\exp -\frac{x^2}{2}$$

Except for the normalization constant, the wave functions are

$$\psi_1 = x \exp -\frac{x^2}{2}, \qquad \psi_3 = (2x^3 - 3x) \exp -\frac{x^2}{2}$$

4.48 In harmonic oscillator problem, the creation ($a^\dagger$) and annihilation (a) operators obey the relation

$$a^\dagger a = \frac{H}{\hbar\omega} - \frac{1}{2}$$

Hence prove that the energy of the ground state $E_0 = 1/2\ \hbar\omega$ and the ground state wave function is $\psi_0 = N_0 \exp(-\text{max}^2/2\hbar)$.

Solution. Given

$$a^\dagger a = \frac{H}{\hbar\omega} - \frac{1}{2}$$

The annihilation operator a annihilates a state and it is known from (Eq. 4. 21) that

$$a|n\rangle = \sqrt{n}\,|n-1\rangle \tag{i}$$

Hence,

$$a|0\rangle = 0 \quad \text{or} \quad a^\dagger a|0\rangle = 0 \tag{ii}$$

Substituting the value of $a^\dagger a$, we get

$$\left(\frac{H}{\hbar\omega} - \frac{1}{2}\right)|0\rangle = 0 \quad \text{or} \quad \left(\frac{E_0}{\hbar\omega} - \frac{1}{2}\right)|0\rangle = 0 \tag{iii}$$

Since $|0\rangle \neq 0$,

$$\frac{E_0}{\hbar\omega} - \frac{1}{2} = 0 \quad \text{or} \quad E_0 = \frac{1}{2}\hbar\omega \tag{iv}$$

Substituting the value of a in $a|0\rangle = 0$, we get

$$\left\{\left(\frac{m\omega}{2\hbar}\right)^{1/2} x + i\left(\frac{1}{2m\hbar\omega}\right)^{1/2} p\right\}|0\rangle = 0$$

$$\left\{\left(\frac{m\omega}{2\hbar}\right)^{1/2} x + \left(\frac{1}{2m\hbar\omega}\right)^{1/2} \hbar\frac{d}{dx}\right\}|0\rangle = 0$$

Multiplying by $(m\omega/2\hbar)^{1/2}$, we obtain

$$\left(\frac{m\omega x}{2\hbar} + \frac{1}{2\hbar}\hbar\frac{d}{dx}\right)|0\rangle = 0$$

$$\left(m\omega x + \hbar\frac{d}{dx}\right)|0\rangle = 0$$

$$\frac{d\psi_0}{dx} = -\frac{m\omega x\psi_0}{\hbar}$$

$$\frac{d\psi_0}{\psi_0} = -\frac{m\omega x}{\hbar}$$

Integrating and taking the exponential, we get

$$\psi_0 = N_0 \exp\left(-\frac{m\omega x^2}{2\hbar}\right)$$

4.49 Consider the infinite square well of width a. Let $u_1(x)$ and $u_2(x)$ be its orthonormal eigenfunctions in the first two states. If $\psi(x) = Au_1(x) + Bu_2(x)$, where A and B are constants, show that (i) $|A|^2 + |B|^2 = 1$; (ii) $\langle E\rangle = |A|^2E_1 + |B|^2E_2$, where E_1 and E_2 are the energy eigenvalues of the $n = 1$ and $n = 2$ states, respectively.

Solution. The energy eigenfunctions and energy eigenvalues of the infinite square well are

$$u_n(x) = \sqrt{\frac{2}{a}}\sin\frac{n\pi x}{a}, \qquad E_n = \frac{\pi^2\hbar^2n^2}{2ma^2}, \qquad n = 1, 2, 3, \ldots \tag{i}$$

(i) The normalizaiton condition gives

$$\langle\psi_n|\psi_n\rangle = 1 \tag{ii}$$

$$\langle(Au_1 + Bu_2)|(Au_1 + Bu_2)\rangle = 1 \tag{iii}$$

Since the eigenfunctions are orthonormal, Eq. (iii) becomes

$$|A|^2\langle u_1|u_1\rangle + |B|^2\langle u_2|u_2\rangle = 1$$

$$|A|^2 + |B|^2 = 1$$

(ii)
$$\langle E\rangle = \langle(Au_1 + Bu_2)|E_{op}|(Au_1 + Bu_2)\rangle$$

$$= \langle(Au_1 + Bu_2)|(AE_1u_1 + BE_2u_2)\rangle$$

$$= |A|^2E_1 + |B|^2E_2$$

4.50 Electrons with energies 1 eV are incident on a barrier 5 eV high 0.4 nm wide. (i) Evaluate the transmission probability. What would be the probability (ii) if the height is doubled, (iii) if the width is doubled, and (iv) comment on the result.

Solution. The transmission probability T is given by

$$T = e^{2\alpha a}, \quad \alpha^2 = \frac{2m(V_0 - E)}{\hbar^2}$$

(i)
$$\alpha^2 = \frac{2(9.1\times10^{-31}\text{ kg})(4\text{ eV})(1.6\times10^{-19}\text{ J/eV})}{(1.054\times10^{-34}\text{ J s})^2}$$

$$\alpha = 10.24\times10^9\text{ m}^{-1}$$

$$\alpha a = (10.24\times10^9\text{ m}^{-1})(0.4\times10^{-9}\text{ m}) = 4.096$$

$$T = \frac{1}{e^{2\alpha a}} = \frac{1}{e^{8.192}} = 2.77\times10^{-4}$$

(ii)
$$\alpha = 15.359\times109\text{ m}^{-1}$$

$$2\alpha a = 2(15.359\times10^9\text{ m}^{-1})(0.4\times10^{-9}\text{ m}) = 12.287$$

$$T = \frac{1}{e^{2\alpha a}} = \frac{1}{e^{12.287}} = 4.6\times10^{-6}$$

(iii)
$$\alpha = 15.359 \times 10^9 \text{ m}^{-1}$$
$$2\alpha a = 2(10.24 \times 10^9 \text{ m}^{-1})(0.8 \times 10^{-9} \text{ m}) = 16.384$$
$$T = \frac{1}{e^{16.384}} = 7.69 \times 10^{-8}$$

(iv) When the barrier height is doubled, the transmission probability decreases by a factor of about 100. However, when the width of the barrier is doubled, the value decreases by a factor of about 10^4. Hence, the transmission probability is more sensitive to the width of the barrier than the height. In the same manner we can easily show that T is more sensitive to the width than the energy of the incident particle.

4.51 Consider two identical linear oscialltors having a spring constant k. The interaction potential is $H = Ax_1x_2$, where x_1 and x_2 are the coordinates of the oscillators. Obtain the energy eigenvalues.

Solution. The Hamiltonian of the system is
$$H = -\frac{\hbar^2}{2m}\frac{\partial^2}{\partial x_1^2} - \frac{\hbar^2}{2m}\frac{\partial^2}{\partial x_2^2} + \frac{1}{2}m\omega^2 x_1^2 + \frac{1}{2}m\omega^2 x_2^2 + Ax_1x_2$$

Writing
$$x_1 = \frac{1}{\sqrt{2}}(y_1 + y_2), \qquad x_2 = \frac{1}{\sqrt{2}}(y_1 - y_2)$$

We have
$$H = -\frac{\hbar^2}{2m}\frac{\partial^2}{\partial y_1^2} - \frac{\hbar^2}{2m}\frac{\partial^2}{\partial y_2^2} + \frac{1}{2}m\omega^2(y_1^2 + y_2^2) + \frac{A}{2}(y_1^2 - y_2^2)$$
$$= -\frac{\hbar^2}{2m}\frac{\partial^2}{\partial y_1^2} - \frac{1}{2}m\left(\omega^2 + \frac{A}{m}\right)y_1^2 - \frac{\hbar^2}{2m}\frac{\partial^2}{\partial y_2^2} + \frac{1}{2}m\left(\omega^2 - \frac{A}{m}\right)y_2^2$$

Hence the system can be regarded as two independent harmonic oscillators having coordinates y_1 and y_2. The energy levels are
$$E_{nn'} = \left(n + \frac{1}{2}\right)\hbar\sqrt{\left(\omega^2 + \frac{A}{m}\right)} + \left(n' + \frac{1}{2}\right)\hbar\sqrt{\left(\omega^2 - \frac{A}{m}\right)}$$

4.52 The energy eigenvalue and the corresponding eigenfunction for a particle of mass m in a one-dimensional potential $V(x)$ are
$$E = 0, \qquad \psi = \frac{A}{x^2 + a^2}$$

Deduce the potential $V(x)$.

Solution.
$$\psi(x) = \frac{A}{x^2 + a^2}$$
$$\frac{d\psi}{dx} = -\frac{2Ax}{(x^2 + a^2)^2}$$

$$\frac{d^2\psi}{dx^2} = -2A\left[\frac{1}{(x^2+a^2)^2} + \frac{x(-2)\,2x}{(x^2+a^2)^3}\right]$$

$$= -2A\frac{[x^2+a^2-4x^2]}{(x^2+a^2)^3} = -2A\frac{a^2-3x^2}{(x^2+a^2)^3}$$

Substituting in the Schrödinger equation, we get

$$\frac{\hbar^2}{2m}\,2A\frac{(a^2-3x^2)}{(x^2+a^2)^3} = \frac{VA}{x^2+a^2} = 0$$

$$V(x) = \frac{\hbar^2}{m}\frac{(3x^2-a^2)}{(x^2+a^2)^2}$$

4.53 A beam of particles having energy 2 eV is incident on a potential barrier of 0.1 nm width and 10 eV height. Show that the electron beam has a probability of 14% to tunnel through the barrier.

Solution. The transmission probability

$$T \cong \frac{16E(V_0-E)e^{-2\alpha a}}{V_0^2}, \qquad \alpha^2 \cong \frac{2m(V_0-E)}{\hbar^2}$$

where a is the width of the barrier, V_0 is the height of the barrier, and E is the energy of the electron.

$$\alpha^2 = \frac{2(9.1\times10^{-31}\text{ kg})(8\text{ eV}\times1.6\times10^{-19}\text{ J/eV})}{(1.05\times10^{-34}\text{ J s})^2}$$

$$= 211.3\times10^{18}\text{ m}^{-2}$$

$$\alpha = 14.536\times10^{9}\text{ m}^{-1}$$

$$\alpha a = (14.536\times10^{9}\text{ m}^{-1})(0.1\times10^{-9}\text{ m}) = 1.4536$$

$$T = \frac{16\times2\text{ eV}\times8\text{ eV}}{(10\text{ eV})^2}\,e^{-29072} = 0.14$$

The percentage probability to tunnel through the barrier is 14.

4.54 For the ground state of a particle of mass m moving in a potential,

$$V(x) = 0,\ 0 < x < a \text{ and } V(x) = \infty \text{ otherwise}$$

Estimate the uncertainty product $(\Delta x)(\Delta p)$.

Solution. The energy of the ground state

$$E = \frac{\pi^2\hbar^2}{2ma^2}$$

This must be equal to $p^2/2m$. Hence,

$$\frac{p^2}{2m} = \frac{\pi^2\hbar^2}{2ma^2} \quad\text{or}\quad p^2 = \frac{\pi^2\hbar^2}{a^2}$$

$$(\Delta p^2) = \langle p\rangle^2 - \langle p\rangle^2$$

Since the box is symmetric, $\langle p\rangle$ will be zero and, therefore,

$$(\Delta p^2) = \langle p^2\rangle = \frac{\pi^2\hbar^2}{a^2}$$

For the particle in the box Δx is not larger than a.
Hence,

$$(\Delta p^2)(\Delta x)^2 = \frac{\pi^2\hbar^2}{a^2}a^2 = \pi^2\hbar^2$$

$$(\Delta p)(\Delta x) = \frac{h}{2}$$

4.55 Let ψ_0 and ψ_2 denote, respectively, the ground state and second excited state energy eigenfunctions of a particle moving in a harmonic oscillator potential with frequency ω. At $t = 0$, if the particle has the wave function

$$\psi(x) = \frac{1}{\sqrt{3}}\,\psi_0(x) + \sqrt{\frac{2}{3}}\,\psi_2(x)$$

(i) Find $\psi(x, t)$ for $t \neq 0$, (ii) Determine the expectation value of energy as a function of time, (iii) Determine momentum and position expectation values as functions of time.

Solution. Including the time dependence, the wave function of a system is

$$\Psi_n(\mathbf{r}, t) = \Psi_n(\mathbf{r}, 0)\exp\left(-\frac{iE_nt}{\hbar}\right)$$

(i) In the present case,

$$\Psi(x, t) = \frac{1}{\sqrt{3}}\,\Psi_0(x)\exp\left(\frac{-iE_0t}{\hbar}\right) + \sqrt{\frac{2}{3}}\,\psi_2(x)\exp\left(\frac{-iE_2t}{\hbar}\right)$$

(ii) $$\langle E\rangle = \left\langle \Psi(x, t)\left|i\hbar\frac{\partial}{\partial t}\right|\Psi(x, t)\right\rangle$$

$$= i\hbar\int\left[\frac{1}{\sqrt{3}}\psi_0(x, t) + \sqrt{\frac{2}{3}}\,\psi_2(x, t)\right]\left[-\frac{iE_0}{\hbar\sqrt{3}}\psi_0(x, t) - \frac{iE_2}{\hbar}\sqrt{\frac{2}{3}}\,\psi_2(x, t)\right]dx$$

$$= \frac{E_0}{3} + \frac{2}{3}E_2 = \frac{1}{3}\hbar\omega + \frac{5}{3}\hbar\omega$$

$$= 2\hbar\omega$$

The cross-terms are zero since $\langle\psi_0(x)\,|\,\psi_2(x)\rangle = 0$.

(iii) The momentum expectation value is

$$\langle p\rangle = \left\langle \Psi(x, t)\left|-i\hbar\frac{d}{dt}\right|\Psi(x, t)\right\rangle$$

The functions $\psi_0(x)$ and $\psi_2(x)$ are even functions of x. When differentiated with respect to x, the resulting function will be odd. Consequently, the integrand will be odd. This makes the integral to vanish. Hence, $\langle p\rangle = 0$.

The position expectation value is

$$\langle x\rangle = \langle \Psi(x, t)|\, x\, |\Psi(x, t)\rangle$$

Again, $\psi_0(x)$ and $\psi_2(x)$ are even. This makes the integrand of the above integral odd, leading to zero. Hence, $\langle x\rangle = 0$.

4.56 For a harmonic oscillator, the Hamiltonian in dimensionless units ($m = \hbar = \omega = 1$) is

$$H = aa^\dagger - \frac{1}{2}$$

where the annihilation (a) and creation ($a^\dagger$) operators are defined by

$$a = \frac{x + ip}{\sqrt{2}}, \qquad a^\dagger = \frac{x - ip}{\sqrt{2}}$$

The energy eigenfunction of a state is

$$\psi_n = (2x^3 - 3x)\exp\left(\frac{-x^2}{2}\right)$$

What is its state? Find the eigenfunctions corresponding to the adjacent states.

Solution. We have the relations

$$a\,|n\rangle = \sqrt{n}\,|n-1\rangle, \qquad a^\dagger|n\rangle = \sqrt{n+1}\,|n+1\rangle$$

$$aa^\dagger\,|n\rangle = \frac{1}{\sqrt{2}}\left(x + \frac{d}{dx}\right)\frac{1}{\sqrt{2}}\left(x - \frac{d}{dx}\right)(2x^3 - 3x)\exp\left(\frac{-x^2}{2}\right)$$

$$= \frac{1}{2}\left(x + \frac{d}{dx}\right)(4x^4 - 12x^2 + 3)\exp\left(\frac{-x^2}{2}\right)$$

$$= (8x^3 - 12x)\exp\left(\frac{-x^2}{2}\right) = 4\,(2x^3 - 3)\exp\left(\frac{-x^2}{2}\right)$$

$$= (3 + 1)|\,n\rangle$$

We have $aa^\dagger = \text{H} + \dfrac{1}{2}$ and $H|\,n\rangle = n + \dfrac{1}{2}$. Then,

$$aa^\dagger\,|\,n\rangle = \left(H + \frac{1}{2}\right)|n\rangle = \left(n + \frac{1}{2} + \frac{1}{2}\right)|n\rangle$$

$$= (n + 1)|n\rangle$$

Hence, the involved state is $n = 3$. The adjacent states are $n = 2$ and $n = 4$. consequently,

$$\psi_2 = \frac{1}{\sqrt{3}}\,a|3\rangle = \frac{1}{\sqrt{3}}\frac{1}{\sqrt{2}}\left(x + \frac{d}{dx}\right)(2x^3 - 3x)\exp\left(\frac{-x^2}{2}\right)$$

$$= \frac{1}{\sqrt{6}}(6x^2 - 3)\exp\left(\frac{-x^2}{2}\right) = \sqrt{\frac{3}{2}}(2x^2 - 1)\exp\left(\frac{-x^2}{2}\right)$$

4.57 A beam of particles, each with energy E approaches a step potential of V_0.

(i) Show that the fraction of the beam reflected and transmitted are independent of the mass of the particle.

(ii) If $E = 40$ MeV and $V_0 = 30$ MeV, what fraction of the beam is reflected and transmitted?

Solution. Details of particles approaching a potential step are discussed in Problem 4.19. We have the relations:

$$\text{Incident flux of particles} = \frac{k_0\hbar}{m} \tag{i}$$

$$\text{Reflected flux of particles} = \frac{k_0\hbar}{m}|A|^2 \tag{ii}$$

$$\text{Transmitted flux of particles} = \frac{k_0\hbar}{m}|B|^2 \tag{iii}$$

where

$$k_0^2 = \frac{2mE}{\hbar^2}, \qquad k^2 = \frac{2m(E - V_0)}{\hbar^2} \tag{iv}$$

$$A = \frac{k_0 - k}{k_0 + k}, \qquad B = \frac{2k_0}{k_0 + k} \tag{v}$$

$$\text{(i) Fraction reflected} = \frac{k_0\hbar\,|A|^2/m}{k_0\hbar/m} = |A|^2$$

$$= \frac{(k_0 - k)^2}{(k_0 + k)^2} = \frac{k_0^2 + k^2 - 2kk_0}{k_0^2 + k^2 + 2kk_0}$$

$$= \frac{(2mE/\hbar^2) + [2m(E - V_0)/\hbar^2] - 2(2m/\hbar^2)\sqrt{E(E - V_0)}}{(2mE/\hbar^2) + [2m(E - V_0)/\hbar^2] + 2(2m/\hbar^2)\sqrt{E(E - V_0)}}$$

$$= \frac{E + (E - V_0) - 2\sqrt{E(E - V_0)}}{E + (E - V_0) + 2\sqrt{E(E - V_0)}} \tag{vi}$$

That is, the fraction reflected is independent of mass.

$$\text{Fraction transmitted} = \frac{k\hbar|B|^2/m}{k_0\hbar/m} = \frac{k}{k_0}|B|^2$$

$$= \frac{k}{k_0}\frac{4k_0^2}{(k_0 + k)^2} = \frac{4kk_0}{(k_0 + k)^2}$$

$$= \frac{4(2m/\hbar^2)\sqrt{(E - V_0)E}}{(2m/\hbar^2)\,[E + (E - V_0) + 2\sqrt{E(E - V_0)}]}$$

$$= \frac{4\sqrt{(E - V_0)E}}{E + (E - V_0) + 2\sqrt{E(E - V_0)}]} \tag{vii}$$

i.e., the fraction transmitted is independent of mass.

(ii) Fraction reflected $= \dfrac{40 + 10 - 2\sqrt{40 \times 10}}{40 + 10 + 2\sqrt{40 \times 10}}$

$$= \frac{10 \text{ meV}}{9 \text{ meV}} = 0.111$$

Fraction transmitted $= \dfrac{4 \times 20}{40 + 10 + 40} = \dfrac{80}{90}$

$$= 0.889$$

4.58 A simple pendulum of length l swings in a vertical plane under the influence of gravity. In the small angle approximation, find the energy levels of the system.

Solution. Taking the mean position of the oscillator as the zero of potential energy, the potential energy in the displaced position (Fig. 4.6) is

$$V = mg\,(l - l\cos\theta) = mgl\,(1 - \cos\theta)$$

When θ is small,

$$\cos\theta = 1 - \frac{\theta^2}{2}, \qquad \sin\theta \cong \theta = \frac{x}{l}$$

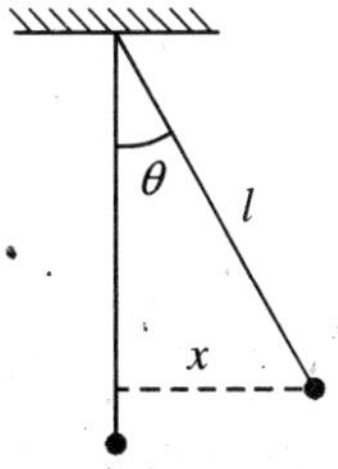

Fig. 4.6 Simple pendulum in the displaced position.

Substituting the value of cos θ and replacing $\theta = x/l$, we get

$$V = \frac{1}{2} mgl\theta^2 = \frac{1}{2} mg\frac{x^2}{l}$$

$$= \frac{1}{2} m\omega^2 x^2, \qquad \omega = \sqrt{\frac{g}{l}}$$

In plane polar coordinates,

$$v_\theta = l\frac{d\theta}{dt} = l\dot{\theta}$$

$$\text{Kinetic energy} = \frac{1}{2} ml^2\dot{\theta}^2 = \frac{1}{2} ml^2\frac{\dot{x}^2}{l^2} = \frac{1}{2} m\dot{x}^2$$

$$= \frac{p_x^2}{2m}$$

The Hamiltonian

$$H = \frac{p_x^2}{2m} + \frac{1}{2} m\omega^2 x^2$$

which is the same as the one-dimensional harmonic oscillator Hamiltonian. The energy eigenvalues are

$$E_n = \left(n + \frac{1}{2}\right)\hbar\omega, \qquad \omega = \sqrt{\frac{g}{l}}, \qquad n = 0, 1, 2, \ldots$$

CHAPTER 5

Three-Dimensional Energy Eigenvalue Problems

In this chapter, we apply the basic ideas developed earlier to some of the important three-dimensional potentials.

5.1 Particle Moving in a Spherically Symmetric Potential

In a spherically symmetric potential $V(r)$, the Schrödinger equation is

$$\nabla^2\psi(\mathbf{r}) + \frac{2m}{\hbar^2}(E - V)\psi(\mathbf{r}) = 0 \tag{5.1}$$

Expressing Eq. (5.1) in the spherical polar coordinates and writing

$$\psi(r, \theta, \phi) = R(r)\,\Theta(\theta)\,\Phi(\phi) \tag{5.2}$$

the Schrödinger equation can be divided into three equations:

$$\frac{d^2\Phi}{d\phi^2} = -m^2\Phi \tag{5.3}$$

$$\frac{1}{\sin\theta}\frac{d}{d\theta}\left(\sin\theta\frac{d\Theta}{d\theta}\right) + \left(\lambda - \frac{m^2}{\sin^2\theta}\right)\Theta = 0 \tag{5.4}$$

$$\frac{1}{r^2}\frac{d}{dr}\left(r^2\frac{dR}{dr}\right) + \frac{2m}{\hbar^2}(E - V)R - \frac{\lambda}{r^2}R = 0 \tag{5.5}$$

where m and λ are the constants to be determined. The normalized solution of the first two equations are

$$\Phi(\phi) = \frac{1}{\sqrt{2\pi}}e^{im\phi}, \qquad m = 0, \pm 1, \pm 2, \ldots \tag{5.6}$$

$$\Theta_l^m(\theta) = \varepsilon \sqrt{\frac{(2l+1)(l-|m|)!}{2(l+|m|)!}}\, P_l^m(\cos\theta), \qquad l = 0, 1, 2, \ldots \tag{5.7}$$

where $P_l^m(\cos\theta)$ are the associated Legendre polynomials and the constant λ in Eq. (5.4) = $l(l+1)$. The spherical harmonics $Y_{lm}(\theta, \phi)$ are the product of these two functions. Hence,

$$Y_{lm}(\theta, \phi) = \varepsilon \sqrt{\frac{(2l+1)(l-|m|)!}{4\pi(l+|m|)!}}\, P_l^m(\cos\theta)\, e^{im\phi} \tag{5.8}$$

where

$$\varepsilon = (-1)^m \quad \text{for } m \geq 0; \qquad \varepsilon = 1 \quad \text{for } m \leq 0$$

5.2 System of Two Interacting Particles

The wave equation of a system of two interacting particles can be reduced into two one particle equations: one representing the translational motion of the centre of mass and the other the representing relative motion of the two particles. In the coordinate system in which the centre of mass is at rest, the second equation is given by

$$-\frac{\hbar^2}{2\mu}\nabla^2\psi(\mathbf{r}) + V(r)\psi(\mathbf{r}) = E\psi(\mathbf{r}), \qquad \mu = \frac{m_1 m_2}{m_1 + m_2} \tag{5.9}$$

5.3 Rigid Rotator

For free rotation, $V(r) = 0$. As the rotator is rigid, the wave function will depend only on the angles θ and ϕ. The rigid rotator wave functions are the spherical harmonics $Y_{lm}(\theta, \phi)$. The energy eigenvalues are

$$E_l = \frac{l(l+1)\hbar^2}{2I}, \qquad l = 0, 1, 2, \ldots \tag{5.10}$$

5.4 Hydrogen Atom

The potential energy of a hydrogen-like atom is given by

$$V(r) = -\frac{Ze^2}{4\pi\varepsilon_0 r}$$

where Z is the atomic number of the nucleus. The Schrödinger equation to be solved is

$$\left(-\frac{\hbar^2}{2\mu}\nabla^2 - \frac{Ze^2}{4\pi\varepsilon_0 r}\right)\psi(\mathbf{r}) = E\psi(\mathbf{r}) \tag{5.11}$$

In spherical polar coordinates, the angular part of the wave function are the spherical harmonics $Y_{lm}(\theta, \phi)$; the radial equation to be solved is

$$\frac{1}{r^2}\frac{d}{dr}\left(r^2\frac{dR}{dr}\right) + \frac{2\mu}{\hbar^2}\left[E - \frac{l(l+1)\hbar^2}{2\mu r^2} + \frac{Ze^2}{4\pi\varepsilon_0 r}\right]R = 0 \tag{5.12}$$

The energy eigenvalues are

$$E_n = -\frac{\mu Z^2 e^4}{32\pi^2 \epsilon_0^2 \hbar^2}\frac{1}{n^2}, \qquad n = 1, 2, 3, \ldots \tag{5.13}$$

The normalized radial wave functions are

$$R_{nl}(r) = -\left\{\left(\frac{2Z}{na_0}\right)^3 \frac{(n-l-1)!}{2n[(n+l)!]^3}\right\}^{1/2} e^{-\rho/2}\rho^l L_{n+l}^{2l+1}(\rho) \tag{5.14}$$

$$\rho = \sqrt{-\frac{8\mu E}{\hbar^2}}\, r, \qquad l = 0, 1, 2, \ldots, (n-1) \tag{5.15}$$

$L_{n+l}^{2l+1}(\rho)$ are the associated Laguerre polynomials. The wave function is given by

$$\psi_{nlm}(r, \theta, \phi) = R_{nl}(r)\, Y_{lm}(\theta, \phi) \tag{5.16}$$

$$n = 1, 2, 3, \ldots; \quad l = 0, 1, 2, \ldots, (n-1); \quad m = 0, \pm 1, \pm 2, \ldots, \pm l$$

The explict form of the ground state wave function is

$$\psi_{100} = \frac{1}{\pi^{1/2}}\left(\frac{Z}{a_0}\right)^{3/2} e^{-Zr/a_0} \tag{5.17}$$

The **radial probability density**, $P_{nl}(r)$ is the probability of finding the electron of the hydrogen atom at a distance r from the nucleus. Thus,

$$P_{nl}(r) = r^2 |R_{nl}|^2 \tag{5.18}$$

PROBLEMS

5.1 A particle of mass m moves in a three-dimensional box of sides a, b, c. If the potential is zero inside and infinity outside the box, find the energy eigenvalues and eigenfunctions.

Solution. As the potential is infinity, the wave function ψ outside the box must be zero. Inside the box, the Schrödinger equation is given by

$$\frac{\partial^2\psi}{\partial x^2}+\frac{\partial^2\psi}{\partial y^2}+\frac{\partial^2\psi}{\partial z^2}+\frac{2mE}{\hbar^2}\psi(x, y, z)=0$$

The equation can be separated into three equations by writing

$$\psi(x, y, z) = X(x)\ Y(y)\ Z(z)$$

Substituting this value of ψ and simplifying, we get

$$\frac{d^2X(x)}{dx^2}+\frac{2m}{\hbar^2}E_xX(x)=0$$

$$\frac{d^2Y(y)}{dy^2}+\frac{2m}{\hbar^2}E_yY(y)=0$$

$$\frac{d^2Z(z)}{dz^2}+\frac{2m}{\hbar^2}E_zZ(z)=0$$

where $E = E_x + E_y + E_z$. Use of the boundary condition $X(x) = 0$ at $x = 0$ and at $x = a$ and the normalization condition give

$$E_x=\frac{n_x^2\pi^2\hbar^2}{2ma^2},\quad n_x = 1, 2, 3, \ldots$$

$$X(x)=\sqrt{\frac{2}{a}}\sin\frac{n_x\pi x}{a}$$

where $n_x = 0$ is left out, which makes $X(x)$ zero everywhere. Similar relations result for the other two equations. Combining the three, we get

$$E=\frac{\pi^2\hbar^2}{2m}\left(\frac{n_x^2}{a^2}+\frac{n_y^2}{b^2}+\frac{n_z^2}{c^2}\right),\quad n_x, n_y, n_z = 1, 2, 3, \ldots$$

$$\psi(x, y, z)=\sqrt{\frac{8}{abc}}\sin\frac{n_x\pi x}{a}\sin\frac{n_y\pi y}{b}\sin\frac{n_z\pi z}{c}$$

5.2 In Problem 5.1, if the box is a cubical one of side a, derive the expression for energy eigenvalues and eigenfunctions. What is the zero point energy of the system? What is the degeneracy of the first and second excited states?

Solution. The energy eigenvalues and eigenfunctions are

$$E_{n_xn_yn_z}=\frac{\pi^2\hbar^2}{2ma^2}(n_x^2+n_y^2+n_z^2)$$

$$\psi_{n_x n_y n_z}(x, y, z) = \sqrt{\frac{8}{a^3}} \sin \frac{n_x \pi x}{a} \sin \frac{n_y \pi x}{a} \sin \frac{n_z \pi x}{a}$$

$$\text{Zero point energy} = E_{111} = \frac{3\pi^2 \hbar^2}{2ma^2}$$

The three independent states having quantum numbers (1,1,2), (1,2,1), (2,1,1) for (n_x, n_y, n_z) have the energy

$$E_{112} = E_{121} = E_{211} = \frac{5\pi^2 \hbar^2}{2ma^2}$$

which is the first excited state and is three-fold degenerate. The energy of the second excited state is

$$E_{122} = E_{212} = E_{221} = \frac{9\pi^2 \hbar^2}{2ma^2}$$

It is also three-fold degenerate.

5.3 A rigid rotator is constrained to rotate about a fixed axis. Find out its normalized eigenfunctions and eigenvalues.

Solution. As the axis of rotation is always along a fixed direction, the rotator moves in a particular plane. If this plane is taken as the *x-y* plane, θ is always 90°, and the wave function ψ is a function of ϕ only. The Schrödinger equation now reduces to

$$-\frac{\hbar^2}{2\mu}\left(\frac{1}{r^2}\frac{d^2\psi(\phi)}{d\phi^2}\right) = E\psi(\phi)$$

$$\frac{d^2\psi(\phi)}{d\phi^2} = -\frac{2\mu r^2 E\psi}{\hbar^2} = -\frac{2IE\psi}{\hbar^2}$$

$$\frac{d^2\psi(\phi)}{d\phi^2} = -m^2\psi(\phi), \qquad m^2 = \frac{2IE}{\hbar^2}$$

The solution of this equation is

$$\psi(\phi) = A \exp(im\phi), \qquad m = 0, \pm 1, \pm 2, \ldots$$

The energy eigenvalues are given by

$$E_m = \frac{\hbar^2 m^2}{2I}, \qquad m = 0, \pm 1, \pm 2, \ldots$$

The normalized eigenfunctions are

$$\psi(\phi) = \frac{1}{\sqrt{2\pi}} \exp(im\phi), \qquad m = 0, \pm 1, \pm 2, \ldots$$

5.4 Calculate the energy difference between the stationary states $l = 1$ and $l = 2$ of the rigid molecule H_2. Use the Bohr frequency rule to estimate the frequency of radiation involved during transition between these two states. Suggest a method for determining the bond length of hydrogen molecule.

Solution. The energy of a rigid rotator is given by

$$E_l = \frac{l(l+1)\hbar^2}{2I}, \qquad l = 0, 1, 2, \ldots$$

$$E_1 = \frac{\hbar^2}{I}, \qquad E_2 = \frac{3\hbar^2}{I}$$

According to Bohr's frequency rule,

$$\nu = \frac{E_2 - E_1}{h} = \frac{2\hbar^2}{Ih} = \frac{h}{2\pi^2 I}$$

$$\text{Moment of inertia } I = \mu r^2 = \frac{m \cdot m}{m + m} r^2 = \frac{m}{2} r^2$$

Here, m is the mass of hydrogen atom and r is the bond length of hydrogen molecule. Substituting this value of I, we get

$$\nu = \frac{h}{\pi^2 m r^2} \quad \text{or} \quad r = \left(\frac{h}{\pi^2 m \nu}\right)^{1/2}$$

5.5 Solve the time independent Schrödinger equation for a three-dimensional harmonic oscillator whose potential energy is

$$V = \frac{1}{2}(k_1 x^2 + k_2 y^2 + k_3 z^2)$$

Solution. The theory we developed for a linear harmonic oscillator can easily be extended to the case of three-dimensional oscillator. The Schrödinger equation for the system is

$$\frac{-\hbar^2}{2m}\nabla^2\psi(x, y, z) + V\psi(x, y, z) = E\psi(x, y, z)$$

This equation can be separated into three equations by writing the wave function

$$\psi(x, y, z) = X(x)\, Y(y)\, Z(z)$$

The Schrödinger equation now separates into three equations of the form

$$\frac{d^2X(x)}{dx^2} + \frac{2m}{\hbar^2}\left(E_x - \frac{1}{2}m\omega_x^2 x^2\right)X(x) = 0$$

$$\frac{d^2Y(x)}{dy^2} + \frac{2m}{\hbar^2}\left(E_y - \frac{1}{2}m\omega_y^2 y^2\right)Y(y) = 0$$

$$\frac{d^2Z(z)}{dz^2} + \frac{2m}{\hbar^2}\left(E_z - \frac{1}{2}m\omega_z^2 z^2\right)Z(z) = 0$$

where $E_x + E_y + E_z = E$, the total energy of the system and

$$\omega_x = \sqrt{\frac{k_1}{m}}, \qquad \omega_y = \sqrt{\frac{k_2}{m}}, \qquad \omega_z = \sqrt{\frac{k_3}{m}}$$

Using the results of linear harmonic oscillator (Eq. 4.13), we get

$$E_x = \left(n_x + \frac{1}{2}\right)\omega_x, \qquad n_x = 0, 1, 2, \ldots$$

$$E_y = \left(n_y + \frac{1}{2}\right)\omega_y, \qquad n_y = 0, 1, 2, \ldots$$

$$E_z = \left(n_z + \frac{1}{2}\right)\omega_z, \qquad n_z = 0, 1, 2, \ldots$$

The eigenfunctions are given by Eq. (4.14), and so

$$\psi_{n_x n_y n_z} = NH_{n_x}(\alpha x)\, H_{n_y}(\beta y) H_{n_z}(\gamma z) \exp\left[-\frac{1}{2}(\alpha^2 x^2 + \beta^2 x^2 + \gamma^2 x^2)\right]$$

where N is the normalization constant and

$$\alpha = \left(\frac{m\omega_x}{\hbar}\right)^{1/2}, \qquad \beta = \left(\frac{m\omega_y}{\hbar}\right)^{1/2}, \qquad \gamma = \left(\frac{m\omega_z}{\hbar}\right)^{1/2}$$

Normalization gives

$$N = \frac{\alpha^{1/2}\beta^{1/2}\gamma^{1/2}}{\pi^{3/4}(2^{n_x+n_y+n_z}\, n_x!\, n_y!\, n_z!)^{1/2}}$$

5.6 For the ground state of the hydrogen atom, evaluate the expectation value of the radius vector r of the electron.

Solution. The wave function of the ground state is given by

$$\psi_{100} = \frac{1}{\sqrt{\pi}}\left(\frac{1}{a_0}\right)^{3/2} \exp\left(\frac{-r}{a_0}\right)$$

$$\langle r\rangle = \int \psi_{100}^* r \psi_{100}\, d\tau = \frac{1}{\pi a_0^3}\int_0^\infty r^3 \exp\left(-\frac{2r}{a_0}\right) dr \int_0^\pi\int_0^{2\pi} \sin\theta\, d\theta\, d\phi$$

The integration over the angular coordinates gives 4π. Using the relation in the Appendix, the r-integral can be evaluated. Thus,

$$\langle r\rangle = \frac{4}{a_0^3}\,\frac{3!}{(2/a_0)^4} = \frac{3}{2}a_0$$

The expectation value of r in the ground state of hydrogen atom is $3a_0/2$.

5.7 Neglelcting electron spin degeneracy, prove that the hydrogen atom energy levels are n^2 fold degenerate.

Solution. In a hydrogen atom, the allowed values of the quantum numbers are n = 1, 2, 3, ...; l = 0, 1, 2, ..., $(n - 1)$; m = 0, ±1, ±2, ..., $\pm l$. For a given value of n, l can have the values 0, 1, 2, ..., $(n - 1)$, and for a given value of l, m can have $(2l + 1)$ values. Therefore, the degeneracy of the nth state is

$$\sum_{l=0}^{n-1}(2l+1) = \sum_{l=0}^{n-1} 2l + n = \frac{2(n-1)n}{2} + n = n^2$$

5.8 Calculate the expectation value of the potential energy V of the electron in the 1s state of hydrogen atom. Using this result, evaluate the expectation value of kinetic energy T.

Solution. Substituting the ground state wave function from Eq. (5.17) and carrying out the angular integration, we get

$$\langle V\rangle = \int \psi_{100}^{*}\left(\frac{-ke^2}{r}\right)\psi_{100}\,d\tau = -ke^2\,\frac{4\pi}{\pi a_0^3}\int_0^\infty \exp\left(-\frac{2r}{a_0}\right) r\,dr$$

Using the standard integral (see appendix), we obtain

$$\langle V\rangle = \frac{-ke^2}{a_0} = \frac{-k^2me^4}{\hbar^2} = 2E_1$$

where E_1, the ground state energy, is equal to $\langle T\rangle + \langle V\rangle$ and, therefore,

$$E_1 = \langle T\rangle + 2E_1$$

or

$$\langle T\rangle = -E_1$$

$$= \frac{me^4}{32\pi^2\varepsilon_0^2\hbar^2}$$

5.9 Evaluate the most probable distance of the electron of the hydrogen atom in its 3d state.

Solution. From Eq. (5.18), the radial probability density

$$P_{nl}(r) = |R_{nl}|^2\, r^2$$

$$R_{32} = \frac{4}{27\sqrt{10}}\left(\frac{1}{3a_0}\right)^{3/2}\left(\frac{r}{a_0}\right)^2 \exp\left(-\frac{r}{3a_0}\right)$$

$$= \text{constant } r^2 \exp\left(-\frac{r}{3a_0}\right)$$

$$P_{32} = \text{constant } r^6 \exp\left(-\frac{2r}{3a_0}\right)$$

To find the most probable distance, we have to set $dP_{32}/dr = 0$, and

$$\frac{dP_{32}}{dr} = 0 = 6r^5 \exp\left(-\frac{2r}{3a_0}\right) - \frac{2r^6}{3a_0}\left(-\frac{2r}{3a_0}\right)$$

where

$$r = 9a_0$$

The most probable distance of a 3d electron in a hydrogen atom is $9a_0$.

5.10 In a stationary state of the rigid rotator, show that the probability density is independent of the angle ϕ.

Solution. In stationary states, the wave functions of a rigid rotator are the spherical harmonics $y_{lm}(\theta, \phi)$ given by

$$y_{lm}(\theta, \phi) = \text{constant } P_l^m(\cos\theta)\, e^{im\phi}$$

$$\text{Probability density} = |Y_{lm}|^2 = \text{constant } |P_l^m(\cos\theta)|^2$$

which is independent of the angle ϕ.

5.11 Calculate the energy difference between the first two rotational energy levels of the CO molecule if the intermolecular separation is 1.131 Å. The mass of the carbon atom is 19.9217×10^{-27} kg are the mass of oxygen atom is 26.5614×10^{-27} kg. Assume the molecule to be rigid.

Solution. The energy of a rigid rotator is given by

$$E_l = \frac{l(l+1)\hbar^2}{2I}$$

$$E_0 = 0, \qquad E_1 = \frac{\hbar^2}{I}, \qquad \Delta E = E_1 - E_0 = \frac{\hbar^2}{I}$$

The reduced mass

$$\mu = \frac{19.9217 \times 26.5614 \times 10^{-27}}{19.9217 + 26.5614} = 11.3837 \times 10^{-27}\ \text{kg}$$

$$I = \mu r^2 = (11.3837 \times 10^{-27}\ \text{kg})\ (1.131 \times 10^{-10}\ \text{m})^2$$

$$= 14.5616 \times 10^{-47}\ \text{kg m}^2$$

$$\Delta E = \frac{\hbar^2}{I} = \frac{(1.054)^2 \times 10^{-68}\ \text{J}^2\text{s}^2}{14.5616 \times 10^{-47}\ \text{kg m}^2} = 7.63 \times 10^{-23}\ \text{J}$$

5.12 What is the probability of finding the 1s-electron of the hydrogen atom at distances (i) 0.5 a_0, (ii) 0.9 a_0, (iii) a_0, and (iv) 1.2 a_0 from the nucleus? Comment on the result.

Solution. The radial probability density $P_{nl}(r) = |R_{nl}|^2 r^2$. Then,

$$R_{10} = \frac{2}{a_0^{3/2}} \exp\left(-\frac{r}{a_0}\right), \qquad P_{10}(r) = \frac{4r^2}{a_0^3} \exp\left(-\frac{2r}{a_0}\right)$$

(i) $$P_{10}(0.5a_0) = \frac{e^{-1}}{a_0} = \frac{0.37}{a_0}.$$

(ii) $$P_{10}(0.9a_0) = \frac{4(0.9)^2}{a_0} e^{-1.8} = \frac{0.536}{a_0}.$$

(iii) $$P_{10}(a_0) = \frac{4e^{-2}}{a_0} = \frac{0.541}{a_0}.$$

(iv) $$P_{10}(1.2a_0) = \frac{4(1.2)^2}{a_0} = \frac{0.523}{a_0}.$$

$P_{10}(r)$ increases as r increases from 0 to a_0 and then decreases, indicating a maximum at $r = a_0$. This is in conformity with Bohr's picture of the hydrogen atom.

5.13 What is the probability of finding the 2s-electron of hydrogen atom at a distance of (i) a_0 from the nucleus, and (ii) $2a_0$ from the nucleus?

Solution.

$$R_{20} = \left(\frac{1}{2a_0}\right)^{3/2}\left(2 - \frac{r}{a_0}\right)\exp\left(-\frac{r}{2a_0}\right)$$

$$P_{20}(r) = \left(\frac{1}{2a_0}\right)^{3}\left(2 - \frac{r}{a_0}\right)^{2} r^2 \exp\left(-\frac{r}{a_0}\right)$$

$$P_{20}(a_0) = \frac{e^{-1}}{8a_0} = \frac{0.37}{8a_0}$$

$$P_{20}\,(2a_0) = 0$$

5.14 For hydrogen atom in a stationary state defined by quantum numbers n, l and m, prove that

$$\langle r\rangle = \int_0^\infty r^3 \,|\,R_{nl}\,|^2\, dr$$

Solution. In a stationary state,

$$\langle r\rangle = \iiint \psi_{nlm}^* r\, \psi_{nlm}\, d\tau = \int_0^\infty |R_{nl}|^2\, r^3\, dr \int_0^\pi \int_0^{2\pi} |Y_{lm}|^2 \sin\theta\, d\theta\, d\phi$$

Since the spherical harmonics are normalized, the value of angular integral is unity, i.e.

$$\langle r\rangle = \int_0^\infty |R_{nl}|^2\, r^3\, dr$$

5.15 Calculate the size, i.e., $\langle r^2\rangle^{1/2}$, for the hydrogen atom in its ground state.

Solution.

$$\psi_{100} = \left(\frac{1}{\pi a_0^3}\right)^{1/2} e^{-r/a_0}$$

$$\langle r^2\rangle = \frac{1}{\pi a_0^3}\iiint \exp\left(-\frac{2r}{a_0}\right) r^4 \sin\theta\, d\theta\, d\phi\, dr$$

The angular integration gives 4π. Use of the integrals in the Appendix gives

$$\langle r^2\rangle = \frac{4}{a_0^3}\int_0^\infty r^4 \exp\left(-\frac{2r}{a_0}\right) dr = \frac{4}{a_0^3}\,\frac{4!}{(2/a_0)^5} = 3a_0^2$$

$$\langle r^2\rangle^{1/2} = \sqrt{3}a_0$$

5.16 Estimate the value of $(\Delta r)^2$ for the ground state of hydrogen atom.

$$(\Delta r)^2 = \langle r^2\rangle - \langle r\rangle^2, \qquad \langle r\rangle = \int_0^\infty |R_{nl}|^2 r^3\, dr$$

Solution. From Problem 5.6, for the ground state,

$$\langle r\rangle = \frac{4}{a_0^3}\int_0^\infty r^3 \exp\left(\frac{-2r}{a_0}\right) dr = \frac{3a_0}{2}$$

We now have (Problem 5.15)

$$\langle r^2\rangle = 3a_0^2$$

$$(\Delta r)^2 = 3a_0^2 - \frac{9}{4}a_0^2 = \frac{3a_0^2}{4}$$

5.17 Calculate the number of revolutions per second which a rigid diatomic molecule makes when it is in the (i) $l = 2$ state, (ii) $l = 5$ state, given that the moment of inertia of the molecule is I.

Solution. Rotational energy of a molecule is

$$E_l = \frac{l(l+1)\hbar^2}{2I}$$

Classically

$$\text{Rotational energy} = \frac{1}{2}I\omega^2 = 2\pi^2 I\nu^2$$

Equating the two expressions for energy, we get

$$\frac{l(l+1)\hbar^2}{2I} = 2\pi^2 I\nu^2 \quad \text{or} \quad \nu = \frac{\sqrt{l(l+1)}\,\hbar}{2\pi I}$$

(i) $l = 2$ state: $\nu = \dfrac{\sqrt{6}\,\hbar}{2\pi I}$

(ii) $l = 5$ state: $\nu = \dfrac{\sqrt{30}\,\hbar}{2\pi I}$

Note: The result can also be obtained by equating the expressions for angular momentum.

5.18 In Problem 5.5, if the oscillator is isotropic: (i) What would be the energy eigenvalues? (ii) What is the degeneracy of the state n?

Solution.

(i) For an isotropic oscillator $k_1 = k_2 = k_3$ and $n_x, n_y, n_z = 0, 1, 2, \ldots$ Hence, the energy expression becomes

$$E = E_x + E_y + E_z = \left(n + \frac{3}{2}\right)\hbar\omega, \qquad n = n_x + n_y + n_z = 0, 1, 2, \ldots$$

(ii) Degeneracy of the state n : The various possibilities are tabulated:

n_x	n_y	n_z	
n	0	0	1 way
$n-1$	1	0	2 ways
$n-1$	0	1	
$n-2$	1	1	
$n-2$	0	2	3 ways
$n-2$	2	0	
⋮	⋮	⋮	
1	$n-1$	0	n ways
1	⋮	⋮	
⋮	⋮	⋮	
0	n	0	$(n+1)$ ways
0	0	n	
⋮	⋮	⋮	

Total no. of ways $= 1 + 2 + 3 + \cdots + (n+1)$
$= (n+1)(n+2)/2$

Degeneracy of the state $(n) = (n+1)(n+2)/2$

5.19 Find the number of energy states and energy levels in the range $E < [15R^2/(8\ ma^2)]$ of a cubical box of side a.

Solution. For a particle in a cubic box of side a, the energy is given by (refer Problem 5.2)

$$E = \frac{\pi^2 \hbar^2}{2ma^2}(n_x^2 + n_y^2 + n_z^2) = \frac{h^2}{8ma^2}(n_x^2 + n_y^2 + n_z^2)$$

Comparing with the given expression, we get

$$n_x^2 + n_y^2 + n_z^2 < 15$$

The number of possible combinations of $(n_x\, n_y\, n_z)$ is

(1 1 1)	1 way
(1 1 2), (1 2 1), (2 1 1)	3 ways
(1 1 3), (1 3 1), (3 1 1)	3 ways
(1 2 2), (2 1 2), (2 2 1)	3 ways
(2 2 2)	1 way
(1 2 3), (1 3 2), (2 1 3), (2 3 1), (3 2 1), (3 1 2)	6 ways
Total	17 ways

Hence the No. of possible states = 17. The No. of energy levels = 6.

5.20 Show that the three $2p$ eigenfunctions of hydrogen atom are orthogonal to each other.

$\psi_{210} = c_1 r e^{-r/2a_0} \cos\theta$, c_1 being constant

$\psi_{21,\pm1} = c_2 r e^{-r/2a_0} \sin\theta\, e^{\pm i\phi}$, c_2 being constant

Solution. The ϕ-dependent part of the product $\psi_{21,1}^* \, \psi_{21,-1}$ gives $e^{-2i\phi}$ The corresponding ϕ integral becomes

$$\int_0^{2\pi} e^{-2i\phi} d\phi = \frac{1}{-2i}\left[e^{-2i\phi}\right]_0^{2\pi} = 0$$

The ϕ integral of

$$\int \psi_{210}^* \, \psi_{211} \, d\tau = \int_0^{2\pi} e^{i\phi} d\phi = 0$$

The ϕ integral of

$$\int \psi_{210}^* \, \psi_{21,-1} = \int_0^{2\pi} e^{-i\phi} \, d\phi = 0$$

Thus, the three 2p eigenfunctions of hydrogen atom are orthogonal to each other.

5.21 Prove that the 1s, 2p and 3d orbitals of a hydrogen-like atom show a single maximum in the radial probability curves. Obtain the values at which these maxima occur.

Solution. The radial probability density $P_{nl} = r^2|R_{nl}|^2$. Then,

$$R_{10} = \text{constant} \times \exp\left(-\frac{Zr}{a_0}\right)$$

$$R_{21} = \text{constant} \times r \exp\left(-\frac{Zr}{2a_0}\right)$$

$$R_{32} = \text{constant} \times r \exp\left(-\frac{Zr}{3a_0}\right)$$

P_{nl} will be maximum when $dP_{nl}/dr = 0$, and hence

$$\frac{dP_{10}}{dr} = 0 = \text{constant}\left(2r - \frac{2Zr^2}{a_0}\right)\exp\left(-\frac{2Zr}{a_0}\right), \qquad r = \frac{a_0}{Z}$$

$$\frac{dP_{21}}{dr} = 0 = \text{constant}\left(4r^3 - \frac{Zr^4}{a_0}\right)\exp\left(-\frac{Zr}{a_0}\right), \qquad r = \frac{4a_0}{Z}$$

Similarly, $dP_{32}/dr = 0$ gives $r = 9a_0/Z$.
In general, $r_{max} = n^2 a_0/Z$.

Note: The result $r_{max} = a_0/Z$ suggests that the 1s-orbital of other atoms shrinks in proportion to the increase in atomic number.

5.22 If the interelectronic repulsion in helium is ignored, what would be its ground state energy and wave function?

Solution. Helium atom has two electrons and $Z = 2$. The ground state energy and wave function of hydrogen-like atom are

$$E_1 = -\frac{k^2 Z^2 me^4}{2\hbar^2} = -13.6\ Z^2\ \text{eV}, \qquad k^2 = \frac{1}{4\pi\varepsilon_0}$$

$$\psi_{100} = \frac{1}{\pi^{1/2}}\left(\frac{Z}{a_0}\right)^{3/2} \exp\left(-\frac{Zr}{a_0}\right)$$

When the interelectronic repulsion is neglected, the energy of the system is the sum of the energies of the two electrons and the wave function is the product of the two functions, i.e.

$$\text{Energy } E = -13.6\ Z^2 - 13.6\ Z^2 = -108.8 \text{ eV}$$

Wave function
$$\psi = \psi_1(r_1)\ \psi_2(r_2) = \frac{1}{\pi}\left(\frac{Z}{a_0}\right)^3 \exp\left(\frac{-Z(r_1 + r_2)}{a_0}\right)$$

where r_1 and r_2 are the radius vector of electrons 1 and 2, respectively.

5.23 Evaluate the most probable distance of the electron of the hydrogen atom in its 2p state. What is the radial probability at that distance?

Solution. The radial probability density

$$P_{nl}(r) = r^2 |R_{nl}|^2$$

and

$$R_{21} = \left(\frac{1}{2a_0}\right)^{3/2} \frac{1}{a_0\sqrt{3}}\, r \exp\left(-\frac{r}{2a_0}\right)$$

$$P_{21}(r) = r^2 R_{21}^2 = \frac{1}{24a_0^5}\, r^4 \exp\left(-\frac{r}{a_0}\right)$$

For P_{21} to be maximum, it is necessary that

$$\frac{dP_{21}}{dr} = \frac{1}{24a_0^5}\left(4r^3 - \frac{r^4}{a_0}\right)\exp\left(-\frac{r}{a_0}\right) = 0$$

$$r = 4a_0$$

The most probable distance is four times the Bohr radius, i.e.

$$P_{21}(4a_0) = \frac{32}{3a_0} \exp(-4)$$

5.24 A positron and an electron form a shortlived atom called *positronium* before the two annihilate to produce gamma rays. Calculate, in electron volts, the ground state energy of positronium.

Solution. The positron has a charge $+e$ and mass equal to the electron mass. The mass μ in the energy expression of hydrogen atom is the reduced mass which, for the positronium atom, is

$$\frac{m_e \cdot m_e}{2m_e} = \frac{m_e}{2}$$

where m_e is the electron mass.

Hence the energy of the positronium atom is half the energy of hydrogen atom.

$$E_n = -\frac{k^2 m_e e^4}{4\hbar^2 n^2}, \quad n = 1, 2, 3, \ldots$$

Then the ground state energy is

$$-\frac{13.6}{2} \text{ eV} = -6.8 \text{ eV}$$

5.25 A mesic atom is formed by a muon of mass 207 times the electron mass, charge –e, and the hydrogen nucleus. Calculate: (i) the energy levels of the mesic atom; (ii) radius of the mesic atom; and (iii) wavelength of the 2p → 1s transition.

Solution.

(i) The system is similar to that of hydrogen atom. Hence the energy levels are given by

$$E_n = -\frac{\mu e^4}{(4\pi\varepsilon_0)^2 2\hbar^2}\frac{1}{n^2}, \qquad n = 1, 2, 3, \ldots$$

where μ is the proton-muon reduced mass

$$\mu = \frac{207m_e \times 1836m_e}{207m_e + 1836m_e} = 186m_e$$

(ii) The radius of the mesic atom will also be similar to that of Bohr atom, see Eq. (1.9).

Radius of the nth orbit $r_n = \dfrac{n^2\hbar^2}{k\mu e^2}$

$$r_1 = \frac{\hbar^2}{k\mu e^2}, \qquad k = 8.984 \times 10^9 \text{ N m}^2\text{ C}^{-2}$$

$$= \frac{(1.05\times 10^{-34}\text{ J s})^2}{(8.984\times 10^9\text{ N m}^2\text{ C}^{-2})} \times \frac{1}{(186\times 9.1\times 10^{-31}\text{ kg})} \times \frac{1}{(1.6\times 10^{-19}\text{ C})^2}$$

$$= 2.832 \times 10^{-13}\text{ m} = 283.2 \times 10^{-15}\text{ m} = 283.2\text{ fm}$$

(iii) $$E_2 - E_1 = \frac{k^2\mu e^4}{2\hbar^2}\left(\frac{1}{1^2} - \frac{1}{2^2}\right)$$

$$= \frac{(8.984\times 10^9\text{ Nm}^2\text{C}^{-2})^2(186\times 9.1\times 10^{-31}\text{ kg})(1.6\times 10^{-19}\text{ C})^4}{2(1.05\times 10^{-34}\text{ J s})^2} \times \frac{3}{4}$$

$$= 304527.4 \times 10^{-21}\text{ J} = 1903.3\text{ eV}$$

$$\lambda = \frac{hc}{E_2 - E_1} = \frac{(6.626\times 10^{-34}\text{ J s})(3\times 10^8\text{ m/s})}{304527.4\times 10^{-21}\text{ J}}$$

$$= 0.65275 \times 10^{-9}\text{ m} = 0.653\text{ nm}$$

5.26 Calculate the value of $\langle 1/r\rangle$ for the electron of the hydrogen atom in the ground state. Use the result to calculate the average kinetic energy $\langle p^2/2m\rangle$ in the ground state. Given

$$\int_0^\infty x^n e^{-ax}\,dx = \frac{n!}{a^{n+1}}$$

Solution. For the ground state,

$$\psi_{100} = \frac{1}{\pi^{1/2}a_0^{3/2}}e^{-r/a_0}, \qquad E_1 = -\frac{\mu e^4}{32\pi^2\varepsilon_0^2\hbar^2}$$

$$\left\langle \frac{1}{r} \right\rangle = \int \psi_{100} \frac{1}{r} \psi_{100}\, d\tau = \frac{1}{\pi a_0^3} \int_0^{\infty} r \exp\left(-\frac{2r}{a_0}\right) dr \int_0^{\pi} \int_0^{2\pi} \sin\theta\, d\theta\, d\phi$$

The angular part of the integral gives 4π. The r-integral gives $a_0^2/4$. Hence,

$$\left\langle \frac{1}{r} \right\rangle = \frac{4\pi}{\pi a_0^3} \frac{a_0^2}{4} = \frac{1}{a_0}$$

$$\langle V(r) \rangle = \left\langle -\frac{e^2}{4\pi\varepsilon_0 r} \right\rangle = -\frac{e^2}{4\pi\varepsilon_0} \left\langle \frac{1}{r} \right\rangle = -\frac{e^2}{4\pi\varepsilon_0 a_0}$$

Therefore,

$$\left\langle \frac{p^2}{2m} \right\rangle = E - \langle V(r) \rangle = -\frac{\mu e^4}{32\pi^2 \varepsilon_0^2 \hbar^2} + \frac{e^2}{4\pi\varepsilon_0 a_0}$$

Since

$$a_0 = \frac{4\pi\varepsilon_0 \hbar^2}{\mu e^2}$$

We have

$$\left\langle \frac{p^2}{2m} \right\rangle = -\frac{\mu e^4}{32\pi^2 \varepsilon_0^2 \hbar^2} + \frac{\mu e^4}{16\pi^2 \varepsilon_0^2 \hbar^2}$$

$$= \frac{\mu e^4}{32\pi^2 \varepsilon_0^2 \hbar^2}$$

In other words, the average value of kinetic energy $\langle KE \rangle = -\langle V \rangle/2$. In fact, this condition is true for all states (see Problem ...)

5.27 A rigid rotator having moments of inertia I rotates freely in the x-y plane. If ϕ is the angle between the x-axis and the rotator axis, (i) find: the energy eigenvalues and eigenfunctions, (ii) the angular speed; and (iii) $\psi(t)$ for $t > 0$ if $\psi(0) = A \cos^2 \phi$.

Solution.

(i) The energy eigenvalues and eigenfunctions (refer Problem 5.3) are

$$E_m = \frac{m^2 \hbar^2}{2I}, \qquad \psi = \frac{1}{\sqrt{2\pi}} \exp(im\phi), \qquad m = 0, \pm 1, \pm 2, \ldots$$

At $t = 0$,

$$\psi(0) = A \cos^2 \phi = \frac{A}{2}(1 + \cos 2\phi)$$

$$\psi(0) = \frac{A}{2} + \frac{A}{4}(e^{i2\phi} + e^{-i2\phi})$$

The first term corresponds to $m = 0$. In the second term, one quantity corresponds to $m = 2$ and the other to $m = -2$.

(ii) The angular speed $\dot{\phi}$ is given by

$$E_m = \frac{1}{2} I \dot{\phi}^2 \quad \text{or} \quad \frac{m^2 \hbar^2}{2I} = \frac{1}{2} I \dot{\phi}^2$$

$$\dot{\phi} = \frac{m\hbar}{I}$$

(iii) $$\psi(t) = \frac{A}{2} + \frac{A}{4} e^{2i\phi} \cdot \exp\left(-\frac{iE_2 t}{\hbar}\right) + \frac{A}{4} e^{-2i\phi} \cdot \exp\left(-\frac{iE_{-2} t}{\hbar}\right)$$

$$= \frac{A}{2} + \frac{A}{4} \exp\left[2i\left(\phi - \frac{\hbar t}{I}\right)\right] + \frac{A}{4} \exp\left[-2i\left(\phi + \frac{\hbar t}{I}\right)\right]$$

5.28 A particle of mass m is confined to the interior of a hollow spherical cavity of radius R_1 with impenetrable walls. Find the pressure exerted on the walls of the cavity by the particle in its ground state.

Solution. The radial wave equation (5.5), with $V(r) = 0$, is

$$\frac{1}{r^2} \frac{d}{dr}\left(r^2 \frac{dR}{dr}\right) + \left[\frac{2mE}{\hbar^2} - \frac{l(l+1)}{r^2}\right] R = 0$$

For the ground state, $l = 0$. Writing

$$R(r) = \frac{\chi(r)}{r}$$

the radial equation reduces to [refer Eq. (5.17)]

$$\frac{d^2\chi}{dr^2} + k^2 \chi = 0, \qquad k^2 = \frac{2mE}{\hbar^2}, \qquad r < R$$

whose solution is

$$\chi = A \sin kr + B \cos kr, \qquad A \text{ and } B \text{ are constants.}$$

R is finite at $r = 0$, i.e., at $r = 0$, $\chi = Rr = 0$. This leads to $B = 0$. Hence,

$$\chi = A \sin kr$$

The condition that $R = 0$ at $r = R_1$ gives

$$0 = A \sin kR_1$$

As A cannot be zero,

$$kR_1 = n\pi \quad \text{or} \quad k = \frac{n\pi}{R_1}, \qquad n = 1, 2, 3, \ldots$$

Hence the solution is

$$\chi = A \sin \frac{n\pi r}{R_1}, \qquad n = 1, 2, 3, \ldots$$

Normalization gives

$$\chi_n = \sqrt{\frac{2}{R_1}} \sin \frac{n\pi r}{R_1}, \qquad n = 1, 2, 3, \ldots$$

with the condition that

$$k = \frac{n\pi}{R_1} \quad \text{or} \quad E_n = \frac{\pi^2 \hbar^2 n^2}{2mR_1^2}$$

The average force F exerted radially on the walls by the particle is given by

$$F = \left\langle -\frac{\partial V}{\partial R} \right\rangle = -\left\langle \frac{\partial H}{\partial R} \right\rangle = -\frac{\partial \langle H \rangle}{\partial R} = -\frac{\partial E}{\partial R}$$

The particle is in its ground state. Hence, $n = 1$ and

$$F = -\frac{\partial E_1}{\partial R} = \frac{\pi^2 \hbar^2}{mR_1^3}$$

The pressure exerted on the walls is

$$p = \frac{F}{4\pi R_1^2} = \frac{\pi \hbar^2}{4mR_1^5}$$

5.29 At time $t = 0$, the wave function for the hydrogen atom is

$$\Psi(\boldsymbol{r}, 0) = \frac{1}{\sqrt{10}}(2\Psi_{100} + \Psi_{210} + \sqrt{2}\Psi_{211} + \sqrt{3}\Psi_{21,-1})$$

where the subscripts are values of the quantum numbers n, l, m. (i) What is the expectation value for the energy of the system? (ii) What is the probability of finding the system with $l = 1$, $m = 1$?

Solution.

(i) The expectation value of the energy of the system

$$\begin{aligned}
\langle E \rangle &= \langle \Psi | H | \Psi \rangle \\
&= \frac{1}{10}\langle (2\Psi_{100} + \Psi_{210} + \sqrt{2}\Psi_{211} + \sqrt{3}\Psi_{21,-1}) | H | (2\Psi_{100} + \Psi_{210} + \sqrt{2}\Psi_{211} + \sqrt{3}\Psi_{21,-1}) \rangle \\
&= \frac{1}{10}\langle (2\Psi_{100} + \Psi_{210} + \sqrt{2}\Psi_{211} + \sqrt{3}\Psi_{21,-1}) | (2E_1\Psi_{100} + E_2\Psi_{210} + \sqrt{2}E_2\Psi_{211} + \sqrt{3}E_2\Psi_{21,-1}) \rangle \\
&= \frac{1}{10}(4E_1 + E_2 + 2E_2 + 3E_2) = \frac{1}{10}(4E_1 + 6E_2)
\end{aligned}$$

Since $E_1 = -13.8$ eV and $E_2 = -3.4$ eV,

$$\langle E \rangle = \frac{1}{10}(-54.4 \text{ eV} - 20.4 \text{ eV}) = -7.48 \text{ eV}$$

(ii) The required probability is given by

$$P = \frac{2}{10}\langle 211 | 211 \rangle = \frac{2}{10} = \frac{1}{5}$$

5.30 Evaluate the radius for which the radial probability distribution $P(r)$ is maximum for the 1s, 2p, 3d orbitals of hydrogen atom. Compare your result with that of Bohr theory. Prove that, in general, when $l = n - 1$, $P(r)$ peaks at the Bohr atom value for circular orbits.

Solution. Evaluation of $P(r)$ for these orbitals is done in Problem 5.21. For 1s, 2p and 3d orbitals, the values are a_0, $4a_0$, $9a_0$, respectively. According to Bohr's theory, the radiis of the Bohr orbits are given by (see Eq. 1.9)

$$r_n = \frac{n^2\hbar^2}{kme^2}, \qquad k = \frac{1}{4\pi\varepsilon_0}$$

From Eq. (1.10),

$$a_0 = \frac{\hbar^2}{kme^2}$$

This gives

$$r_1 = a_0, \qquad r_2 = 4a_0, \qquad r_3 = 9a_0$$

which is in agreement with the quantum mechanical results. Hence, the maximum radial probability peaks at

$$r_{\max} = n^2 a_0$$

The above values are for s ($l = 0$), p ($l = 1$), and d ($l = 2$) orbitals. Generalizing, when $l = n - 1$, $P(r)$ peaks at the Bohr atom value.

5.31 Evaluate the difference in wavelength $\Delta\lambda = \lambda_H - \lambda_D$ between the first line of Balmer series for a hydrogen atom (λ_H) and the corresponding line for a deuterium atom (λ_D).

Solution. The first line of the Balmer series is the tranisition $n = 3 \to n = 2$. Then,

$$\nu = \frac{2\pi^2k^2\mu e^4}{h^3}\left(\frac{1}{2^2} - \frac{1}{3^2}\right) = \frac{2\pi^2k^2\mu e^4}{h^3} \times \frac{5}{36}$$

$$\lambda_{\text{H}} = \frac{c}{\nu_{\text{H}}} = \frac{36ch^3}{5 \times 2\pi^2k^2\mu_{\text{H}}e^4}$$

$$\lambda_{\text{D}} = \frac{36ch^3}{5 \times 2\pi^2k^2\mu_{\text{D}}e^4}$$

$$\Delta\lambda = \lambda_{\text{H}} - \lambda_{\text{D}} = \frac{36ch^3}{10\pi^2k^2e^4}\left(\frac{1}{\mu_{\text{H}}} - \frac{1}{\mu_{\text{D}}}\right)$$

$$\mu_{\text{H}} = \frac{m_pm_e}{m_p + m_e}, \qquad \mu_{\text{D}} = \frac{2m_pm_e}{2m_p + m_e}$$

$$\frac{1}{\mu_{\text{H}}} - \frac{1}{\mu_{\text{D}}} = \frac{\mu_{\text{D}} - \mu_{\text{H}}}{\mu_{\text{H}}\mu_{\text{D}}} = \frac{1}{2m_p}$$

$$\Delta\lambda = \lambda_{\text{H}} - \lambda_{\text{D}} = \frac{36ch^3}{10\pi^2k^2e^4}\,\frac{1}{2m_p}$$

$$= \frac{36\,(3 \times 10^8 \text{ m/s})\,(6.626 \times 10^{-34} \text{ J s})^3}{10\pi^2\,(8.984 \times 10^9 \text{ N m}^2\text{C}^{-2})^2\,(1.6 \times 10^{-19}\text{ C})^4\;2(1836 \times 9.1 \times 10^{-31}\text{ kg})}$$

$$= 0.18 \times 10^{-9} \text{ m} = 0.18 \text{ nm}$$

5.32 A quark having one-third the mass of a proton is confined in a cubical box of side 1.8×10^{-15} m. Find the excitation energy in MeV from the first excited state to the second excited state.

Solution. The energy eigenvalue for a particle of mass m in a cubical box of side a is given by (refer Problem 5.2)

$$E_{n_1 n_2 n_3} = \frac{\pi^2 \hbar^2}{2ma^2}(n_1^2 + n_2^2 + n_3^2)$$

First excited state: $E_{211} = E_{121} = E_{112} = \dfrac{6\pi^2\hbar^2}{2ma^2}$

Second excited state: $E_{221} = E_{212} = E_{122} = \dfrac{9\pi^2\hbar^2}{2ma^2}$

$$m = \frac{1.67262 \times 10^{-27}\ \text{kg}}{3} = 0.55754 \times 10^{-27}\ \text{kg}$$

$$\Delta E = \frac{3\pi^2\hbar^2}{2ma^2}$$

$$= \frac{3\pi^2 (1.05 \times 10^{-34}\ \text{J s})^2}{2(0.55754 \times 10^{-27}\ \text{kg})(1.8 \times 10^{-15}\ \text{m})^2}$$

$$= 9.0435 \times 10^{-11}\ \text{J} = \frac{9.0435 \times 10^{-11}\ \text{J}}{1.6 \times 10^{-19}\ \text{J/eV}}$$

$$= 565.2\ \text{MeV}$$

5.33 A system consisting of HCl molecules is at a temperature of 300 K. In the vibrational ground state, what is the ratio of number of molecules in the ground rotational state to the number in the first excited state? The moment of inertia of the HCl molecule is 2.3×10^{-47} kg m^2.

Solution. The factors that decide the number of molecules in a state are the Boltzmann factor and the degeneracy of the state. The degeneracy of a rotational level is $(2J + 1)$. If N_0 is the number of molecules in the $J = 0$ state, the number in the Jth state is

$$N_j = (2J + 1)\, N_0 \exp\left(-\frac{E_J}{kT}\right)$$

Hence,

$$\frac{N_0}{N_1} = \frac{1}{3}\exp\left(\frac{E_1}{kT}\right)$$

Rotational energy $\qquad E_J = \dfrac{J(J+1)\hbar^2}{2I}, \qquad J = 0, 1, 2, \ldots$

$$E_1 = \frac{2\hbar^2}{2I} = \frac{\hbar^2}{I}$$

$$\frac{E_1}{kT} = \frac{\hbar^2 \times 1}{IkT} = \frac{(1.054 \times 10^{-34}\ \text{J s})^2}{(2.3 \times 10^{-47}\ \text{kg} \cdot \text{m}^2)(1.38 \times 10^{-23}\ \text{JK}^{-1})\ 300\ \text{K}}$$

$$= 0.117$$

$$\frac{N_0}{N_1} = \frac{1}{3} e^{0.117} \cong 0.375$$

Note: Due to the factor $(2J + 1)$ in the expression for N_J, the level $J = 0$ need not be the one having the maximum number.

5.34 An electron of mass m and charge $-e$ moves in a region where a uniform magnetic field $\boldsymbol{B} = \nabla \times \boldsymbol{A}$ exists in the z-direction.

(i) Write the Hamiltonian operator of the system.
(ii) Prove that p_y and p_z are constants of motion.
(iii) Obtain the Schrödinger equation in cartestian coordinates and solve the same to obtain the energy values.

Solution.

(i) Given $\boldsymbol{B} = \nabla \times A$. We have

$$\boldsymbol{B} = \hat{i}\left(\frac{\partial A_z}{\partial y} - \frac{\partial A_y}{\partial z}\right) + \hat{j}\left(\frac{\partial A_z}{\partial z} - \frac{\partial A_z}{\partial x}\right) + \hat{k}\left(\frac{\partial A_y}{\partial x} - \frac{\partial A_x}{\partial y}\right)$$

Since the field is in the z-direction,

$$\frac{\partial A_z}{\partial y} - \frac{\partial A_y}{\partial z} = 0$$

$$\frac{\partial A_x}{\partial z} - \frac{\partial A_z}{\partial x} = 0$$

$$\frac{\partial A_y}{\partial x} - \frac{\partial A_x}{\partial y} = 0$$

On the basis of these equations, we can take

$$A_x = A_z = 0, \qquad A_y = Bx \quad \text{or} \quad \boldsymbol{A} = Bx\hat{j}$$

The Hamiltonian operator

$$H = \frac{1}{2m}\left(\boldsymbol{p} + \frac{e}{c}\boldsymbol{A}\right)^2, \qquad p = -i\hbar\nabla$$

$$= \frac{1}{2m}\left(p_x^2 + p_y^2 + p_z^2 + \frac{e^2}{c^2}A^2 + \frac{e}{c}\boldsymbol{p}\cdot\boldsymbol{A} + \frac{e}{c}\boldsymbol{A}\cdot\boldsymbol{p}\right)$$

$$= \frac{1}{2m}\left(p_x^2 + p_y^2 + p_z^2 + \frac{e^2B^2x^2}{c^2} + \frac{e}{c}p_yBx + \frac{e}{c}Bxp_y\right)$$

$$= \frac{1}{2m}\left[p_x^2 + \left(p_y + \frac{eBx}{c}\right)^2 + p_z^2\right]$$

where p_x, p_y, p_z are operators.

(ii) Since the operator p_y commutes with p_x, p_z and x,

$$[p_y, H] = [p_z, H] = 0$$

Hence p_y and p_z are constants.

(iii) The Schrödinger equation is

$$\frac{1}{2m}\left[p_x^2 + \left(p_y + \frac{eBx}{c}\right)^2 + p_z^2\right]\psi = E\psi$$

$$\frac{1}{2m}\left[p_x^2 + \left(p_y + \frac{eBx}{c}\right)^2 \psi = \left(E - \frac{p_z^2}{2m}\right)\psi\right]$$

Let us change the variable by defining

$$\chi = x + \frac{cp_y}{eB}, \qquad p_\chi = p_x$$

$$p_y + \frac{eBx}{c} = p_y + \frac{eB}{c}\left(\chi - \frac{cp_y}{eB}\right) = \frac{eB\chi}{c}$$

In terms of the new variables, $[\chi, p_\chi] = i\hbar$. Hence, the above equation reduces to

$$\left[\frac{p_\chi^2}{2m} + \frac{m}{2}\left(\frac{eB}{mc}\right)^2 \chi^2\right]\psi = \left(E - \frac{p_z^2}{2m}\right)\psi$$

Since p_z is constant, this equation is the same as the Schrödinger equation of a simple harmonic oscillator of angular frequency $\omega = eB/mc$ and energy eigenvalue $E - (p_z^2/2m)$. Therefore,

$$E - \frac{p_z^2}{2m} = \left(n + \frac{1}{2}\right)\hbar\omega, \qquad n = 0, 1, 2, \ldots$$

$$E = \left(n + \frac{1}{2}\right)\hbar\omega + \frac{p_z^2}{2m}, \qquad n = 0, 1, 2, \ldots$$

5.35 Consider the free motion of a particle of mass M constrained to a circle of radius r. Find the energy eigenvalues and eigenfunctions.

Solution. The system has only one variable, viz. the azimuthal angle ϕ. The classical energy equation is

$$E = \frac{p^2}{2m}$$

where p is the momentum perpendicular to the radius vector of the particle. Since the z-component of angular momentum $L_z = pr$,

$$E = \frac{L_z^2}{2Mr^2}$$

The operator for L_z is $-i\hbar\,(\partial/\partial\phi)$.

Replacing E and L_z by their operators and allowing the operator equation to operate on the eigenfunction $\Psi(\phi, t)$, we have

$$i\hbar\frac{\partial}{\partial t}\Psi = \frac{1}{2Mr^2}\left(-i\hbar\frac{\partial}{\partial\phi}\right)^2\Psi$$

$$= \frac{-\hbar^2}{2Mr^2}\frac{\partial^2\Psi}{\partial\phi^2}$$

A stationary state solution with energy eigenvalue E has the form

$$\Psi(\phi, t) = \psi(\phi)\, e^{-iEt/\hbar}$$

where $\psi(\phi)$ is the solution of

$$-\frac{\hbar^2}{2Mr^2}\frac{d^2\psi(\phi)}{d\phi^2} = E\psi(\phi)$$

$$\frac{d^2\psi(\phi)}{d\phi^2} = \frac{2Mr^2E\psi(\phi)}{\hbar^2}$$

$$= -k^2\psi, \qquad k^2 = \frac{2Mr^2E}{\hbar^2}$$

This equation has the solution

$$\psi(\phi) = Ae^{\pm ik\phi}$$

For ψ to be single valued,

$$\psi(\phi + 2\pi) = \psi(\phi)$$

This requirement leads to the condition

$$k = m, \qquad m = 0, 1, 2, \ldots$$

$$\frac{2Mr^2E_m}{\hbar^2} = m^2$$

$$E_m = \frac{\hbar^2 m^2}{2Mr^2}, \qquad m = 0, 1, 2, \ldots$$

The normalization of the eigenfunction leads to

$$\psi(\phi) = \frac{1}{\sqrt{2\pi}}\, e^{im\phi}, \qquad m = 0, 1, 2, \ldots$$

5.36 A particle of mass m is subjected to the spherically symmetric attractive square well potential defined by

$$V(r) = \begin{cases} -V_0, & 0 < r < a \\ 0, & r > a \end{cases}$$

Find the minimum depth of the potential well needed to have (i) one bound state of zero angular momentum, and (ii) two bound states of zero angular momentum.

Solution. The radial equation for a state with zero angular momentum, $\lambda = l(l+1) = 0$ in Eq. (5.5) is

$$\frac{1}{r^2}\frac{d}{dr}\left(r^2\frac{dR}{dr}\right) + \frac{2m}{\hbar^2}(E - V)R = 0$$

Since the potential is attractive, E must be negative. Hence,

$$\frac{1}{r^2}\frac{d}{dr}\left(r^2\frac{dR}{dr}\right) + \frac{2m}{\hbar^2}(V_0 - |E|)R = 0, \qquad 0 < r < a \tag{i}$$

$$\frac{1}{r^2}\frac{d}{dr}\left(r^2\frac{dR}{dr}\right) - \frac{2m|E|}{\hbar^2}R = 0, \qquad r > 0 \tag{ii}$$

To solve Eqs. (i) and (ii), we write

$$R = \frac{u(r)}{r}, \qquad k_1^2 = \frac{2m}{\hbar^2}(V_0 - |E|), \qquad k_2^2 = \frac{2m|E|}{\hbar^2} \tag{iii}$$

In terms of these quantities, equations (iii) reduce to

$$\frac{d^2u}{dr^2} + k_1^2 u = 0, \qquad 0 < r < a \tag{iv}$$

$$\frac{d^2u}{dr^2} - k_2^2 u = 0, \qquad r > 0 \tag{v}$$

The solutions of these equations are

$$u(r) = A \sin k_1 r + B \cos k_1 r \tag{vi}$$

$$u(r) = C \exp(-k_2 r) + D \exp(k_2 r) \tag{vii}$$

As $r \to 0$, $u(r)$ must tend to zero. This makes B zero. The solution $\exp(k_2 r)$ is not finite as $r \to \infty$. Hence, $D = 0$, and the solutions are

$$u(r) = A \sin k_1 r, \qquad 0 < r < a \tag{viii}$$

$$u(r) = C \exp(-k_2 r), \qquad r > 0 \tag{ix}$$

Applying the continuity conditions on $u(r)$ and du/dr at $r = a$, we get

$$A \sin(k_1 a) = C \exp(-k_2 a)$$

$$Ak_1 \cos k_1 a = -k_2 C \exp(-k_2 a)$$

Dividing one by the other and multiplying throughout by a, we obtain

$$k_1 a \cot k_1 a = -k_2 a \tag{x}$$

Writing

$$k_1 a = \beta, \qquad k_2 a = \gamma$$

we have

$$\beta^2 + \gamma^2 = \frac{2mV_0 a^2}{\hbar^2} \tag{xi}$$

which is the equation of a circle in the $\beta\gamma$-plane with radius $(2mV_0a^2/\hbar^2)^{1/2}$. Equation (x) becomes

$$\beta \cot \beta = -\gamma$$

To get the solution, $\beta \cot \beta$ against β is plotted along with circles of radii $(2mV_0a^2/\hbar^2)^{1/2}$ for different values of V_0a^2 (Fig. 5. 1). As β and γ can have only positive values, the intersection of the two curves in the first quadrant gives the energy levels.

(i) From Fig. 5.1, it follows that there will be one intersection if $\pi/2$ < radius < $3\pi/2$

$$\frac{\pi^2}{4} < \frac{2mV_0a^2}{\hbar^2} < \frac{9\pi^2}{4}$$

$$\frac{\pi^2\hbar^2}{8ma^2} < V_0 < \frac{9\pi^2\hbar^2}{8ma^2}$$

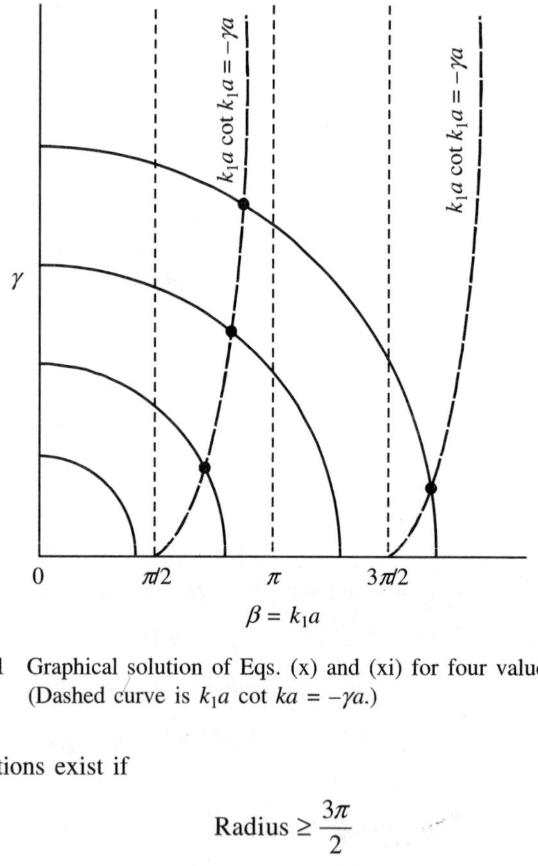

Fig. 5.1 Graphical solution of Eqs. (x) and (xi) for four values of V_0a^2. (Dashed curve is $k_1a \cot ka = -\gamma a$.)

(ii) Two intersections exist if

$$\text{Radius} \geq \frac{3\pi}{2}$$

$$\frac{2mV_0a^2}{\hbar^2} \geq \frac{9\pi^2}{4}$$

$$V_0 \geq \frac{9\pi^2\hbar^2}{8ma^2}$$

5.37 Write the radial part of the Schrödinger equation for hydrogen atom. Neglect the terms in $1/r$ and $1/r^2$ in the equation. Find the solution under these conditions in terms of the energy eigenvalues and hence the radial probability density. For the ground state, when is the probability density maximum? Comment on the result. Use the energy expression for the ground state.

Solution. The radial part of the equation is

$$\frac{1}{r^2}\frac{d}{dr}\left(r^2\frac{dR}{dr}\right)+\frac{2\mu}{\hbar^2}\left[E-\frac{l(l+1)\hbar^2}{2\mu r^2}+\frac{ke^2}{r}\right]R=0 \tag{i}$$

where $k = 1/4\pi\varepsilon_0$, $l = 0, 1, 2, \ldots$. Simplifying, we get

$$\frac{d^2R}{dr^2}+\frac{2}{r}\frac{dR}{dr}+\frac{2\mu}{\hbar^2}\left[E-\frac{l(l+1)\hbar^2}{2\mu r^2}+\frac{ke^2}{r}\right]R=0$$

Neglecting the terms in $1/r$ and $1/r^2$, we obtain

$$\frac{d^2R}{dr^2}+\frac{2\mu ER}{\hbar^2}=0 \tag{ii}$$

For bound states, E is negative. Hence,

$$\frac{d^2R}{dr^2}-A^2R=0, \qquad A^2=\frac{2\mu|E|}{\hbar^2} \tag{iii}$$

where solution is

$$R(r)=C_1e^{Ar}+C_2e^{-Ar}$$

where C_1 and C_2 are constants.
The physically acceptable solution is

$$R(r)=C_2e^{-Ar} \tag{iv}$$

The radial probability density

$$P=R^2r^2=C_2^2rh2e^{-2Ar}$$

For P to be maximum, it is necessary that

$$\frac{dP}{dr}=C_2^2\,(2re^{-2Ar}-2Ar^2e^{-2Ar})=0$$

$$1-Ar=0 \quad \text{or} \quad r=\frac{1}{A}=\frac{\hbar}{\sqrt{2m|E|}} \tag{v}$$

For the ground state, we have

$$|E|=\frac{k^2me^4}{2\hbar^2}$$

Substituting this value of $|E|$ in the expression for r, we get

$$r=\frac{\hbar^2}{k\mu e^2}=\frac{4\pi\varepsilon_0\hbar^2}{\mu e^2}=a_0$$

where a_0 is the Bohr radius, i.e., for the ground state, the radial probability density is maximum at the Bohr radius. The Bohr theory stipulates that the electron will be revolving at a distance a_0 from the origin. Here, the probability density is maximum at the Bohr radius with the possibility for a spherical distribution.

5.38 A crystal has some negative ion vacancies, each containing one electron. Treat these electrons as moving freely inside a volume whose dimensions are of the order of lattice constant. Assuming the value of lattice constant, estimate the longest wavelength of electromagnetic radiation absorbed by these electrons.

Solution. The energy levels of an electron in a cubical box of side a is (refer Problem 5.2)

$$E_{n_x,n_y,n_z} = \frac{\pi^2\hbar^2}{2ma^2}(n_x^2 + n_y^2 + n_z^2), \qquad n_x, n_y, n_z = 1, 2, 3, \ldots$$

Lattice constant $a \cong 1\text{Å} = 10^{-10}$ m.
The energy of the ground state is given by

$$E_{111} = \frac{\pi^2\hbar^2}{2ma^2} \times 3 = \frac{\pi^2(1.05\times 10^{-34}\ \text{J s})^2 \times 3}{2(9.1\times 10^{-31}\ \text{kg})(10^{-10}\ \text{m})^2}$$

$$= 1.795 \times 10^{-17}\ \text{J}$$

The longest wavelength corresponds to the transition from energy E_{111} to E_{211}, and hence

$$E_{211} = \frac{\pi^2\hbar^2 \times 6}{2ma^2} = 3.59 \times 10^{-17}\ \text{J}$$

$$\text{Longest wavelength } \lambda = \frac{c}{\nu} = \frac{ch}{E_{211} - E_{111}}$$

$$\lambda = \frac{(3\times 10^8\ \text{ms}^{-1})(6.626\times 10^{-34}\ \text{J s})}{1.795\times 10^{-17}\ \text{J}}$$

$$= 11.07 \times 10^{-9}\ \text{m} = 11.07\ \text{nm}$$

5.39 A particle of mass m is constrained to move between two concentric spheres of radii a and b $(b > a)$. If the potential inside is zero, find the ground state energy and the form of the wave function.

Solution. When the system is in the ground state and when $V = 0$, the radial wave equation (5.5) takes the form

$$\frac{1}{r^2}\frac{d}{dr}\left(r^2\frac{dR}{dr}\right) + k^2 R = 0, \qquad k^2 = \frac{2mE}{\hbar^2} \tag{i}$$

Writing $R(r) = \chi(r)/r$, Eq. (i) takes the form

$$\frac{d^2\chi}{dr^2} + k^2\chi = 0, \qquad a < r < b \tag{ii}$$

The solution of this equation is

$$\chi = A \sin kr + B \cos kr \tag{iii}$$

where A and B are constants.

The function $\chi(r)$ must be zero at $r = a$ and at $r = b$. For χ to be zero at $r = a$, Eq. (iii) must be of the form

$$\chi(r) = A \sin k(r - a) \tag{iv}$$

$\chi(r) = 0$ at $r = (b)$ gives

$$0 = A \sin k(b - a)$$

This is possible only if

$$k(b - a) = n\pi \quad \text{or} \quad k = \frac{n\pi}{b - a}$$

Substituting the value of k, we get

$$\frac{2mE}{\hbar^2} = \frac{n^2\pi^2}{(b - a)^2}, \qquad n = 1, 2, 3, \ldots$$

$$E_n = \frac{\pi^2\hbar^2 n^2}{2m(b - a)^2} \tag{v}$$

The ground state energy

$$E_1 = \frac{\pi^2\hbar^2}{2m(b - a)^2} \tag{vi}$$

Substituting the value of k in Eq. (iv), for the ground state,

$$\chi(r) = A \sin \frac{\pi(r - a)}{b - a}$$

$$R(r) = \frac{\chi(r)}{r} = \frac{A}{r} \sin \frac{\pi(r - a)}{b - a}$$

5.40 What are atomic orbitals? Explain in detail the p-orbitals and represent them graphically.

Solution. The wave function $\psi_{nlm}(r, \theta, \phi)$, which describes the motion of an electron in a hydrogen atom is called an atomic orbital. When $l = 0, 1, 2, \ldots$, the corresponding wave functions are s-orbital, p-orbital, d-orbital, and so on, respectively. For a given value of l, m can have the values $0, \pm 1, \pm 2, \ldots, \pm l$, and the radial part is the same for all the $(2l + 1)$ wave functions. Hence, the wave functions are usually represented by the angular part $Y_{lm}(\theta, \phi)$ only. Thus, the states having $n = 2$, $l = 1$ have $m = 1, 0, -1$, and the states are denoted by $2p_1$, $2p_0$, and $2p_{-1}$. The $Y_{lm}(\theta, \phi)$ values for these three states are

$$Y_{11} = -\left(\frac{3}{8\pi}\right)^{1/2} \sin\theta\, e^{-i\phi}, \qquad Y_{1,0} = \left(\frac{3}{4\pi}\right)^{1/2} \cos\theta$$

$$Y_{1,-1} = \left(\frac{3}{8\pi}\right)^{1/2} \sin\theta\, e^{-i\phi}$$

For $m \neq 0$, the orbitals are imaginary functions. It is convenient to deal with real functions obtained by linear combination of these functions. For the p-orbitals,

$$\psi(p_x) = \frac{\psi(p = 1) + \psi(p = -1)}{\sqrt{2}} = \left(\frac{3}{4\pi}\right)^{1/2} \sin\theta \cos\phi$$

$$\psi(p_y) = \frac{-i[\psi(p=1) - \psi(p=-1)]}{\sqrt{2}} = \left(\frac{3}{4\pi}\right)^{1/2} \sin\theta \sin\phi$$

$$\psi(p_z) = \psi(p_0) = \left(\frac{3}{4\pi}\right)^{1/2} \cos\theta$$

The representations of orbitals are usually done in two ways: in one method, the graphs of $\psi(p_x)$, $\psi(p_y)$ and $\psi(p_z)$ are plotted and, in the second approach, contour surfaces of constant probability density are drawn. The representations of the angular part for the p-orbitals are shown in Fig. 5.2. The plot of probability density has the cross-section of numeral 8.

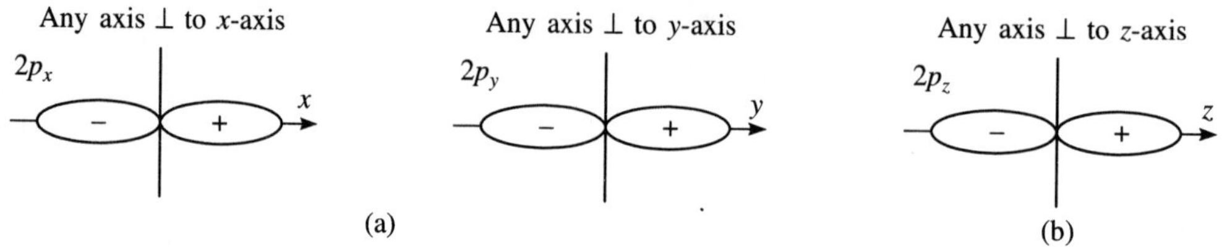

Fig. 5.2 Representation of the angular part of wave function for p-orbitals; (a) Plot of $Y_{lm}(\theta, \phi)$; (b) Plot of $|Y_{lm}(\theta, \phi)|^2$.

Each p-orbital is made of two lobes touching at the origin. The p_x-orbital is aligned along the x-axis, the p_y-orbital along the y-axis, and the p_z-orbital along the z-axis. The two lobes are separated by a plane called **nodal plane**.

5.41 The first line in the rotation spectrum of CO molecule has a wave number of 3.8424 cm^{-1}. Calculate the C$=$O bond length in CO molecule. The Avagadro number is 6.022×10^{23}/mole.

Solution. The first line corresponds to the $l = 0$ to $l = 1$ transition. From Eq. (5.10),

$$E_1 - E_0 = \frac{\hbar^2}{\mu r^2} \quad \text{or} \quad h\nu = \frac{\hbar^2}{\mu r^2}$$

$$r^2 = \frac{h}{4\pi^2 \mu \nu} = \frac{h}{4\pi^2 \mu \bar{\nu} c}$$

$$\mu = \frac{(12 \text{ g/mol})(15.9949 \text{ g/mol})}{(27.9949 \text{ g/mol})(6.022 \times 10^{23}/\text{mol})} = 1.1385 \times 10^{-23} \text{ g}$$

$$= 1.1385 \times 10^{-26} \text{ kg}$$

$$r^2 = \frac{6.626 \times 10^{-34} \text{ J s}}{4\pi^2 (1.1385 \times 10^{-26} \text{ kg})(384.24 \text{ m}^{-1})(3 \times 10^8 \text{ m/s})}$$

$$= 1.2778 \times 10^{-20} \text{ m}^2$$

$$r = 1.13 \times 10^{-10} \text{ m}$$

5.42 The $l = 0$ to $l = 1$ rotational absorption line of $^{13}C^{16}O$ molecule occurs at 1.102×10^{11} Hz and that of $C^{16}O$ at 1.153×10^{11} Hz. Find the mass number of the carbon isotope in $C^{16}O$.

Solution. For a diatomic molecule from Eq. (5.10),

$$E_1 - E_0 = \frac{\hbar^2}{I} = \frac{\hbar^2}{\mu r^2}$$

where μ is the reduced mass.
Writing $E_1 - E_0 = h\nu_1$ for the first molecule and $h\nu_2$ for the second one, we obtain

$$\frac{\nu_1}{\nu_2} = \frac{\mu_2}{\mu_1}$$

$$\mu_1 = \frac{13 \times 16}{29 \times N}, \qquad \mu_2 = \frac{m \times 16}{(m + 16)\mathrm{N}}$$

where N is Avagadro's number. Substituting the above values, we get

$$\frac{1.102 \times 10^{11}}{1.153 \times 10^{11}} = \frac{29m}{13(m + 16)}$$

Solving, we get

$$m \cong 12.07 \cong 12$$

The mass of the carbon in $C^{16}O$ is 12.

5.43 An electron is subjected to a potential $V(z) = -e^2/4z$. Write the Schrödinger equation and obtain the ground state energy.

Solution. The Hamiltonian operator

$$H = -\frac{\hbar^2}{2m}\left(\frac{\partial^2}{\partial x^2} + \frac{\partial^2}{\partial y^2} + \frac{\partial^2}{\partial z^2}\right) - \frac{e^2}{4z}$$

The Schrödinger equation is

$$-\frac{\hbar^2}{2m}\left(\frac{\partial^2}{\partial x^2} + \frac{\partial^2}{\partial y^2} + \frac{\partial^2}{\partial z^2}\right)\psi - \frac{e^2}{4z}\psi = E\psi(x, y, z) \tag{i}$$

Writing

$$\psi(x, y, z) = \phi_x(x)\, p_y(y)\, \phi_z(z) \tag{ii}$$

and substituting it in Eq. (i), we get the following equations:

$$-\frac{\hbar^2}{2m}\frac{d^2}{dx^2}\phi_x(x) = E_x\phi_x(x)$$

$$\frac{d^2}{dx^2}\phi_x(x) = -k_x^2\phi_x(x), \qquad k_x^2 = \frac{2mE_x}{\hbar^2} \tag{iii}$$

$$\frac{d^2}{dy^2}\phi_y(y) = -k_y^2\phi_y(y), \qquad k_y^2 = \frac{2mE_y}{\hbar^2} \tag{iv}$$

$$-\frac{\hbar^2}{2m}\frac{d^2\phi_z}{dz^2} - \frac{e^2}{4z}\phi_z = E_z\phi_z \tag{v}$$

where $E = E_x + E_y + E_z$. Since the potential depends only on z, k_x^2 and k_y^2 are constants. Hence,

$$E_x = \frac{k_x^2\hbar^2}{2m} = \frac{p_x^2}{2m}$$

$$E_y = \frac{p_y^2}{2m}$$

Therefore,

$$E_z = E - \frac{p_x^2}{2m} - \frac{p_y^2}{2m} \tag{vi}$$

For hydrogen atom with zero angular momentum, the radial equation is

$$\frac{1}{r^2}\frac{d}{dr}\left(r^2\frac{dR}{dr}\right) + \frac{2m}{\hbar^2}\left[E + \frac{ke^2}{r}\right]R = 0, \qquad k = \frac{1}{4\pi\varepsilon_0}$$

Writing

$$R = \frac{\chi(r)}{r}$$

we have

$$\frac{d^2\chi}{dr^2} + \frac{2m}{\hbar^2}\left(E + \frac{ke^2}{r}\right)\chi = 0 \tag{vii}$$

$$-\frac{\hbar^2}{2m}\frac{d^2\chi}{dr^2} - \frac{ke^2}{r}\chi = E\chi \tag{viii}$$

Equation (v) is of the same form as Eq. (viii) with 1/4 in place of k. The hydrogen atom ground state energy is

$$E_1 = -\frac{k^2me^4}{2\hbar^2} \tag{ix}$$

Hence,

$$E_z = -\frac{me^4}{32\hbar^2} \tag{x}$$

From Eqs. (x) and (vi),

$$E = \frac{p_x^2}{2m} + \frac{p_y^2}{2m} - \frac{me^4}{32\hbar^2} \tag{xi}$$

5.44 Write the radial part of the Schrödinger equation of a particle of mass m moving in a central potential $V(r)$. Identify the effective potential for nonzero angular momentum.

Solution. The radial equation for the particle moving in a central potential is

$$\frac{1}{r^2}\frac{d}{dr}\left(r^2\frac{dR}{dr}\right) + \frac{2m}{\hbar^2}\left[E - V(r) - \frac{l(l+1)\hbar^2}{2mr^2}\right]R = 0$$

Writing

$$R(r) = \frac{\chi(r)}{r}$$

the above equation reduces to

$$\frac{d^2x}{dr^2} + \frac{2m}{\hbar^2}\left[E - V(r) - \frac{l(l+1)\hbar^2}{2mr^2}\right]x = 0$$

This equation has the form of a one-dimensional Schrödinger equation of a particle of mass m moving in a field of effective potential

$$V_{\text{eff}} = V(r) + \frac{l(l+1)\hbar^2}{2mr^2}$$

The additional potential $l(l + 1)\hbar^2/(2mr^2)$ is a repulsive one and corresponds to a force $l(l + 1)\hbar^2/mr^3$, called the *centrifugal force.*

5.45 A particle of mass m moves on a ring of radius a on which the potential is constant.
(i) Find the allowed energies and eigenfunctions
(ii) If the ring has two turns, each having a radius a, what are the energies and eigenfunctions?

Solution.

(i) The particle always moves in a particular plane which can be taken as the xy-palne. Hence, $\theta = 90°$, and the three-dimensional Schrödinger equation reduces to a one-dimensional equation in the angle ϕ. (refer Problem 5.3). Thus, the Schrödinger equation takes the form

$$-\frac{\hbar^2}{2m}\left(\frac{1}{a^2}\frac{d^2\psi(\phi)}{d\phi^2}\right) = E\psi(\phi)$$

Since $ma^2 = I$, the moment of inertia is

$$\frac{d^2\psi(\phi)}{d\phi^2} = -\frac{2IE\psi}{\hbar^2}$$

The solution and energy eigenvalues (see Problem 5.3) are

$$E_n = \frac{\hbar^2 n^2}{2I}, \qquad n = 0, \pm 1, \pm 2, \ldots$$

$$\psi_n(\phi) = \frac{1}{\sqrt{2\pi}} \exp\,(in\phi), \qquad n = 0, \pm 1, \pm 2, \ldots$$

(ii) The Schrödinger equation will be the same. However, the wave function must be the same at angles ϕ and 4π, i.e.,

$$\psi(\phi) = \psi(\phi + 4\pi)$$

$$e^{in\phi} = e^{in(\phi + 4\pi)}$$

$$e^{in4\pi} = 1 \quad \text{or} \quad \cos\,(n\,4\pi) = 1$$

$$n = 0, \ \pm\frac{1}{2}, \ \pm 1, \ \pm\frac{3}{2}$$

Hence, the energy and wave function are

$$E_n = \frac{\hbar^2 n^2}{2I}, \qquad n = 0, \pm\frac{1}{2}, \pm 1, \ldots$$

$$\psi_n = Ae^{in\phi}, \qquad n = 0, \pm\frac{1}{2}, \pm 1, \ldots$$

Defining $m = 2n$, we get

$$E_m = \frac{\hbar^2 m^2}{8I}, \qquad m = 0, \pm 1, \pm 2, \ldots$$

$$\psi_m = A \exp\,[i(m/2)\phi], \qquad m = 0, \pm 1, \pm 2, \ldots$$

Normalization gives

$$|A|^2 \int_0^{4\pi} \Psi^* \Psi \, d\phi = 1 \quad \text{or} \quad A = \frac{1}{\sqrt{4\pi}}$$

$$\psi_m = \frac{1}{\sqrt{4\pi}} \exp\,[i(m/2)\phi]$$

CHAPTER 6

Matrix Formulation and Symmetry

6.1 Matrix Representation of Operators and Wave Functions

In this approach, the observables are represented by matrices in a suitable function space defined by a set of orthonormal functions $u_1, u_2, u_3, \ldots, u_n$. The matrix element of an operator A is defined as

$$A_{ij} = \langle u_i | A | u_j \rangle \tag{6.1}$$

The diagonal matrix elements are real and for the offdiagonal elements, $A_{ji} = A_{ij}^*$. The matrix representation with respect to its own eigenfunctions is diagonal and the diagonal elements are the eigenvalues of the operator. According to the expansion theorem, the wave function

$$|\psi(x)\rangle = \sum_i c_i |u_i\rangle, \qquad c_i = \langle u_i | \psi \rangle \tag{6.2}$$

The matrix representation of the wave function is given by a column matrix formed by the expansion coefficients $c_1, c_2, c_3, \ldots, c_n$. If one uses the eigenfunctions of the Hamiltonian for a representation, then

$$\Psi_n (x, t) = \psi_n(x) \exp\left(-\frac{iE_n t}{\hbar}\right)$$

$$A_{mn}(t) = A_{mn}(0) \exp\left(\frac{i\omega_{mn} t}{\hbar}\right), \qquad \omega_{mn} = \frac{E_m - E_n}{\hbar} \tag{6.3}$$

6.2 Unitary Transformation

The transformation of a state vector ψ into another state vector ψ' can be done by the unitary transformation

$$\psi' = U\psi \tag{6.4}$$

where U is a unitary matrix obeying $UU^\dagger = U^\dagger U = 1$. Then the linear Hermitian operator A transforms as

$$A' = UAU^\dagger \quad \text{or} \quad A = U^\dagger A' U \tag{6.5}$$

The Schrödinger equation in matrix form constitutes a system of simultaneous differential equations for the time-dependent expansion coefficients $c_i(t)$ of the form

$$i\hbar \frac{\partial c_i(t)}{\partial t} = \sum_j H_{ij} c_j(t), \qquad i = 1, 2, 3, \ldots \tag{6.6}$$

where H_{ij} are the matrix elements of the Hamiltonian.

6.3 Symmetry

Symmetry plays an important role in understanding number of phenomena in Physics. A transformation that leaves the Hamiltonian invariant is called a **symmetry transformation**. The existence of a symmetry transformation implies the conservation of a dynamical variable of the system.

6.3.1 Translation in Space

Consider reference frames S and S' with S' shifted from S by ρ and x and x' being the coordinates of a point P on the common x-axis. Let the functions ψ and ψ' be the wave functions in S and S'. For the point P,

$$\psi(x) = \psi'(x'), \qquad x' = x - \rho \tag{6.7}$$

The wave function $\psi(x)$ is transformed into $\psi'(x)$ by the action of the operator $i\rho p_x/\hbar$, i.e.,

$$\psi'(x) = \left(1 + \frac{i\rho p_x}{\hbar}\right)\psi(x) \tag{6.8}$$

Let $|x\rangle$ and $|x'\rangle$ be the position eigenstates for a particle at the coordinate x measured from O and O', respectively. It can be proved that

$$|x\rangle' = \left(1 - \frac{i\rho p_x}{\hbar}\right)|x\rangle \tag{6.9}$$

From a generalization of this equation, the unitary operator that effects the transformation is given by

$$U_T = I - \frac{i\boldsymbol{\rho}\cdot\boldsymbol{p}}{\hbar} \tag{6.10}$$

The invariance of the Hamiltonian under translation in space requires that $\boldsymbol{p}$ must commute with H. Then the linear momentum of the system is conserved.

6.3.2 Translation in Time

For an infinitesimal time translation τ,

$$\Psi'(x,t) = \left[1 + i\tau\left(\frac{-H}{\hbar}\right)\right]\Psi(x, t) \tag{6.11}$$

The unitary operator that effects the transformation is

$$U = 1 - \frac{i\tau H}{\hbar} \tag{6.12}$$

From the form of U, it is obvious that it commutes with H. Hence the total energy of the system is conserved if the system is invariant under translation in time.

6.3.3 Rotation in Space

Let $oxyz$ and $ox'y'z'$ be two coordinate systems. The system $ox'y'z'$ is rotated anticlockwise through an angle θ about the z-axis. The wavefunction at a point P has a definite value independent of the system of coordinates. Hence,

$$\psi'(\boldsymbol{r}') = \psi(\boldsymbol{r}) \tag{6.13}$$

It can be proved that

$$\psi'(\boldsymbol{r}) = \left(1 + \frac{i\theta L_z}{\hbar}\right)\psi(\boldsymbol{r}) \tag{6.14}$$

where L_z is the z-component of angular momentum. For rotation about an arbitrary axis,

$$\psi'(\boldsymbol{r}) = \left(1 + \frac{i\theta \boldsymbol{n}\cdot\boldsymbol{L}}{\hbar}\right)\psi(\boldsymbol{r}) \tag{6.15}$$

where $\boldsymbol{n}$ is the unit vector along the arbitrary axis. The unitary operator for an infinitesimal rotation θ is given by

$$U_R(\boldsymbol{n}, \theta) = \left(1 + \frac{i\theta \boldsymbol{n}\cdot\boldsymbol{J}}{\hbar}\right) \tag{6.16}$$

where $\boldsymbol{J}$ is the total angular momentum. This leads to the statement that the conservation of total angular momentum is a consequence of the rotational invariance of the system.

6.3.4 Space Inversion

Reflection through the origin is space inversion or parity operation. Associated with such an operation, there is a unitary operator, called the **parity operator** P. For a wave function $\psi(\boldsymbol{r})$, the parity operator P is defined by

$$P\psi(\boldsymbol{r}) = \psi(-\boldsymbol{r}) \tag{6.17}$$

$$P^2\psi(\boldsymbol{r}) = P\psi(-\boldsymbol{r})\psi(\boldsymbol{r}) \tag{6.18}$$

Hence, the eigenvalues of P are +1 or −1, i.e., the eigenfunctions either change sign (odd parity) or remains the same (even parity) under inversion. The parity operator is Hermitian. The effect of parity operation on observables $\boldsymbol{r}$, $\boldsymbol{\phi}$ and $\boldsymbol{L}$ is given by

$$P\boldsymbol{r}P^\dagger = -\boldsymbol{r}, \qquad P\boldsymbol{p}P^\dagger = -\boldsymbol{p}, \qquad P\boldsymbol{L}P^\dagger = \boldsymbol{L} \tag{6.19}$$

If $PHP^\dagger = H$, then the system has space inversion symmetry and the operator P commutes with the Hamiltonian.

6.3.5 Time Reversal

Another important transformation is time reversal, $t' = -t$. Denoting the wave function after time reversal by $\Psi'(\mathbf{r}, t')$, we get

$$\Psi'(\mathbf{r}, t') = T\Psi(\mathbf{r}, t), \qquad t' = -t \tag{6.20}$$

where T is the time reversal operator. If A is a time-independent operator and A' its transform, then

$$A' = T A T^{-1} \tag{6.21}$$

To be in conformity with the time reversal invariance in classical mechanics, it is necessary that

$$\mathbf{r}' = T\mathbf{r}T^{-1} = \mathbf{r}, \qquad \mathbf{p}' = T\mathbf{p}T^{-1} = -\mathbf{p}, \quad \mathbf{L}' = T\mathbf{L}T^{-1} = -\mathbf{L} \tag{6.22}$$

The operator T commutes with the Hamiltonian operator H.

Another interesting result is that T operating on any number changes it into its complex conjugate.

PROBLEMS

6.1 The base vectors of a representation are $\begin{pmatrix}1\\0\end{pmatrix}$ and $\begin{pmatrix}0\\1\end{pmatrix}$. Construct a transformation matrix U for transformation to another representation having the base vectors

$$\begin{pmatrix}1/\sqrt{2}\\1/\sqrt{2}\end{pmatrix} \quad \text{and} \quad \begin{pmatrix}-1/\sqrt{2}\\1/\sqrt{2}\end{pmatrix}$$

Solution. The transformation matrix U must be such that

$$\begin{pmatrix}1/\sqrt{2}\\1/\sqrt{2}\end{pmatrix} = \begin{pmatrix}U_{11} & U_{12}\\U_{21} & U_{22}\end{pmatrix}\begin{pmatrix}0\\1\end{pmatrix}, \qquad \begin{pmatrix}-1/\sqrt{2}\\1/\sqrt{2}\end{pmatrix} = \begin{pmatrix}U_{11} & U_{12}\\U_{21} & U_{22}\end{pmatrix}\begin{pmatrix}0\\1\end{pmatrix}$$

Solving we get

$$U_{11} = 1/\sqrt{2}, \qquad U_{21} = 1/\sqrt{2}, \qquad U_{12} = -1/\sqrt{2}, \qquad U_{22} = 1/\sqrt{2}$$

$$U = \begin{pmatrix}1/\sqrt{2} & -1/\sqrt{2}\\1/\sqrt{2} & 1/\sqrt{2}\end{pmatrix}, \qquad U^{\dagger} = \begin{pmatrix}1/\sqrt{2} & 1/\sqrt{2}\\-1/\sqrt{2} & 1/\sqrt{2}\end{pmatrix}$$

It follows that $UU^{\dagger} = 1$. Hence U is unitary.

6.2 Prove that the fundamental commutation relation $[x, p_x] = i\hbar$ remains unchanged under unitary transformation.

Solution. Let U be the unitary operator that effects the transformation. Then,

$$x' = UxU^{\dagger}, \qquad p_x' = Up_xU^{\dagger}$$

$$\begin{aligned}
[x', p_x'] &= x'p_x' - p_x'x' \\
&= (UxU^{\dagger})\,(Up_xU^{\dagger}) - (Up_xU^{\dagger})\,(UxU^{\dagger}) \\
&= Uxp_xU^{\dagger} - Up_xxU^{\dagger} = U(xp_x - p_xx)\,U^{\dagger} \\
&= Ui\hbar U^{\dagger} = i\hbar UU^{\dagger} = i\hbar
\end{aligned}$$

Hence the result.

6.3 The raising $(a^{\dagger})$ and lowering (a) operators of harmonic oscillator satisfy the relations

$$a\,|n\rangle = \sqrt{n}\,|\,n-1\rangle,\; a^{\dagger}|n\rangle = \sqrt{n+1}\,|\,n+1\rangle, \qquad n = 0, 1, 2, \ldots$$

Obtain the matrices for a and $a^{\dagger}$.

Solution. Multiplying the first equation from left by $\langle n'|$, we get

$$\langle n'\,|\,a\,|\,n\rangle = \sqrt{n}\,\langle n'\,|\,n-1\rangle = \sqrt{n}\,\delta_{n',n-1}$$

This equation gives the matrix elements of a. Hence,

$$\langle 0\,|\,a\,|\,1\rangle = 1, \qquad \langle 1\,|\,a\,|\,2\rangle = \sqrt{2}, \qquad \langle 2\,|\,a\,|\,3\rangle = \sqrt{3}, \ldots$$

Multiplying the second equation from left by $\langle n'|$, we obtain

$$\langle n'|a^\dagger|n\rangle = \sqrt{n+1}\,\langle n|n+1\rangle = \sqrt{n+1}\,\delta_{n',n+1}$$

The matrix elements are

$$\langle 1|a^\dagger|0\rangle = 1, \qquad \langle 2|a^\dagger|1\rangle = \sqrt{2}, \qquad \langle 3|a^\dagger|2\rangle = \sqrt{3}; \ldots$$

The complete matrices are

$$a = \begin{pmatrix} 0 & 1 & 0 & 0 & 0 & \ldots \\ 0 & 0 & \sqrt{2} & 0 & 0 & \ldots \\ 0 & 0 & 0 & \sqrt{3} & 0 & \ldots \\ \vdots & \vdots & \vdots & \vdots & \vdots & \vdots \end{pmatrix}, \qquad a^\dagger = \begin{pmatrix} 0 & 0 & 0 & 0 & \ldots \\ 1 & 0 & 0 & 0 & \ldots \\ 0 & \sqrt{2} & 0 & 0 & \ldots \\ 0 & 0 & \sqrt{3} & 0 & \ldots \\ \vdots & \vdots & \vdots & \vdots & \vdots \end{pmatrix}$$

6.4 Show that the expectation values of operators do not change with unitary transformation.

Solution. Let A and A' be an operator before and after unitary transformation. Then,

$$A' = UAU^\dagger, \qquad U^\dagger U = UU^\dagger = 1$$

$$\begin{aligned}\langle A\rangle &= \langle\psi|A|\psi\rangle = \langle\psi|U^\dagger UAU^\dagger U|\psi\rangle \\ &= \langle U\psi|\,UAU^\dagger\,|U\psi\rangle \\ &= \langle\psi'|A'|\psi'\rangle = \langle A'\rangle\end{aligned}$$

That is, the expectation value does not change with unitary transformation.

6.5 A representation is given by the base vectors $\begin{pmatrix}1\\0\end{pmatrix}$ and $\begin{pmatrix}0\\1\end{pmatrix}$. Construct the transformation matrix U for transformation to another representation consisting of basis vectors

$$\begin{pmatrix}1/\sqrt{2}\\ i/\sqrt{2}\end{pmatrix} \quad\text{and}\quad \begin{pmatrix}-1/\sqrt{2}\\ -i/\sqrt{2}\end{pmatrix}$$

Also show that the matrix is unitary.

Solution. The transformation matrix U must satisfy the conditions:

$$\begin{pmatrix}1/\sqrt{2}\\ i/\sqrt{2}\end{pmatrix} = \begin{pmatrix}U_{11} & U_{12}\\ U_{21} & U_{22}\end{pmatrix}\begin{pmatrix}1\\0\end{pmatrix}, \qquad \begin{pmatrix}1/\sqrt{2}\\ -i/\sqrt{2}\end{pmatrix} = \begin{pmatrix}U_{11} & U_{12}\\ U_{21} & U_{22}\end{pmatrix}\begin{pmatrix}0\\1\end{pmatrix}$$

$$U_{11} = \frac{1}{\sqrt{2}}, \qquad U_{21} = \frac{i}{\sqrt{2}}, \qquad U_{12} = \frac{1}{\sqrt{2}}, \qquad U_{22} = \frac{-i}{\sqrt{2}}$$

$$U = \begin{pmatrix}1/\sqrt{2} & 1/\sqrt{2}\\ i/\sqrt{2} & -i/\sqrt{2}\end{pmatrix}, \qquad U^\dagger = \begin{pmatrix}1/\sqrt{2} & -i/\sqrt{2}\\ 1/\sqrt{2} & i/\sqrt{2}\end{pmatrix}$$

$$UU^{\dagger} = \begin{pmatrix} 1/\sqrt{2} & 1/\sqrt{2} \\ i/\sqrt{2} & -i/\sqrt{2} \end{pmatrix} \begin{pmatrix} 1/\sqrt{2} & -i/\sqrt{2} \\ 1/\sqrt{2} & i/\sqrt{2} \end{pmatrix} = \begin{pmatrix} 1 & 0 \\ 0 & 1 \end{pmatrix}$$

Thus, U is unitary.

6.6 For 2×2 matrices A and B, show that the eigenvalues of AB are the same as those of BA.

Solution.

$$A = \begin{pmatrix} a_{11} & a_{12} \\ a_{21} & a_{22} \end{pmatrix}, \qquad B = \begin{pmatrix} b_{11} & b_{12} \\ b_{21} & b_{22} \end{pmatrix}$$

$$AB = \begin{pmatrix} a_{11}b_{11} + a_{12}b_{21} & a_{11}b_{12} + a_{12}b_{22} \\ a_{21}b_{11} + a_{22}b_{21} & a_{21}b_{12} + a_{22}b_{22} \end{pmatrix}$$

The characteristic equation of AB is given by

$$\begin{vmatrix} a_{11}b_{11} + a_{12}b_{21} - \lambda & a_{11}b_{12} + a_{12}b_{22} \\ a_{21}b_{11} + a_{22}b_{21} & a_{21}b_{12} + a_{22}b_{22} - \lambda \end{vmatrix} = 0$$

$$\lambda^2 - \lambda \,\mathrm{Tr}\,(AB) + |AB| = 0$$

Since $|AB| = |A||B|$, $|AB| = |BA|$. As Tr (AB) = Tr (BA), the characteristic equation for AB is the same as the characteristic equation for BA. Hence, the eigenvalues of AB are the same these of BA.

6.7 Prove the following: (i) the scalar product is invariant under a unitary transformation; (ii) the trace of a matrix is invariant under unitary transformation; and (iii) if $[A, B]$ vanishes in one representation, it vanishes in any other representation.

Solution.

(i) $\langle \phi | A | \psi \rangle = \langle \phi | U^{\dagger} U A U^{\dagger} U | \psi \rangle = \langle U\phi | U A U^{\dagger} | U\psi \rangle = \langle \phi' | A' | \psi' \rangle$

Setting $A = I$, the above equation reduces to

$$\langle \phi | \psi \rangle = \langle \phi' | \psi' \rangle$$

i.e., the scalar product is invariant under unitary transformation.

(ii) $A_{mm} = \langle \psi_m | A | \psi_m \rangle = \langle \psi_m | U^{\dagger} U A U^{\dagger} U | \psi_m \rangle = \langle U\psi_m | U A U^{\dagger} | U\psi_m \rangle$

$= \langle \psi'_m | A' | \psi'_m \rangle = A'_{mn}$

Thus,

$$\sum_m A_{mm} = \sum_m A'_{mn}$$

In other words, the trace is invariant under a unitary transformation.

(iii) $A'B' - B'A' = UAU^{\dagger}\, UBU^{\dagger} - UBU^{\dagger}\, UAU^{\dagger} = UABU^{\dagger} - UBAU^{\dagger}$

$= U(AB - BA)U^{\dagger}$

If $AB - BA = 0$, then $A'B' - B'A'$. Hence the result.

6.8 Show that a linear transformation which preserves length of vectors is represented by an orthogonal matrix.

Solution. Let x and x' be the n-dimensional and transformed vectors, respectively. Then,

$$x' = Ax, \qquad \sum_{i=1}^{n} x_i'^2 = \sum_{i=1}^{n} x_i^2$$

where A is the $n \times n$ transformation matrix. Substituting the value of x_i', we get

$$\sum_{i=1}^{n} \left(\sum_j A_{ij} x_j \right) \left(\sum_k A_{ik} x_k \right) = \sum_{i=1}^{n} x_i^2$$

$$\sum_{i=1}^{n} \sum_{j=1}^{n} \sum_{k=1}^{n} A_{ij} A_{ik} x_j x_k = \sum_{i=1}^{n} x_i^2$$

This equation, to be valid, it is necessary that

$$\sum_{i=1}^{n} A_{ij} A_{ik} = \delta_{jk} \quad \text{or} \quad (A'A)_{jk} = \delta_{jk}$$

where A' is the transpose of the matrix A. Therefore, A is an orthogonal matrix.

6.9 Prove that the parity of spherical harmonics $Y_{l,m}(\theta, \phi)$ is $(-1)^l$.

Solution. When a vector $\mathbf{r}$ is reflected through the origin, we get the vector $-\mathbf{r}$. In spherical polar coordinates, this operation corresponds to the following changes in the angles θ and ϕ, leaving r unchanged:

$$\theta \rightarrow (\pi - \theta) \quad \text{and} \quad \phi \rightarrow (\phi + \pi)$$

We have

$$Y_{l,m}(\theta, \phi) = CP_l^m(\cos\theta)\exp(im\phi), \quad C \text{ being constant}$$

$$\begin{aligned} Y_{l,m}(\pi - \theta, \phi + \pi) &= CP_l^m[\cos(\pi - \theta)]\exp[im(\phi + \pi)] \\ &= CP_l^m(-\cos\theta)\exp(im\phi)\exp(im\pi) \\ &= CP_l^m(\cos\theta)(-1)^{l+m}\exp(im\phi)(-1)^m \\ &= (-1)^l\, Y_{l,m}(\theta, \phi) \end{aligned}$$

During simplification we have used the result $P_n^m(-x) = (-1)^{n+m} P_n^m(x)$. That is, the parity of spherical harmonics is given by $(-1)^l$.

6.10 If $\psi_+(\mathbf{r})$ and $\psi_-(\mathbf{r})$ are the eigenfunctions of the parity operator belonging to even and odd eigenstates, show that they are orthogonal.

Solution. From definition we have

$$P\psi_+(\mathbf{r}) = \psi_+(\mathbf{r}), \qquad P\psi_-(\mathbf{r}) = -\psi_-(\mathbf{r})$$

$$\langle \psi_+(\mathbf{r}) | \psi_-(\mathbf{r}) \rangle = \langle \psi_+(\mathbf{r}) | PP | \psi_-(\mathbf{r}) \rangle$$

Here, we have used the result $P^2 = 1$. Since P is Hermitian,

$$\langle \psi_+(\mathbf{r}) | \psi_-(\mathbf{r}) \rangle = \langle P\psi_+(\mathbf{r}) | P | \psi_-(\mathbf{r}) \rangle = -\langle \psi_+(\mathbf{r}) | \psi_-(\mathbf{r}) \rangle$$

This is possible only when

$$\langle \psi_+(r) \mid \psi_-(r) \rangle = 0$$

Here, $\psi_+(\boldsymbol{r})$ and $\psi_-(\boldsymbol{r})$ are orthogonal.

6.11 Use the concept of parity to find which of the following integrals are nonzero. (i) $\langle 2s|x|2p_y\rangle$; (ii) $\langle 2p_x|x|2p_y\rangle$. The functions in the integrals are hydrogen-like wave functions.

Solution. We have the result that the integral $\int_{-\infty}^{\infty} f(x)\,dx$ is zero if $f(x)$ is an odd function and finite if it is an even function. In $\langle 2s\,|\,x\,|\,2p_y\rangle$, the parity of the function $\langle 2s|$ is $(-1)^0 = 1$. Hence the parity is even. The parity of the function $|\,2p_y\rangle$ is $(-1)^l = -1$, which is odd. Hence the parity of the given integral is even × odd × odd, which is even. The value of the integral is therefore finite. The parity of the integrand in $\langle 2p_x\,|\,x\,|\,2p_y\rangle$ is odd × odd × odd, which is odd. The integral therefore vanishes.

6.12 Obtain the generators G_z, G_x and G_y for infinitesimal rotation of a vector about z, x and y axes respectively.

Solution. The generator for infinitesimal rotation about the z-axis (Eq. 6.14) is the coefficient of $i\theta$ in $(1 + i\theta G_z)$, where θ is the infinitesimal rotation angle. Let A be a vector with components A_x, A_y, A_z. If the vector rotates about the z-axis through θ, then

$$A_x' = A_x \cos\theta + A_y \sin\theta$$

$$A_y' = -A_x \sin\theta + A_y \cos\theta$$

$$A_z' = A_z$$

Since rotation is infinitesimal, $\cos\theta \cong 1$ and $\sin\theta \cong \theta$, and the above equation can be put in matrix form as

$$\begin{pmatrix} A_x' \\ A_y' \\ A_z' \end{pmatrix} = \begin{pmatrix} 1 & \theta & 0 \\ -\theta & 1 & 0 \\ 0 & 0 & 1 \end{pmatrix}\begin{pmatrix} A_x \\ A_y \\ A_z \end{pmatrix} = \left[\begin{pmatrix} 1 & \theta & 0 \\ 0 & 1 & 0 \\ 0 & 0 & 1 \end{pmatrix} + \begin{pmatrix} 0 & \theta & 0 \\ -\theta & 0 & 0 \\ 0 & 0 & 0 \end{pmatrix}\right]\begin{pmatrix} A_x \\ A_y \\ A_z \end{pmatrix}$$

Comparing the coefficient on RHS with $1 + i\theta G_z$, we get

$$i\theta G_z = \begin{pmatrix} 1 & \theta & 0 \\ -\theta & 1 & 0 \\ 0 & 0 & 1 \end{pmatrix} = i\theta \begin{pmatrix} 0 & -i & 0 \\ i & 0 & 0 \\ 0 & 0 & 0 \end{pmatrix}$$

Hence,

$$G_z = \begin{pmatrix} 0 & -i & 0 \\ i & 0 & 0 \\ 0 & 0 & 0 \end{pmatrix}$$

Proceeding on similar lines, the generators G_x and G_y for rotation about the x and y-axes are given by

$$G_x = \begin{pmatrix} 0 & 0 & 0 \\ 0 & 0 & -i \\ 0 & i & 0 \end{pmatrix}, \qquad G_y = \begin{pmatrix} 0 & 0 & i \\ 0 & 0 & 0 \\ -i & 0 & 0 \end{pmatrix}$$

6.13 Prove that the parity operator is Hermitian and unitary.

Solution. For any two wave functions $\psi_1(\mathbf{r})$ and $\psi_2(\mathbf{r})$, we have

$$\int_{-\infty}^{\infty} \psi_1^*(\mathbf{r})\, P\psi_2(\mathbf{r})\, d\mathbf{r} = \int_{-\infty}^{\infty} \psi_1^*(\mathbf{r})\, \psi_2(\mathbf{r})\, d(-\mathbf{r})\, d\mathbf{r}$$

On the RHS, changing the variable $\mathbf{r}$ to $-\mathbf{r}$, we get

$$\int_{-\infty}^{\infty} \psi_1^*(\mathbf{r})\, P\psi_2(\mathbf{r})\, d\mathbf{r} = \int_{\infty}^{-\infty} \psi_1^*(-\mathbf{r})\, \psi_2(\mathbf{r})\, d(-\mathbf{r})$$

$$= \int_{-\infty}^{\infty} \psi_1^*(-r)\, \psi_2(r)\, dr$$

$$= \int_{-\infty}^{\infty} [P\psi_1(r)]^*\, \psi_2(r)\, dr$$

Hence the operator P is Hermitian, i.e., $P = P^\dagger$. We have $P^2 = 1$ or $PP^\dagger = 1$. Thus, P is unitary.

6.14 Use the concept of parity to find which of the following integrals are nonzero: (i) $\langle 2s|x^2|2p_x\rangle$; (ii) $\langle 2p_x|x^2|2p_x\rangle$; and (iii) $\langle 2p|x|3d\rangle$. The functions in the integrals are hydrogen-like wave functions.

Solution.

(i) $\langle 2s|x^2|2p_x\rangle$.
The parity of the integrand is even × even × odd = odd. Hence the integral vanishes

(ii) $\langle 2p_x|x^2|2p_x\rangle$.
The parity of the integrand is odd × even × odd = even. Hence the integral is finite.

(iii) $\langle 2p|x|3d\rangle$.
The parity of the integrand is odd × odd × even = even. Hence the integral is finite.

6.15 For a spinless particle moving in a potential $V(r)$, show that the time reversal operator T commutes with the Hamiltonian.

Solution.

$$H = \frac{p^2}{2m} + V(r)$$

From Eq. (6.22),

$$TrT^{-1} = r$$

Multiplying by T from RHS, we get

$$TrT^{-1}T = rT \quad \text{or} \quad Tr = rT$$

Using the relations $Tr = rT$ and $Tp = -pT$, we obtain

$$TH = T\frac{p^2}{2m} + TV(r) = \frac{-pTp}{2m} + VT$$

$$= \frac{p^2}{2m}T + VT = HT$$

$$[T, H] = 0$$

6.16 Show that the time reversal operator operating on any number changes it into its complex conjugate.

Solution. From Eq. (6.22),

$$x' = TxT^{-1} = x, \qquad p_x' = Tp_xT^{-1} = -p \tag{i}$$

We now evaluate the fundamental commutation relation $[x', p_x']$:

$$[x', p_x'] = [TxT^{-1}, Tp_xT^{-1}] = [x, -p_x] = -i\hbar \tag{ii}$$

The value of $[x', p_x']$ can also be written as

$$[x', p_x'] = T[x, p_x]T^{-1} = T(i\hbar)T^{-1} \tag{iii}$$

From Eqs. (ii) and (iii),

$$T(i\hbar)T^{-1} = -i\hbar$$

which is possible only if T operating on any number changes it into its complex conjugate.

6.17 For a simple harmonic oscillator, ω is the angular frequency and $x_{nl}(0)$ is the nlth matrix element of the displacement x at time $t = 0$. Show that all matrix elements $x_{nl}(0)$ vanish except those for which the transition frequency $\omega_{nl} = \pm\omega$, where $\omega_{nl} = (E_n - E_l)/\hbar$.

Solution. The Hamiltonian of a simple harmonic oscillator is

$$H = \frac{p^2}{2m} + \frac{1}{2}m\omega^2x^2 \tag{i}$$

The equation of motion for the operator x in the Heisenberg picture is

$$\begin{aligned} i\hbar\frac{dx}{dt} &= [x, H] = \frac{1}{2m}[x, p^2] + \frac{1}{2m}m\omega^2[x, x^2] \\ &= \frac{1}{2m}(p[x, p] + [x, p]p) \\ &= \frac{i\hbar}{2m}(p + p) = i\hbar\frac{p}{m} \end{aligned}$$

$$\dot{x} = \frac{p}{m} \tag{ii}$$

Similarly,

$$\dot{p} = -m\omega^2x \tag{iii}$$

Differentiating Eq. (i) with respect to t and substituting the value of $\dot{p}$ from Eq. (ii), we obtain

$$\ddot{x} + \omega^2x = 0 \tag{iv}$$

In matrix form,

$$\ddot{x}_{nl} + \omega^2x_{nl} = 0 \tag{v}$$

From Eq. (6.3),

$$x_{nl}(t) = x_{nl}(0)\exp(i\omega_{nl}t) \tag{vi}$$

Differentiaing twice with respect to t, we get

$$\ddot{x}_{nl}(t) = -\omega_{nl}^2x_{nl}(0)\exp(i\omega_{nl}t) = -\omega_{nl}^2x_{nl}(t) \tag{vii}$$

Combining Eqs. (v) and (vii), we obtain

$$(\omega_{nl}^2 - \omega^2)\, x_{nl}(t) = 0$$

When $t = 0$,

$$(\omega_{nl}^2 - \omega^2)\, x_{nl}(0) = 0$$

That is, if $\omega_{nl}^2 - \omega^2 = 0$ or $\omega_{nl} = \pm\omega$, then $x_{nl}(0) \neq 0$. Thus, $x_{nl}(0)$ matix elements vanish except those for which the transition frequency $\omega_{nl} = \pm\omega$.

6.18 When a state vector ψ transforms into another state vector ψ' by a unitary transformation, an operator A transforms as A'. Show that (i) if A is Hermitian, then A' is Hermitian; (ii) the eigenvalues of A' are the same as those of A.

Solution.

(i) We have

$$A' = UAU^{\dagger}$$

$$(A')^{\dagger} = (UAU^{\dagger})^{\dagger} = UA^{\dagger}U^{\dagger}$$

where we have used the rule $(ABC)^{\dagger} = C^{\dagger}B^{\dagger}A^{\dagger}$. Since A is Hermitian, $A = A^{\dagger}$. Then,

$$(A')^{\dagger} = UAU^{\dagger} = A'$$

i.e., $A^{\dagger}$ is Hermitian.

(ii) The eigenvalue equation of A is

$$A\psi_n = a_n\psi_n$$

where a_n is the eigenvalue. Since $U^{\dagger}U = 1$,

$$AU^{\dagger}U\psi_n = a_nU^{\dagger}U(U\psi_n)$$

Operating from left by U, we get

$$(UAU^{\dagger})(U\psi_n) = a_nUU^{\dagger}(U\psi_n)$$

$$A'(U\psi_n) = a_n(U\psi_n)$$

Denoting $U\psi_n$ by ψ'_n, we obtain

$$A'\psi'_n = a_n\psi'_n$$

Thus, the eigenvalues of A are also eigenvalues of A'.

6.19 Prove that (i) a unitary transformation transforms one complete set of basis vectors into another, (ii) the same unitary transformation also transforms the matrix representation of an operator with respect to one set into the other.

Solution.

(i) Let the two orthonormal sets of basis functions be $\{\mathbf{u}_i\}$ and $\{\mathbf{v}_i\}$, i = 1, 2, 3, ¼. Since any function can be expanded as a linear combination of an orthonormal set,

$$\mathbf{u}_n = \sum_m U_{mn}\mathbf{v}_m, \qquad m = 1, 2, 3, \ldots$$

where the expansion coefficient

$$U_{mn} = \langle \mathbf{v}_m | \mathbf{u}_n \rangle$$

Next consider the product $UU^\dagger$, i.e.,

$$(UU^\dagger)_{mn} = \sum_k U_{mk}U^\dagger_{kn} = \sum_k U_{mk}U^*_{nk}$$

$$= \sum_k \langle \mathbf{v}_m|\mathbf{u}_k\rangle\langle \mathbf{v}_n|\mathbf{u}_k\rangle^* = \sum_k \langle \mathbf{v}_m|\mathbf{u}_k\rangle\langle \mathbf{u}_k|\mathbf{v}_n\rangle$$

$$= \langle \mathbf{v}_m|\mathbf{v}_n\rangle = \delta_{mn} \tag{ii}$$

Similarly,

$$(U^\dagger U)_{mn} = \delta_{mn} \tag{iii}$$

Hence, U is a unitary matrix. Let a wave function ψ be represented in the basis $\{\mathbf{u}_n\}$ by the coefficients c_n forming a column vector c, and in the basis $\{\mathbf{v}_m\}$ by the coefficients b_m forming a column vector b, i.e.,

$$|\psi\rangle = \sum_n c_n|\mathbf{u}_n\rangle, \qquad c_n = \langle \mathbf{u}_n|\psi\rangle \tag{iv}$$

$$|\psi\rangle = \sum_m b_m|\mathbf{v}_m\rangle, \qquad b_m = \langle \mathbf{v}_m|\psi\rangle \tag{v}$$

Substituting $\langle\psi|$ from Eq. (iv), we get

$$b_m = \sum_n \langle \mathbf{v}_m|\mathbf{u}_n\rangle c_n = \sum_n U_{mn}c_n$$

In matrix form,

$$b = Uc \tag{vi}$$

which is the required result.

(ii) Let A and A' be matrices representing an operator A in the bases $\{\mathbf{u}\}$ and $\{\mathbf{v}\}$, respectively. Then,

$$A_{kl} = \langle \mathbf{u}_k|A|\mathbf{u}_l\rangle, \qquad A'_{mn} = \langle \mathbf{v}_m|A|\mathbf{v}_n\rangle \tag{vii}$$

Expanding $|\mathbf{v}_m\rangle$ and $|\mathbf{v}_n\rangle$ in terms of $|\mathbf{u}\rangle$ and replacing the expansion coefficients, we get

$$|\mathbf{v}_m\rangle = \sum_k d_k|\mathbf{u}_l\rangle = \sum_k \langle \mathbf{u}_k|\mathbf{v}_m\rangle|\mathbf{u}_k\rangle$$

$$|\mathbf{v}_n\rangle = \sum_l f_l|\mathbf{u}_l\rangle = \sum_i \langle \mathbf{u}_l|\mathbf{v}_n\rangle|\mathbf{u}_l\rangle$$

Substituting these values of $|\mathbf{v}_m\rangle$ and $|\mathbf{v}_n\rangle$ in Eq. (vii), we get

$$A'_{mn} = \sum_k\sum_l \langle \mathbf{u}_k|\mathbf{v}_m\rangle^*\langle \mathbf{u}_k|A|\mathbf{u}_l\rangle\langle \mathbf{u}_l|\mathbf{v}_n\rangle$$

$$= \sum_k\sum_l \langle \mathbf{v}_m|\mathbf{u}_k\rangle\langle \mathbf{u}_k|A|\mathbf{u}_l\rangle\langle \mathbf{u}_l|\mathbf{v}_n\rangle$$

$$= \sum_k\sum_l U_{mk}A_{kl}(U^\dagger)_{ln}$$

In matrix form,

$$A' = UAU^\dagger \quad \text{or} \quad A = U^\dagger A'U$$

Hence the result.

6.20 (i) Evaluate the fundamental commutation relation $[x', p'_x]$, where x' and p' are the coordinate and momentum after time reversal. (ii) Find the form of the time-dependent Schrödinger equation after time reversal ($t \to t' = -t$).

Solution.

(i) The commutator is evaluated in Problem 6.16, Hence,

$$[x', p'_x] = [TxT^{-1}, Tp_xT^{-1}]$$

$$= [x, -p_x] = -i\hbar \qquad \text{(i)}$$

(ii) The time-independent Schrödinger equation of a particle moving in a potential $V(r)$ is

$$i\hbar\frac{\partial\Psi(\boldsymbol{r}, t)}{\partial t} = H\Psi(\boldsymbol{r}, t) \qquad \text{(ii)}$$

Since T commutes with the Hamiltonian H,

$$T\left[i\hbar\frac{\partial\Psi(\boldsymbol{r}, t)}{\partial t}\right] = HT\Psi(\boldsymbol{r}, t) \qquad \text{(iii)}$$

T operating on any number changes it into its complex conjugate. Hence, $T(i\hbar)T^{-1} = -i\hbar$, i.e., $T(i\hbar) = -i\hbar T$. Equation (iii) now reduces to

$$-i\hbar\frac{\partial}{\partial t'}\Psi'(\boldsymbol{r}, t') = H\Psi'(\boldsymbol{r}, t')$$

$$i\hbar\frac{\partial}{\partial t}\Psi'(\boldsymbol{r}, t') = H\Psi'(\boldsymbol{r}, t')$$

That is, the Schrödinger equation satisfied by the time reversed function $\Psi'(\boldsymbol{r}, t')$ has the same form as the original one.

6.21 Consider two coordinate systems $oxyz$ and $ox'y'z'$. The system $ox'y'z'$ is rotated anticlockwise through an infinitesimal angle θ about an arbitrary axis. The wave functions $\psi(\boldsymbol{r})$ and $\psi'(\boldsymbol{r})$ are the wave functions of the same physical state referred to $oxyz$ and $ox'y'z'$ and is related by the equation

$$\psi'(r) = \left(I + \frac{i\theta}{\hbar}\boldsymbol{n}\cdot\boldsymbol{J}\right)\psi(\boldsymbol{r})$$

where n is the unit vector along the arbitrary axis and $\boldsymbol{J}$ is the total angular momentum. Find the condition for the Hamiltonian H to be invariant under the transformation.

Solution. The operator that effects the transformation is

$$U = I + \frac{i\theta}{\hbar}\boldsymbol{n}\cdot\boldsymbol{J}$$

$$H' = UHU^{\dagger}$$

$$= \left(I + \frac{i\theta}{\hbar}\boldsymbol{n}\cdot\boldsymbol{J}\right)H\left(I - \frac{i\theta}{\hbar}\boldsymbol{n}\cdot\boldsymbol{J}\right)$$

$$= H + \frac{i\theta}{\hbar}\boldsymbol{n}\cdot(\boldsymbol{J}H - H\boldsymbol{J})$$

$$= H + \frac{i\theta}{\hbar}\boldsymbol{n}\cdot[\boldsymbol{J}, H]$$

For H to be invariant under the transformation, $H' = H$. This is possible only when $[J, H] = 0$, i.e., the total angular momentum must commute with the Hamiltonian. In other words, the total angular momentum must be a constant of motion.

6.22 Show that the parity operator commutes with the orbital angular momentum operator.

Solution. Let P be the parity operator and $\boldsymbol{L} = \boldsymbol{r} \times \boldsymbol{p}$ be the orbital angular momentum operator. Consider an arbitrary wave function $f(r)$. Then,

$$\begin{aligned} PLf(r) &= P(\boldsymbol{r} \times \boldsymbol{p})f(r) \\ &= (-\boldsymbol{r}) \times (-\boldsymbol{p})f(-r) \\ &= (\boldsymbol{r}) \times (\boldsymbol{p})f(-r) \\ &= LPf(r) \end{aligned}$$

$$(PL - LP)\, f(r) = 0$$

Thus, P commutes with L.

6.23 A real operator A satisfies the equation

$$A^2 - 5A + 6 = 0$$

(i) What are the eigenvalues of A?
(ii) What are the eigenvectors of A;
(iii) Is A an observable?

Solution.

(i) As A satisfies a quadratic equation, it will have two eigenvalues. Hence it can be represented by a 2×2 matrix. Its eigenvalues are the roots of the equation

$$\lambda^2 - 5\lambda + 6 = 0$$

Solving, we get

$$(\lambda - 3)(\lambda - 2) = 0 \quad \text{or} \quad \lambda = 2 \text{ or } 3$$

The simplest 2×2 matrrix with eigenvalues 2 and 3 is

$$A = \begin{pmatrix} 2 & 0 \\ 0 & 3 \end{pmatrix}$$

(ii) The eigenvalue equation corresponding to the eigenvalue 2 is

$$\begin{pmatrix} 2 & 0 \\ 0 & 3 \end{pmatrix}\begin{pmatrix} a_1 \\ a_2 \end{pmatrix} = 2\begin{pmatrix} a_1 \\ a_2 \end{pmatrix}$$

which leads to $a_1 = 1$, $a_2 = 0$. The other eigenvalue 3 leads to $a_1 = 0$, $a_2 = 1$, i.e., the eigenvectors are

$$\begin{pmatrix} 1 \\ 0 \end{pmatrix} \quad \text{and} \quad \begin{pmatrix} 0 \\ 1 \end{pmatrix}$$

(iii) Since $A = A^\dagger$, the matrix A is Hermitian. Hence, it is an observable.

6.24 The ground state wave function of a linear harmonic oscillator is

$$\psi_0(x) = A \exp\left(-\frac{m\omega x^2}{2\hbar}\right)$$

where A is a constant. Using the raising and lowering operators, obtain the wave function of the first excited state of the harmonic oscillator.

Solution. The lowering (a) and raising ($a^\dagger$) operators are defined by

$$a = \sqrt{\frac{m\omega}{2\hbar}}x + i\frac{1}{\sqrt{2m\hbar\omega}}p \qquad \text{(i)}$$

$$a^\dagger = \sqrt{\frac{m\omega}{2\hbar}}x - i\frac{1}{\sqrt{2m\hbar\omega}}p \qquad \text{(ii)}$$

From the definition, it is obvious that

$$[a, a^\dagger] = 1, \qquad a^\dagger a = \frac{H}{\hbar\omega} - \frac{1}{2} \qquad \text{(iii)}$$

Allowing the Hamiltonian to operate on $a^\dagger|0\rangle$ and using Eq. (iii), we have

$$\text{H}a^\dagger|0\rangle = \left(a^\dagger a + \frac{1}{2}\right)\hbar\omega a^\dagger|0\rangle$$

$$= \hbar\omega a^\dagger a a^\dagger\,|0\rangle + \frac{1}{2}\hbar\omega a^\dagger|0\rangle$$

Since $[a, a^\dagger] = 1$ or $aa^\dagger = a^\dagger a + 1$,

$$Ha^\dagger|0\rangle = \hbar\omega a^\dagger(a^\dagger a + 1)\,|0\rangle + \frac{1}{2}\hbar\omega a^\dagger|0\rangle$$

$$= \hbar\omega a^\dagger a^\dagger a|0\rangle + \hbar\omega a^\dagger|0\rangle + \frac{1}{2}\hbar\omega a^\dagger|0\rangle$$

$$= 0 + \frac{3}{2}\hbar\omega a^\dagger|0\rangle$$

Hence,

$$|1\rangle = a^\dagger|0\rangle = \left[\sqrt{\frac{m\omega}{2\hbar}}x - i\frac{1}{\sqrt{2m\hbar\omega}}p\right]|0\rangle$$

$$= A\sqrt{\frac{m\omega}{2\hbar}}xe^{-m\omega x^2/2\hbar} - A\frac{\hbar}{\sqrt{2m\hbar\omega}}\frac{d}{dx}\exp(-m\omega x^2/2\hbar)$$

$$= A\sqrt{\frac{2m\omega}{2\hbar}}x\exp\left(-\frac{m\omega x^2}{2\hbar}\right)$$

6.25 If E_m and E_n are the energies corresponding to the eigenstates $|m\rangle$ and $|n\rangle$, respectively, show that

$$\sum_n (E_m - E_n)|\langle m|x|n\rangle|^2 = -\frac{\hbar^2}{2M}$$

where M is the mass of the particle.

Solution.

$$[[H, x], x] = Hx^2 - 2xHx + x^2H$$

$$\begin{aligned}\langle m|[[H, x], x]|m\rangle &= \langle m|Hx^2|m\rangle - 2\langle m|xHx|m\rangle + \langle m|x^2H|m\rangle \\ &= E_m \langle m|x^2|m\rangle - 2\langle m|xHx|m\rangle + E_m \langle m|x^2|m\rangle \\ &= 2E_m \langle m|x^2|m\rangle - 2\langle m|xHx|m\rangle\end{aligned}$$

where the Hermitian property of H is used. Now,

$$\begin{aligned}\langle m|x^2|m\rangle &= \sum_n \langle m|x|n\rangle\langle n|x|m\rangle \\ &= \sum_n |\langle m|x|n\rangle|^2\end{aligned}$$

$$\begin{aligned}\langle m|xHx|m\rangle &= \sum_n \langle m|xH|n\rangle\langle n|x|m\rangle \\ &= \sum_n E_n|\langle m|x|n\rangle|^2\end{aligned}$$

Hence,

$$\langle m|[[H, x], x]|m\rangle = 2\sum_n (E_m - E_n)|\langle m|x|n\rangle|^2$$

For the Hamiltonian,

$$H = \frac{p^2}{2M} + V(x)$$

$$\begin{aligned}[H, x] &= \frac{1}{2M}[p^2, x] + [V(x), x] \\ &= \frac{1}{2M}p[p, x] + \frac{1}{2M}[p, x]p = -\frac{i\hbar p}{M}\end{aligned}$$

$$[[H, x], x] = -\frac{i\hbar}{M}[p, x] = -\frac{\hbar^2}{M}$$

Equating the two relations, we get

$$\sum_n (E_m - E_n)|\langle m|x|n\rangle|^2 = -\frac{\hbar^2}{2M}$$

CHAPTER 7

Angular Momentum and Spin

Angular momentum is an important and interesting property of physical systems, both in classical and quantum mechanics. In this chapter, we consider the operators representing angular momentum, their eigenvalues, eigenvectors and matrix representation, we also discuss the concept of an intrinsic angular momentum, called spin, and the addition of angular momenta.

7.1 Angular Momentum Operators

Replacing p_x, p_y and p_z by the respective operators in angular momentum $\boldsymbol{L} = \boldsymbol{r} \times \boldsymbol{p}$, we can get the operators for the components L_x, L_y and L_z, i.e.,

$$L_x = -i\hbar\left(y\frac{\partial}{dz} - z\frac{\partial}{dy}\right) \tag{7.1}$$

$$L_y = -i\hbar\left(z\frac{\partial}{dx} - x\frac{\partial}{dz}\right) \tag{7.2}$$

$$L_z = -i\hbar\left(x\frac{\partial}{dy} - y\frac{\partial}{dx}\right) \tag{7.3}$$

Instead of working with L_x and L_y, it is found convenient to work with L_+ and L_- defined by

$$L_+ = L_x + iL_y, \qquad L_- = L_x - iL_y \tag{7.4}$$

L_+ and L_- are respectively called **raising and lowering operators** and together referred to as **ladder operators**.

7.2 Angular Momentum Commutation Relations

Some of the important angular momentum commutation relations are

$$[L_x, L_y] = i\hbar L_z, \quad [L_y, L_z] = i\hbar L_x, \quad [L_z, L_x] = i\hbar L_y \tag{7.5}$$

$$[L^2, L_x] = [L^2, L_y] = [L^2, L_z] = 0 \tag{7.6}$$

From the definition of L_+ and L_-, it is evident that they commute with L^2:

$$[L^2, L_+] = 0, \qquad [L^2, L_-] = 0 \tag{7.7}$$

As the components L_x, L_y, L_z are noncommuting among themselves, it is not possible to have simultaneous eigenvectors for L^2, L_x, L_y, L_z. However, there can be simultaneous eigenvectors for L^2, and one of the components, say, L_z. The eigenvalue-eigenvector equations are

$$L^2 Y_{lm}(\theta, \phi) = l(l+1)\hbar^2 Y_{lm}(\theta, \phi), \qquad l = 0, 1, 2, \ldots \tag{7.8}$$

$$L_z Y_{lm}(\theta, \phi) = m\hbar Y_{lm}(\theta, \phi), \qquad m = 0, \pm 1, \pm 2, \ldots, \pm l \tag{7.9}$$

Experimental results such as spectra of alkali metals, anomalous Zeeman effect, Stern-Gerlach experiment, etc., could be explained only by invoking an additional intrinsic angular momentum, called **spin**, for the electron in an atom. Hence the classical definition $\boldsymbol{L} = \boldsymbol{r} \times \boldsymbol{p}$ is not general enough to include spin and we may consider a general angular momentum $\boldsymbol{J}$ obeying the commutation relations

$$[J_x, J_y] = i\hbar J_z, \qquad [J_y, J_z] = i\hbar J_x, \qquad [J_z, J_x] = i\hbar J_y \tag{7.10}$$

as the more appropriate one.

7.3 Eigenvalues of J^2 and J_z

The square of the general angular momentum $\boldsymbol{J}$ commutes with its components. As the components are non-commuting among themselves, J^2 and one of the components, say J_z, can have simultaneous eigenkets at a time. Denoting the simultaneous eigenkets by $|\,jm\rangle$, the eigenvalue-eigenket equations of J^2 and J_z are

$$J^2 |\,jm\rangle = j(j+1)\hbar^2 |\,jm\rangle, \qquad j = 0, \frac{1}{2}, 1, \frac{3}{2}, \ldots \tag{7.11}$$

$$J_z |\,jm\rangle = m\hbar\,|\,jm\rangle, \qquad m = -j, -j+1, \ldots, (j-1), j \tag{7.12}$$

7.4 Spin Angular Momentum

To account for experimental observations, Uhlenbeck and Goudsmit proposed that an electron in an atom should possess an intrinsic angular momentum in addition to orbital angular momentum. This intrinsic angular momentum $\boldsymbol{S}$ is called the **spin angular momentum** whose projection on the z-axis can have the values $S_z = m_s\hbar$, $m_s = \pm 1/2$. The maximum measurable component of $\boldsymbol{S}$ in units of $\hbar$ is called the **spin** of the particle s. The spin angular momentum gives rise to the magnetic moment, which was confirmed by Dirac. Thus,

$$\mu_s = -\frac{e}{m}\boldsymbol{S} \tag{7.13}$$

For spin $-1/2$ system, the matrices representing S_x, S_y, S_z are

$$S_x = \frac{1}{2}\hbar\begin{pmatrix} 0 & 1 \\ 1 & 0 \end{pmatrix}, \qquad S_y = \frac{1}{2}\hbar\begin{pmatrix} 0 & -i \\ i & 0 \end{pmatrix}, \qquad S_z = \frac{1}{2}\hbar\begin{pmatrix} 1 & 0 \\ 0 & -1 \end{pmatrix} \tag{7.14}$$

Another useful matrix is the σ matrix defined by

$$S = \frac{1}{2}\hbar\sigma$$

where

$$\sigma_x = \begin{pmatrix} 0 & 1 \\ 1 & 0 \end{pmatrix}, \qquad \sigma_y = \begin{pmatrix} 0 & -i \\ i & 0 \end{pmatrix}, \qquad \sigma_z = \begin{pmatrix} 1 & 0 \\ 0 & -1 \end{pmatrix} \tag{7.15}$$

The σ_x, σ_y and σ_z matrices are called **Pauli's spin matrices.**

7.5 Addition of Angular Momenta

Consider two noninteracting systems having angular momenta $\boldsymbol{J}_1$ and $\boldsymbol{J}_2$; let their eigenkets be $|j_1m_1\rangle$ and $|j_2m_2\rangle$, respectively, i.e.,

$$J_1^2|j_1m_1\rangle = j_1(j_1+1)\hbar^2|j_1m_1\rangle \tag{7.16}$$

$$J_{1z}^2|j_1m_1\rangle = m_1\hbar|j_1m\rangle \tag{7.17}$$

$$J_2^2|j_2m_2\rangle = j_2(j_2+1)\hbar^2|j_2m_2\rangle \tag{7.18}$$

$$J_{2z}^2|j_2m_2\rangle = m_2\hbar|j_1m_1\rangle \tag{7.19}$$

where

$$m_1 = j_1, j_1-1, \ldots, -j_1; \qquad m_2 = j_2, j_2-1, \ldots, -j_2$$

Since the two systems are noninteracting,

$$[J_1, J_2] = 0, \qquad [J_1^2, J_2^2] = 0 \tag{7.20}$$

Hence the operators $J_1^2, J_{1z}, J_2^2, J_{2x}$ form a complete set with simultaneous eigenkets $|j_1m_1 \quad j_2m_2\rangle$. For the given values of j_1 and j_2,

$$|j_1m_1j_2m_2\rangle = |j_1m_1\rangle|j_2m_2\rangle = |m_1m_2\rangle \tag{7.21}$$

For the total angular momentum vector $\boldsymbol{J} = \boldsymbol{J}_1 + \boldsymbol{J}_2$,

$$[J^2, J_z] = [J^2, J_1^2] = [J^2, J_2^2] = 0 \tag{7.22}$$

Hence, J^2, J_z, J_1^2, J_2^2 will have simultaneous eigenkets and let them be $|jmj_1j_2\rangle$. For given values of j_1 and j_2, this becomes $|jm\rangle$. The unknown kets $|jm\rangle$ can be expressed as a linear combination of the known kets $|m_1m_2\rangle$ as

$$|jm\rangle = \sum_{m_1,m_2} C_{jmm_1m_2}|m_1m_2\rangle \tag{7.23}$$

The coefficients $C_{jmm_1m_2}$ are called the **Clebsh-Gordan coefficients** or **Wigner coefficients.** Multiplying Eq. (7.23) by the bra $\langle m_1m_2|$, we get

$$\langle m_1m_2|jm\rangle = C_{jmm_1m_2} \tag{7.24}$$

With this value in Eq. (7.23), we have

$$|jm\rangle = \sum_{m_1,m_2} |m_1m_2\rangle\langle m_1m_2|jm\rangle \tag{7.25}$$

PROBLEMS

7.1 Prove the following commutation relations for the angular momentum operators L_x, L_y, L_z and $\boldsymbol{L}$:

(i) $[L_x, L_y] = i\hbar L_z$; $[L_y, L_z] = i\hbar L_x$; $[L_z, L_x] = i\hbar L_y$

(ii) $[L^2, L_x] = [L^2, L_y] = [L^2, L_z] = 0$

Solution. The angular momentum L of a particle is defined by

$$\boldsymbol{L} = \boldsymbol{r} \times \boldsymbol{p} = (yp_z - zp_y)\hat{i} + (zp_x - xp_z)\hat{j} + (xp_y - yp_x)\hat{k}$$

(i) $[L_x, L_y] = [yp_z - zp_y, zp_x - xp_z] = [yp_z, zp_x] - [yp_z, xp_z] - [zp_y, zp_x] + [zp_y, xp_z]$

In the second and third terms on RHS, all the variables involved commute with each other. Hence both of them vanish. Since y and p_x commute with z and p_z,

$$[yp_z, zp_x] = yp_x[p_z, z] = -i\hbar yp_x$$

$$[zp_y, xp_z] = xp_y[z, p_z] = i\hbar xp_y$$

Therefore,

$$[L_x, L_y] = i\hbar(xp_y - yp_x) = i\hbar L_z$$

Similarly, we can prove that

$$[L_y, L_z] = i\hbar L_x, \qquad [L_z, L_x] = i\hbar L_y$$

(ii)
$$\begin{aligned}
[L^2, L_x] &= [L_x^2 + L_y^2 + L_z^2, L_x] \\
&= [L_x^2, L_x] + [L_y^2, L_x] + [L_z^2, L_x] \\
&= 0 + L_y[L_y, L_x] + [L_y, L_x]L_y + L_z[L_z, L_x] + [L_z, L_x]L_z \\
&= L_y(-i\hbar L_z) + (-i\hbar L_z)L_y + L_z(i\hbar L_y) + (i\hbar L_y)L_z \\
&= 0
\end{aligned}$$

Thus we can conclude that

$$[L^2, L_x] = [L^2, L_y] = [L^2, L_z] = 0$$

7.2 Express the operators for the angular momentum components L_x, L_y and L_z in the spherical polar coordinates.

Solution. The gradient in the spherical polar coordinates is given by

$$\nabla = \hat{r}\frac{\partial}{\partial r} + \hat{\theta}\frac{1}{r}\frac{\partial}{\partial \theta} + \hat{\phi}\frac{1}{r\sin\theta}\frac{\partial}{\partial \phi}$$

where $\hat{r}$, $\hat{\theta}$ and $\hat{\phi}$ are the unit vectors along the r, θ and ϕ directions. The angular momentum

$$\begin{aligned}
\mathbf{L} = \mathbf{r} \times \mathbf{p} &= -i\hbar(\mathbf{r} \times \nabla) \\
&= -i\hbar\left(\boldsymbol{r} \times \hat{r}\frac{\partial}{\partial r} + \boldsymbol{r} \times \hat{\theta}\frac{1}{r}\frac{\partial}{\partial \theta} + \boldsymbol{r} \times \hat{\phi}\frac{1}{r\sin\theta}\frac{\partial}{\partial \phi}\right)
\end{aligned}$$

Since $\mathbf{r} = r\hat{r}$, $\hat{r} \times \hat{r} = 0$, $\hat{r} \times \hat{\theta} = \hat{\phi}$ and $\hat{r} \times \hat{\phi} = -\hat{\theta}$,

$$L = -i\hbar\left(\hat{\phi}\frac{\partial}{\partial\theta} - \hat{\theta}\frac{1}{\sin\theta}\frac{\partial}{\partial\phi}\right)$$

Resolving the unit vectors $\hat{\theta}$ and $\hat{\phi}$ in cartesian components (see Appendix), we get

$$\hat{\theta} = \cos\theta\cos\phi\hat{i} + \cos\theta\sin\phi\hat{j} - \sin\theta\hat{k}$$

$$\hat{\phi} = -\sin\theta\hat{i} + \cos\phi\hat{j}$$

Substituting the values of $\hat{\theta}$ and $\hat{\phi}$, we obtain

$$L = -i\hbar\left[(-\sin\phi\hat{i} + \cos\phi\hat{j})\frac{\partial}{\partial\theta} - (\cos\theta\cos\phi\hat{i} + \cos\theta\sin\phi\hat{j} - \sin\phi\hat{k})\frac{1}{\sin\theta}\frac{\partial}{\partial\phi}\right]$$

Collecting the coefficients of $\hat{i}$, $\hat{j}$ and $\hat{k}$, we get

$$L_x = i\hbar\left(\sin\phi\frac{\partial}{\partial\theta} + \cos\phi\cot\theta\frac{\partial}{\partial\phi}\right)$$

$$L_y = -i\hbar\left(\cos\phi\frac{\partial}{\partial\theta} - \sin\phi\cot\theta\frac{\partial}{\partial\phi}\right)$$

$$L_z = -i\hbar\frac{\partial}{\partial\phi}$$

7.3 Obtain the expressions for L_+, L_- and L^2 in the spherical polar coordinates.

Solution. To evaluate L_+ in the spherical polar coordinate system, substitue the values of L_x and L_y from Problem 7.2 in $L_+ = L_x + iL_y$. Then,

$$L_+ = -i\hbar\left(\sin\phi\frac{\partial}{\partial\phi} + \cot\theta\cos\phi\frac{\partial}{\partial\phi}\right) + \hbar\left(\cos\phi\frac{\partial}{\partial\theta} - \cot\theta\sin\phi\frac{\partial}{\partial\phi}\right)$$

$$= \hbar(\cos\phi + i\sin\phi)\frac{\partial}{\partial\theta} + i\hbar\cot\theta(\cos\phi + i\sin\phi)\frac{\partial}{\partial\phi}$$

$$= \hbar e^{i\phi}\left(\frac{\partial}{\partial\phi} + i\cot\theta\frac{\partial}{\partial\phi}\right)$$

$$L_- = L_x - iL_y = -\hbar e^{-i\phi}\left(\frac{\partial}{\partial\theta} - i\cot\theta\frac{\partial}{\partial\phi}\right)$$

$$L_+L_- = -\hbar^2 e^{i\phi}\left(\frac{\partial}{\partial\theta} + i\cot\theta\frac{\partial}{\partial\phi}\right)e^{i\phi}\left(\frac{\partial}{\partial\theta} - i\cot\theta\frac{\partial}{\partial\phi}\right)$$

$$= -\hbar^2\left[\frac{\partial^2}{\partial\theta^2} + \cot\theta\frac{\partial}{\partial\phi} + \cot^2\theta\frac{\partial^2}{\partial\phi^2} + i(\text{cosec}^2\theta - \cot^2\theta)\frac{\partial}{\partial\phi}\right]$$

$$= -\hbar^2\left(\frac{\partial^2}{\partial\theta^2} + \cot\theta\frac{\partial}{\partial\theta} + \cot^2\theta\frac{\partial^2}{\partial\phi^2} + i\frac{\partial}{\partial\phi}\right)$$

$$L_- L_+ = -\hbar^2 \left(\frac{\partial^2}{\partial \theta^2} + \cot\theta \frac{\partial}{\partial \theta} + \cot^2\theta \frac{\partial^2}{\partial \phi^2} - i \frac{\partial}{\partial \phi} \right)$$

$$L^2 = L_x^2 + L_y^2 + L_z^2 = \frac{1}{2}(L_+ L_- + L_- L_+) + L_z^2$$

$$= -\hbar^2 \left(\frac{\partial^2}{\partial \theta^2} + \cot\theta \frac{\partial}{\partial \theta} + \cot^2\theta \frac{\partial^2}{\partial \phi^2} + \frac{\partial^2}{\partial \phi^2} \right)$$

$$= -\hbar^2 \left(\frac{\partial^2}{\partial \theta^2} + \frac{\cos\theta}{\sin\theta} \frac{\partial}{\partial \theta} + \frac{1}{\sin^2\theta} \frac{\partial^2}{\partial \phi^2} \right)$$

$$= -\hbar^2 \left[\frac{1}{\sin\theta} \frac{\partial}{\partial \theta} \left(\sin\theta \frac{\partial}{\partial \theta} \right) + \frac{1}{\sin^2\theta} \frac{\partial^2}{\partial \phi^2} \right]$$

7.4 What is the value of the uncertainty product (ΔL_x) (ΔL_y) in a representation in which L^2 and L_z have simultaneous eigenfunctions? Comment on the value of this product when $l = 0$.

Solution. If the commutator of operators A and B obey the relation $[A, B] = iC$, then

$$(\Delta A)(\Delta B) \geq \frac{|\langle C \rangle|}{2}$$

In the representation in which L^2 and L_z have simultaneous eigenfunctions,

$$[L_x, L_y] = i\hbar L_Z$$

Therefore, it follows that

$$(\Delta L_x)\ (\Delta L_y) \geq \frac{\hbar}{2} |\langle L_z \rangle| \geq \frac{h}{2} m\hbar$$

$$(\Delta L_x)\ (\Delta L_y) \geq \frac{m\hbar^2}{2}$$

This is understandable as $Y_{lm}(\theta, \phi)$ is not an eigenfunction of L_x and L_y when $l \neq 0$. When $l = 0$, $m = 0$, $Y_{00} = 1/\sqrt{4\pi}$. Hence,

$$(\Delta L_x)(\Delta L_y) \geq 0$$

7.5 Evaluate the following commutators.

Solution.

(i) $[L_x, [L_y, L_z]] = [L_x, i\hbar L_x] = i\hbar[L_x, L_x] = 0$.

(ii) $[L_y^2, L_x] = L_y[L_y, L_x] + [L_y, L_x]L_y = -i\hbar(L_y L_z + L_z L_y)$.

(iii) $[L_x^2, L_y^2] = L_x[L_x, L_y^2] + [L_x, L_y^2]L_x = L_x\{[L_x, L_y]L_y + L_y[L_x, L_y]\}$

$+ \{[L_x, L_y]L_y + L_y[L_x, L_y]\}L_x$

$= i\hbar(L_x L_z L_y + L_x L_y L_z + L_z L_y L_x + L_y L_z L_x)$.

7.6 Evaluate the commutator $[L_x, L_y]$ in the momentum representation.

Solution.

$$L_x = yp_z - zp_y; \quad L_y = zp_x - xp_z; \quad L_z = xp_y - yp_x$$

$$[L_x, L_y] = [yp_z - zp_y, zp_x - xp_z] = [yp_z, zp_x] - [yp_z, xp_z] - [zp_y, zp_x] + [zp_y, xp_z]$$

$$= yp_x [p_z, z] - 0 - 0 + p_y x [z, p_z]$$

In the momentum representation $[z, p_z] = i\hbar$,

$$[L_x, L_y] = i\hbar (xp_y - yp_x) = i\hbar L_z$$

7.7 Show that the raising and lowering operators L_+ and L_- are Hermitian conjugates.

Solution.

$$\langle m|L_+|n\rangle = \langle m|L_x|n\rangle + i\langle m|L_y|n\rangle$$

$$= \langle n|L_x|m\rangle^* + i\langle n|L_y|m\rangle^*$$

$$= \langle n|(L_x - iL_y)|m\rangle^* = \langle n|L_-|m\rangle^*$$

Hence the result.

7.8 Prove that the spin matrices S_x and S_y have $\pm\hbar/2$ eigenvalues, i.e.,

$$S_x = \frac{1}{2}\hbar\begin{pmatrix} 0 & 1 \\ 1 & 0 \end{pmatrix} \qquad S_y = \frac{1}{2}\hbar\begin{pmatrix} 0 & -i \\ i & 0 \end{pmatrix}$$

Solution. The characteristic determinant of the S_x matrix is given by

$$\begin{vmatrix} -\lambda & \hbar/2 \\ \hbar/2 & -\lambda \end{vmatrix} = 0 \quad \text{or} \quad \lambda^2 - \frac{\hbar^2}{4} = 0 \quad \text{or} \quad \lambda = \pm\frac{1}{2}\hbar$$

Similarly, the eigenvalues of S_y are $\pm\dfrac{1}{2}\hbar$.

7.9 The operators J_+ and J_- are defined by $J_+ = J_x + iJ_y$ and $J_- = J_x + iJ_y$, where J_x and J_y are the x- and y-components of the general angular momentum J. Prove that

(i) $j_+|j, m\rangle = [j(j+1) - m(m+1)]^{1/2}\hbar|j, m+1\rangle$

(ii) $j_-|j, m\rangle = [j(j+1) - m(m-1)]^{1/2}\hbar|j, m-1\rangle$

Solution. J_z operating on $|jm\rangle$ gives

$$J_z|jm\rangle = m\hbar|jm\rangle \tag{i}$$

Operating from left by J_+, we get

$$J_+J_z|jm\rangle = m\hbar J_+|jm\rangle$$

Since

$$[J_z, J_+] = \hbar J_+ \quad \text{or} \quad J_+J_z = J_zJ_+ - \hbar J_+$$

we have

$$(J_zJ_+ - \hbar J_+)|jm\rangle = m\hbar J_+|jm\rangle$$

$$J_zJ_+|jm\rangle = (m+1)\hbar J_+|jm\rangle \tag{ii}$$

This implies that $J_+|jm\rangle$ is an eigenket of J_z with eigenvalue $(m + 1)\hbar$. The eigenvalue equation for J_z with eigenvalue $(m + 1)\hbar$ can also be written as

$$J_z\,|\,j, m+1\rangle = (m + 1)\,\hbar\,|\,j, m + 1\rangle \tag{iii}$$

Since the eigenvalues of J_z, see Eqs. (ii) and (iii), are equal, the eigenvectors can differ at the most by a multiplicative constant, say, a_m. Now,

$$J_+|\,jm\rangle = a_m|\,j, m + 1\rangle \tag{iv}$$

Similarly,

$$J_-|\,jm\rangle = b_m|\,j, m - 1\rangle \tag{v}$$

$$a_m = \langle j, m+1\,|J_+|\,jm\rangle \quad \text{or} \quad a_m^* = \langle jm|J_-|\,j, m+1\rangle \tag{vi}$$

$$b_m = \langle j, m-1\,|J_-|\,jm\rangle \quad \text{or} \quad b_{m+1} = \langle jm|J_-|\,j, m+1\rangle \tag{vii}$$

Comparing Eqs. (vi) and (vii), we get

$$a_m^* = b_{m+1} \tag{viii}$$

Operating Eq. (iv) from left by J_-, we obtain

$$J_-J_+|\,jm\rangle = a_m J_-|\,j, m + 1\rangle$$

It is easily seen that

$$J_-J_+ = J^2 - J_z^2 - \hbar J_z$$

Using this result and Eq. (v), we have

$$(J^2 - J_z^2 - \hbar J_z)|\,jm\rangle = a_m b_{m+1}\,|\,jm\rangle$$

$$[j(j+1) - m^2 - m]\,\hbar^2\,|\,jm\rangle = |\,a_n|^2\,|\,jm\rangle$$

$$a_m = [j(j+1) - m(m + 1)]^{1/2}\,\hbar \tag{ix}$$

With this value of a_m,

$$J_+\,|\,jm\rangle = [j(j+1) - m(m + 1)]^{1/2}\,\hbar|\,j, m + 1\rangle \tag{x}$$

$$\langle j'm'|J_+|\,jm\rangle = [j(j+1) - m(m + 1)]^{1/2}\,\hbar\,\delta_{jj'}\delta_{m',m+1} \tag{xi}$$

Similarly,

$$\langle j'm'|J_-|\,jm\rangle = [j(j+1) - m(m - 1)]^{1/2}\,\hbar\,\delta_{jj'}\delta_{m',m-1} \tag{xii}$$

7.10 A particle is in an eigenstate of L_z. Prove that $\langle J_x\rangle = \langle J_y\rangle = 0$. Also find the value of $\langle J_x^2\rangle$ and $\langle J_y^2\rangle$.

Solution. Let the eigenstate of J_z be $|\,jm\rangle$. We have

$$J_x = \frac{J_+ + J_-}{2}, \qquad J_y = \frac{J_+ - J_-}{2i}$$

$$\langle J_x \rangle = \frac{1}{2}\langle jm|J_+|jm\rangle + \frac{1}{2}\langle jm|J_-|jm\rangle$$

$$= \frac{1}{2}\sqrt{j(j+1) - m(m+1)}\,\hbar\langle jm|j,m+1\rangle + \frac{1}{2}\sqrt{j(j+1) - m(m-1)}\,\hbar\langle jm|j,m-1\rangle = 0$$

since $\langle jm|j, m+1\rangle = \langle jm|j, m-1\rangle = 0$. Similarly, $\langle J_y \rangle = 0$. We have the relation

$$J_x^2 + J_y^2 = J^2 - J_z^2$$

In the eigenstate $|jm\rangle$, this relation can be rewritten as

$$\langle jm|(J_x^2 + J_y^2)|jm\rangle = \langle jm|(J^2 - J_z^2)|jm\rangle$$

$$\langle jm|J_x^2|jm\rangle + \langle jm|J_y^2|jm\rangle = j(j+1)\hbar^2 - m^2\hbar^2$$

It is expected that $\langle J_x^2 \rangle = \langle J_y^2 \rangle$ and, therefore,

$$\langle J_x^2 \rangle = \langle J_y^2 \rangle = \frac{1}{2}[j(j+1)\hbar^2 - m^2\hbar^2]$$

7.11 $Y_{lm}(\theta, \phi)$ form a complete set of orthonormal functions of (θ, ϕ). Prove that

$$\sum_l \sum_{m=-l}^{l} |Y_{lm}\rangle\langle Y_{lm}| = 1$$

where 1 is the unit operator.

Solution. On the basis of expansion theorem, any function of θ and ϕ may be expanded in the form

$$\psi(\theta,\phi) = \sum_l \sum_m C_{lm} Y_{lm}(\theta, \phi)$$

In Dirac's notation,

$$|\psi\rangle = \sum_l \sum_m C_{lm}\,|Y_{lm}\rangle$$

Operating from left by $\langle Y_{lm}|$ and using the orthonormality relation

$$\langle Y_{l'm'}|Y_{lm}\rangle = \delta_{ll'}\delta_{mm'}$$

we get

$$C_{lm} = \langle Y_{lm}|\psi\rangle$$

Substituting this value of C_{lm}, we obtain

$$|\psi\rangle = \sum_l \sum_{m=-l}^{l} |Y_{lm}\rangle\langle Y_{lm}|\psi\rangle$$

From this relation it follows that

$$\sum_l \sum_{m=-l}^{l} |Y_{lm}\rangle\langle Y_{lm}| = 1$$

7.12 The vector $\boldsymbol{J}$ gives the sum of angular momenta $\boldsymbol{J}_1$ and $\boldsymbol{J}_2$. Prove that

$$[J_x, J_y] = i\hbar J_z,\ [J_y, J_z] = i\hbar J_x,\ [J_z, J_x] = i\hbar J_y$$

Is $\boldsymbol{J}_1 - \boldsymbol{J}_2$ an angular momentum?

Solution. Given $\boldsymbol{J} = \boldsymbol{J}_1 + \boldsymbol{J}_2$:

$$\begin{aligned}[J_x, J_y] &= [J_{1x} + J_{2x}, J_{1y} + J_{2y}] \\ &= [J_{1x}, J_{1y}] + [J_{1x}, J_{2y}] + [J_{2x}, J_{1y}] + [J_{2x}, J_{2y}] \\ &= i\hbar J_{1z} + 0 + 0 + i\hbar J_{2z} \\ &= i\hbar(J_{1z} + J_{2z}) = i\hbar J_z\end{aligned}$$

By cyclic permutation of the coordinates, we can write the other two commutation relations. Writing

$$\boldsymbol{J}_1 - \boldsymbol{J}_2 = \boldsymbol{J}'$$

$$\begin{aligned}[J'_x, J'_y] &= [J_{1x} - J_{2x}, J_{1y} - J_{2y}] \\ &= [J_{1x}, J_{1y}] - [J_{1x}, J_{2y}] - [J_{2x}, J_{1y}] + [J_{2x}, J_{2y}] \\ &= i\hbar J_{1z} - 0 - 0 + i\hbar J_{2z} = i\hbar\,(J_{1z} + J_{2z})\end{aligned}$$

which is not the operator for J'_z. Hence $\boldsymbol{J}_1 - \boldsymbol{J}_2$ is not an angular momentum.

7.13 Write the operators for the square of angular momentum and its z-component in the spherical polar coordinates. Using the explicit form of the spherical harmonic, verify that $Y_{11}(\theta, \phi)$ is an eigenfunction of L^2 and L_z with the quantum numbers $l = 1$ and $m = 1$.

Solution. The operators for L^2 and L_z are

$$\begin{aligned}L^2 &= -\hbar^2\left[\frac{1}{\sin\theta}\frac{\partial}{\partial\theta}\left(\sin\theta\frac{\partial}{\partial\theta}\right) + \frac{1}{\sin^2\theta}\frac{\partial^2}{\partial\phi^2}\right] \\ &= -\hbar^2\left[\frac{\partial^2}{\partial\theta^2} + \cot\theta\frac{\partial}{\partial\theta} + \frac{1}{\sin^2\theta}\frac{\partial^2}{\partial\phi^2}\right]\end{aligned}$$

$$L_z = -i\hbar\frac{\partial}{\partial\theta}$$

The spherical harmonic $Y_{11} = -\left(\dfrac{3}{8\pi}\right)^{1/2}\sin\theta e^{i\phi}$

$$\begin{aligned}L^2Y_{11} &= \left(\frac{3}{8\pi}\right)^{1/2}\hbar^2\left[\frac{\partial^2}{\partial\theta^2} + \cot\theta\frac{\partial}{\partial\theta} + \frac{1}{\sin^2\theta}\frac{\partial^2}{\partial\theta^2}\right]\sin\theta e^{i\phi} \\ &= \left(\frac{3}{8\pi}\right)^{1/2}\hbar^2\left[-\sin\theta + \cot\theta\cos\theta - \frac{1}{\sin^2\theta}\sin\theta\right]e^{i\phi} \\ &= \left(\frac{3}{8\pi}\right)^{1/2}\hbar^2\left[-\sin\theta + \frac{\cos^2\theta}{\sin\theta} - \frac{1}{\sin\theta}\right]e^{i\phi} \\ &= \left(\frac{3}{8\pi}\right)^{1/2}\hbar^2\left[\frac{-\sin^2\theta + \cos^2\theta - 1}{\sin\theta}\right]e^{i\phi} \\ &= \left(\frac{3}{8\pi}\right)^{1/2}\hbar^2(-2\sin\theta)e^{i\phi} = 2\hbar^2Y_{11}\end{aligned}$$

$$L_z Y_{11} = +i\hbar \frac{\partial}{\partial \phi}\left(\frac{3}{8\pi}\right)^{1/2} \sin\theta\, e^{i\phi}$$

$$= -\left(\frac{3}{8\pi}\right)^{1/2} \sin\theta\, e^{i\phi} = Y_{11}$$

Hence the required result.

7.14 The raising (J_+) and lowering (J_-) operators are defined by $J_+ = J_x + iJ_y$ and $J_- = J_x - iJ_y$. Prove the following identities:

(i) $[J_x, J_\pm] = \mp \hbar J_z$

(ii) $[J_y, J_\pm] = -i\hbar J_z$

(iii) $[J_z, J_\pm] = \pm \hbar J_\pm$

(iv) $J_+J_- = J^2 - J_z^2 + \hbar J_z$

(v) $J_-J_+ = J^2 - J_z^2 - \hbar J_z$

Solution.

(i) $$[J_x, J_\pm] = [J_x, J_x] \pm i[J_x, J_y] = 0 \pm i(i\hbar)\, J_z = \mp\hbar J_z$$

(ii) $$[J_y, J_\pm] = [J_y, J_x] \pm i[J_y, J_y] = -i\hbar J_z$$

(iii) $$[J_x, J_\pm] = [J_z, J_x] \pm i[J_z, J_y] = i\hbar J_y \pm i(-i\hbar J_x) = \hbar(\pm J_x + iJ_y) = \pm\hbar J_\pm$$

(iv) $$J_+J_- = (J_x + iJ_y)(J_x - iJ_y) = J_x^2 + J_y^2 - i(J_xJ_y - J_yJ_x) = J^2 - J_z^2 - i[J_x, J_y] = J^2 - J_z^2 + \hbar J_z$$

(v) $$J_-J_+ = (J_x - iJ_y)(J_x - iJ_y) = J_x^2 + J_y^2 + i(J_xJ_y - J_yJ_x) = J^2 - J_z^2 + i[J_x, J_y] = J^2 - J_z^2 - \hbar J_z$$

7.15 In the $|jm\rangle$ basis formed by the eigenkets of J^2 and J_z, show that

$$\langle jm|J_-J_+|jm\rangle = (j - m)(j + m + 1)\hbar^2$$

where $J_+ = J_x + iJ_y$ and $J_- = J_x - iJ_y$.

Solution. In Problem 7.14, we have proved that

$$J_-J_+ = J^2 - J_z^2 - \hbar J_z$$

$$\langle jm|J_-J_+|jm\rangle = \langle jm|J^2 - J_z^2 - \hbar J_z|jm\rangle = [j(j+1) - m^2 - m]\hbar^2\langle jm|jm\rangle$$

Since $\langle jm|jm\rangle = 1$,

$$\langle jm|J_-J_+|jm\rangle = [j^2 - m^2 + j - m]\hbar^2$$
$$= [(j+m)(j-m) + (j-m)]\hbar^2$$
$$= (j-m) + (j+m+1)\hbar^2$$

7.16 In the $|jm\rangle$ basis formed by the eigenkets of the operators J^2 and J_z, obtain the relations for their matrices. Also obtain the explicit form of the matrices for $j = 1/2$ and $j = 1$.

Solution. As J^2 commutes with J_z, the matrices for J^2 and J_z will be diagonal. The eigenvalue-eigenket equations of the operators J^2 and J_z are

$$J^2|jm\rangle = j(j+1)\hbar^2|jm\rangle \tag{i}$$

$$J_z|jm\rangle = m\hbar|jm\rangle \tag{ii}$$

where

$$j = 0,\ 1/2,\ 1,\ 3/2,\ \ldots;\quad m = j,\ j-1,\ j-2,\ \ldots,\ -j$$

Multiplication of Eqs. (i) and (ii) from left by $\langle j'm'|$ gives the J^2 and J_z matrix elements:

$$\langle j'm'|J^2|jm\rangle = j(j+1)\hbar^2\delta_{jj'}\delta_{mm'}$$
$$\langle j'm'|J_z|jm\rangle = m\hbar\delta_{jj'}\delta_{mm'}$$

The presence of the factors $\delta_{jj'}$ and $\delta_{mm'}$ indicates that the matrices are diagonal as expected. The matrices for J^2 and J_z are:

$$j = \frac{1}{2},\qquad m = \frac{1}{2},\ -\frac{1}{2}$$

$$j = 1,\qquad m = 1,\ 0,\ -1$$

7.17 Using the values of $J_+|jm\rangle$ and $J_-|jm\rangle$, obtain the matrices for J_x and J_y for $j = 1/2$ and $j = 1$.

Solution. In Problem 7.9, we have proved that

$$J_+|jm\rangle = [j(j+1) - m(m+1)]^{1/2}\hbar|j, m+1\rangle \tag{i}$$

$$J_-|jm\rangle = [j(j+1) - m(m-1)]^{1/2}\hbar|j, m-1\rangle \tag{ii}$$

Premultiplying these equations by $\langle j'm'|$, we have

$$\langle j'm'|J_+|jm\rangle = [j(j+1) - m(m+1)]^{1/2}\hbar\delta_{jj'}\delta_{m',m+1} \tag{iii}$$

$$\langle j'm'|J_-|jm\rangle = [j(j+1) - m(m-1)]^{1/2}\hbar\delta_{jj'}\delta_{m',m-1} \tag{iv}$$

Equations (iii) and (iv) give the matrix elements for J_+ and J_- matrices. From these, J_x and J_y can be evaluated using the relations

$$J_x = \frac{1}{2}(J_+ + J_-),\qquad J_y = -\frac{i}{2}(J_+ - J_-)$$

For $j = \frac{1}{2}$:
$$J_+ = \hbar\begin{pmatrix} 0 & 1 \\ 0 & 0 \end{pmatrix}, \qquad J_- = \hbar\begin{pmatrix} 0 & 0 \\ 1 & 0 \end{pmatrix}$$

$$J_x = \frac{\hbar}{2}\begin{pmatrix} 0 & 1 \\ 1 & 0 \end{pmatrix}, \qquad J_y = \frac{\hbar}{2}\begin{pmatrix} 0 & -i \\ i & 0 \end{pmatrix}$$

For $j = 1$:
$$J_+ = \hbar\begin{pmatrix} 0 & \sqrt{2} & 0 \\ 0 & 0 & \sqrt{2} \\ 0 & 0 & 0 \end{pmatrix}, \qquad J_- = \hbar\begin{pmatrix} 0 & 0 & 0 \\ \sqrt{2} & 0 & 0 \\ 0 & \sqrt{2} & 0 \end{pmatrix}$$

$$J_x = \frac{\hbar}{\sqrt{2}}\begin{pmatrix} 0 & 1 & 0 \\ 1 & 0 & 1 \\ 0 & 1 & 0 \end{pmatrix}, \qquad J_y = \frac{\hbar}{\sqrt{2}}\begin{pmatrix} 0 & -i & 0 \\ i & 0 & -i \\ 0 & i & 0 \end{pmatrix}$$

7.18 State the matrices that represent the x, y, z components of the spin angular momentum vector $\boldsymbol{S}$ and obtain their eigenvalues and eigenvectors.

Solution. The matrices for S_x, S_y and S_z are

$$S_x = \frac{\hbar}{2}\begin{pmatrix} 0 & 1 \\ 1 & 0 \end{pmatrix}, \quad S_y = \frac{\hbar}{2}\begin{pmatrix} 0 & -i \\ i & 0 \end{pmatrix}, \quad S_z = \frac{\hbar}{2}\begin{pmatrix} 1 & 0 \\ 0 & -1 \end{pmatrix}$$

Let the eigenvalues of S_z be λ. The values of λ are the solutions of the secular determinant

$$\begin{vmatrix} \frac{1}{2}\hbar - \lambda & 0 \\ 0 & -\frac{1}{2}\hbar - \lambda \end{vmatrix} = 0$$

$$\left(\lambda - \frac{1}{2}\hbar\right)\left(\lambda + \frac{1}{2}\hbar\right) = 0$$

$$\lambda = \frac{1}{2}\hbar \quad \text{or} \quad -\frac{1}{2}\hbar$$

Let the eigenvector of S_z corresponding to the eigenvalue $\frac{1}{2}\hbar$ be $\begin{pmatrix} a_1 \\ a_2 \end{pmatrix}$.

Then,

$$\frac{1}{2}\hbar\begin{pmatrix} 1 & 0 \\ 0 & -1 \end{pmatrix}\begin{pmatrix} a_1 \\ a_2 \end{pmatrix} = \frac{1}{2}\hbar\begin{pmatrix} a_1 \\ a_2 \end{pmatrix}$$

$$\begin{pmatrix} a_1 \\ -a_2 \end{pmatrix} = \begin{pmatrix} a_1 \\ a_2 \end{pmatrix} \quad \text{or} \quad a_2 = 0$$

The normalization condition gives

$$|a_1|^2 = 1 \quad \text{or} \quad a_1 = 1$$

i.e., the eigenvector of S_z corresponding to the eigenvalue $\frac{1}{2}\hbar$ is $\begin{pmatrix}1\\0\end{pmatrix}$. Following the same procedure, the eigenvector of S_z corresponding to the eigenvalue $-\frac{1}{2}\hbar$ is $\begin{pmatrix}0\\1\end{pmatrix}$. The same procedure can be followed for the S_x and S_y matrices. The results are summarized as follows:

Spin matrix S_x: Eigenvalue $\frac{1}{2}\hbar$ Eigenvector $\frac{1}{\sqrt{2}}\begin{pmatrix}1\\1\end{pmatrix}$

Eigenvalue $-\frac{1}{2}\hbar$ Eigenvector $\frac{1}{\sqrt{2}}\begin{pmatrix}1\\-1\end{pmatrix}$

Spin matrix S_y: Eigenvalue $\frac{1}{2}\hbar$ Eigenvector $\frac{1}{\sqrt{2}}\begin{pmatrix}1\\i\end{pmatrix}$

Eigenvalue $-\frac{1}{2}\hbar$ Eigenvector $\frac{1}{\sqrt{2}}\begin{pmatrix}1\\-i\end{pmatrix}$

7.19 Derive matrices for the operators J^2, J_z, J_x and J_y for $j = 3/2$.

Solution. For $j = 3/2$, the allowed values of m are 3/2, 1/2, –1/2 and –3/2. With these values for j and m, matrices for J^2 and J_z are written with the help of Eqs. (7.11) and (7.12). Then,

$$J^2 = \frac{15}{4}\hbar^2\begin{pmatrix}1 & 0 & 0 & 0\\ 0 & 1 & 0 & 0\\ 0 & 0 & 1 & 0\\ 0 & 0 & 0 & 1\end{pmatrix}, \qquad J_z = \frac{1}{2}\hbar\begin{pmatrix}3 & 0 & 0 & 0\\ 0 & 1 & 0 & 0\\ 0 & 0 & -1 & 0\\ 0 & 0 & 0 & -3\end{pmatrix}$$

Equations (8.44) and (8.45) give the matrices for J_+ and J_- as

$$J_+ = \hbar\begin{pmatrix}0 & \sqrt{3} & 0 & 0\\ 0 & 0 & 2 & 0\\ 0 & 0 & 0 & \sqrt{3}\\ 0 & 0 & 0 & 0\end{pmatrix}, \qquad J_- = \hbar\begin{pmatrix}0 & 0 & 0 & 0\\ \sqrt{3} & 0 & 2 & 0\\ 0 & 2 & 0 & 0\\ 0 & 0 & \sqrt{3} & 0\end{pmatrix}$$

The matrices for J_x and J_y follow from the relations

$$J_x = \frac{1}{2}(J_+ + J_-), \quad J_y = \frac{1}{2i}(J_+ + J_-)$$

$$J_x = \frac{1}{2}\hbar\begin{pmatrix}0 & \sqrt{3} & 0 & 0\\ \sqrt{3} & 0 & 2 & 0\\ 0 & 2 & 0 & \sqrt{3}\\ 0 & 0 & \sqrt{3} & 0\end{pmatrix}, \qquad J_- = \frac{\hbar}{2i}\begin{pmatrix}0 & \sqrt{3} & 0 & 0\\ -\sqrt{3} & 0 & 2 & 0\\ 0 & -2 & 0 & \sqrt{3}\\ 0 & 0 & -\sqrt{3} & 0\end{pmatrix}$$

7.20 If the angular momentum operators obey the rule $[J_x, J_y] = -i\hbar J_z$ and similar commutation relations for the other components, evaluate the commutators $[J^2, J_x]$ and $[J^2, J_+]$. What would be the roles of J_+ and J_- in the new situation?

Solution.

$$[J^2, J_x] = [J_x^2, J_x] + [J_y^2, J_x] + [J_z^2, J_x]$$

$$= J_y [J_y, J_x] + [J_y, J_x] J_y + J_z [J_z, J_x] + [J_z, J_x] J_z$$

$$= i\hbar J_y J_z + i\hbar J_z J_y - i\hbar J_z J_y - i\hbar J_y J_z = 0$$

Similarly, $[J^2, J_y] = 0$. Hence,

$$[J^2, J_+] = [J^2, J_x] + i[J^2, J_y] = 0$$

Let us evaluate $[J_z, J_+]$ and $[J_z, J_-]$:

$$[J_z, J_+] = [J_z, J_x] + i[J_z, J_y] = -i\hbar J_y - \hbar J_x = -\hbar J_+$$

Similarly, $[J_z, J_-] = \hbar J_-$.

Thus, with the new definition, J_+ would be a lowering operator and J_- would be a raising operator.

7.21 For Pauli's matrices, prove that (i) $[\sigma_x, \sigma_y] = 2i\sigma_z$, (ii) $\sigma_x\sigma_y\sigma_z = i$.

Solution.

(i) We have

$$\mathbf{S} = \frac{1}{2}\hbar\sigma, \qquad [S_x, S_y] = i\hbar S_z$$

Substituting the values of S_x, S_y and S_z, we get

$$\left[\frac{1}{2}\hbar\sigma_x, \frac{1}{2}\hbar\sigma_y\right] = i\hbar\frac{1}{2}\hbar\sigma_z \quad \text{or} \quad [\sigma_x, \sigma_y] = 2i\sigma_z$$

(ii)
$$\sigma_x\sigma_y\sigma_z = \begin{pmatrix} 0 & 1 \\ 1 & 0 \end{pmatrix}\begin{pmatrix} 0 & -i \\ i & 0 \end{pmatrix}\begin{pmatrix} 1 & 0 \\ 0 & -1 \end{pmatrix}$$

$$= \begin{pmatrix} i & 0 \\ 0 & -i \end{pmatrix}\begin{pmatrix} 1 & 0 \\ 0 & -1 \end{pmatrix} = \begin{pmatrix} i & 0 \\ 0 & i \end{pmatrix} = i$$

7.22 Prove by direct matrix multiplication that the Pauli matrices anticommute and they follow the commutation relations $[\sigma_x, \sigma_y] = 2i\sigma_z$, xyz cyclic.

Solution.

$$\sigma_x\sigma_y + \sigma_y\sigma_x = \begin{pmatrix} 0 & 1 \\ 1 & 0 \end{pmatrix}\begin{pmatrix} 0 & -i \\ i & 0 \end{pmatrix} + \begin{pmatrix} 0 & -i \\ i & 0 \end{pmatrix}\begin{pmatrix} 0 & 1 \\ 1 & 0 \end{pmatrix}$$

$$= \begin{pmatrix} i & 0 \\ 0 & -i \end{pmatrix}\begin{pmatrix} -i & 0 \\ 0 & i \end{pmatrix} = 0$$

$$[\sigma_x, \sigma_y] = \sigma_x\sigma_y - \sigma_y\sigma_x = \begin{pmatrix} i & 0 \\ 0 & -i \end{pmatrix} - \begin{pmatrix} -i & 0 \\ 0 & i \end{pmatrix}$$

$$= \begin{pmatrix} 2i & 0 \\ 0 & -2i \end{pmatrix} = 2i\begin{pmatrix} 1 & 0 \\ 0 & -1 \end{pmatrix} = 2i\sigma_z$$

7.23 The components of arbitrary vectors **A** and **B** commute with those of σ. Show that $(\sigma \cdot \mathbf{A})(\sigma \cdot \mathbf{B}) = \mathbf{A} \cdot \mathbf{B} + i\sigma \cdot (\mathbf{A} \times \mathbf{B})$.

Solution.

$$(\sigma \cdot \mathbf{A})(\sigma \cdot \mathbf{B}) = (\sigma_x A_x + \sigma_y A_y + \sigma_z A_z)\,(\sigma_x B_x + \sigma_y B_y + \sigma_z B_z)$$

$$= \sigma_x^2 A_x B_x + \sigma_y^2 A_y B_y + \sigma_z^2 A_z B_z + \sigma_x\sigma_y\, A_x B_y + \sigma_y\sigma_x\, A_y B_x$$

$$+ \sigma_x\sigma_z\, A_x B_z + \sigma_y\sigma_z\, A_y B_z + \sigma_z\sigma_y\, A_z B_y + \sigma_z\sigma_x\, A_z B_x$$

Using the relations

$$\sigma_x^2 = \sigma_y^2 = \sigma_z^2 = 1$$

$$\sigma_x\sigma_y = i\sigma_z, \qquad \sigma_y\sigma_z = i\sigma_x, \qquad \sigma_z\sigma_x = i\sigma_y$$

$$\sigma_x\sigma_y + \sigma_y\sigma_x = \sigma_y\sigma_z + \sigma_z\sigma_y = \sigma_z\sigma_x + \sigma_x\sigma_z = 0$$

we get

$$(\sigma \cdot \boldsymbol{A})\,(\sigma \cdot \boldsymbol{B}) = (\boldsymbol{A} \cdot \boldsymbol{B}) + i\sigma_z\,(A_x B_y - A_y B_x) + i\sigma_y\,(A_z B_x - A_x B_z) + i\sigma_x\,(A_y B_z - A_z B_y)$$

$$= (\boldsymbol{A} \cdot \boldsymbol{B}) + i\sigma \cdot (\boldsymbol{A} \times \boldsymbol{B})$$

7.24 Obtain the normalized eigenvectors of σ_x and σ_y matrices.

Solution. The eigenvalue equation for the matrix s_x for the eigenvalue +1 is

$$\begin{pmatrix} 0 & 1 \\ 1 & 0 \end{pmatrix}\begin{pmatrix} a_1 \\ a_2 \end{pmatrix} = 1\begin{pmatrix} a_1 \\ a_2 \end{pmatrix}$$

$$\begin{pmatrix} a_2 \\ a_1 \end{pmatrix} = \begin{pmatrix} a_1 \\ a_2 \end{pmatrix} \quad \text{or} \quad a_1 = a_2$$

Normalization gives $|a_1|^2 + |a_2|^2 = 1$ or $a_1 = a_2 = 1/\sqrt{2}$.

The normalized eigenvector of σ_x for the eigenvalue +1 is $\dfrac{1}{\sqrt{2}}\begin{pmatrix} 1 \\ 1 \end{pmatrix}$.

The normalized eigenvector of σ_x for for the eigenvalue –1 is $\dfrac{1}{\sqrt{2}}\begin{pmatrix} 1 \\ -1 \end{pmatrix}$.

The eigenvalue equation for the matrix σ_y for the eigenvalue +1 is

$$\begin{pmatrix} 0 & -i \\ i & 0 \end{pmatrix}\begin{pmatrix} a_1 \\ a_2 \end{pmatrix} = \begin{pmatrix} a_1 \\ a_2 \end{pmatrix} \quad \text{or} \quad a_1 i = a_2$$

Normalization gives

$$|a_1|^2 + |a_1 i|^2 = 1 \quad \text{or} \quad 2a_1^2 = 1, \qquad a_1 = \frac{1}{\sqrt{2}}, \qquad a_2 = \frac{i}{\sqrt{2}}$$

The normalized eigenvector of σ_y for the eigenvalue +1 is $\frac{1}{\sqrt{2}}\begin{pmatrix}1\\ i\end{pmatrix}$.

The normalized eigenvector of σ_y for the eigenvalue −1 is $\frac{1}{\sqrt{2}}\begin{pmatrix}1\\ -i\end{pmatrix}$.

7.25 Using Pauli's spin matrix representation, reduce each of the operators
(i) $S_x^2S_yS_z^2$; (ii) $S_x^2S_y^2S_z^2$; (iii) $S_xS_yS_z^3$

Solution.

$$\text{(i)}\quad S_x^2S_yS_z^2 = \left(\frac{\hbar}{2}\right)^2\sigma_x^2\frac{\hbar}{2}\sigma_y\left(\frac{\hbar}{2}\right)^2\sigma_z^2 = \left(\frac{\hbar}{2}\right)^5\sigma_y.$$

$$\text{(ii)}\quad S_x^2S_y^2S_z^2 = \left(\frac{\hbar}{2}\right)^2\sigma_x^2\left(\frac{\hbar}{2}\right)^2\sigma_y^2\left(\frac{\hbar}{2}\right)^2\sigma_z^2 = \left(\frac{\hbar}{2}\right)^6.$$

$$\text{(iii)}\quad S_xS_yS_z^3 = \frac{\hbar}{2}\sigma_x\frac{\hbar}{2}\sigma_y\left(\frac{\hbar}{2}\right)^3\sigma_z^3 = \left(\frac{\hbar}{2}\right)^5\sigma_x\sigma_y\sigma_z = \left(\frac{\hbar}{2}\right)^5 i.$$

7.26 Determine the total angular momentum that may arise when the following angular momenta are added:
(i) $j_1 = 1, j_2 = 1$; (ii) $j_1 = 3, j_2 = 4$; (iii) $j_1 = 2, j_2 = 1/2$.

Solution. When the angular momenta j_1 and j_2 are combined, the allowed total angular momentum (j) values are given by $(j_1 + j_2), (j_1 + j_2 - 1), \ldots, |j_1 - j_2|$.

(i) For $j_1 = 1, j_2 = 1$, the allowed j values are 2, 1, 0.
(ii) For $j_1 = 3, j_2 = 4$, the allowed j values are 7, 6, 5, 4, 3, 2, 1.
(iii) For $j_1 = 2, j_2 = 1/2$, the allowed j values are 5/2, 3/2.

7.27 Determine the orbital momenta of two electrons:
(i) Both in d-orbitals; (ii) both in p-orbitals; (iii) in the configuration p^1d^1.

Solution.

(i) When the two electrons are in d orbitals, $l_1 = 2, l_2 = 2$. The angular momentum quantum number values are 4, 3, 2, 1, 0. The angular momenta in units of $\hbar$ are

$$\sqrt{l(l+1)} = \sqrt{20}, \sqrt{12}, \sqrt{6}, \sqrt{2}, 0$$

(ii) When both the electrons are in p-orbitals, $l_1 = 1, l_2 = 1$. The possible values of l are 2, 1, 0. The angular momenta are $\sqrt{6}, \sqrt{2}, 0$.

(iii) The configuration p^1d^1 means $l_1 = 1, l_2 = 2$. The possible l values are 3, 2, 1. Hence, the angular momenta are $\sqrt{12}, \sqrt{6}, \sqrt{2}$.

7.28 For any vector **A**, show that $[\sigma, \mathbf{A}\cdot\sigma] = 2i\mathbf{A}\times\sigma$.

Solution. The x-component on LHS is

$$\left[\sigma_x, A_x\sigma_x + A_y\sigma_y + A_z\sigma_z\right] = A_x\left[\sigma_x, \sigma_x\right] + A_y\left[\sigma_x, \sigma_y\right] + A_z\left[\sigma_x, \sigma_z\right]$$
$$= 0 + 2iA_y\sigma_z - 2iA_z\sigma_y$$

Adding all the three components, we get

$$[\sigma, A\cdot\sigma] = \hat{i}\,2i(A_y\sigma_z - A_z\sigma_y) + \hat{j}\,2i(A_z\sigma_x - A_x\sigma_z) + \hat{k}\,2i(A_x\sigma_y - A_y\sigma_x) = 2i\mathbf{A}\times\sigma$$

7.29 The sum of the two angular momenta $\boldsymbol{J}_1$ and $\boldsymbol{J}_2$ are given by $\boldsymbol{J} = \boldsymbol{J}_1 + \boldsymbol{J}_2$. If the eigenkets of J_1^2 and J_2^2 are $|j_1m_1\rangle$ and $|j_2m_2\rangle$, respectively, find the number of eigenstates of J^2.

Solution. Let the orthogonal eigenkets of J^2 and J_z be $|jm\rangle$. The quantum number j can have the values $(j_1 + j_2)$, $(j_1 + j_2 - 1)$, ..., $|j_1 - j_2|$. We can have $(2j + 1)$ independent kets for each of the values of j. Hence the total number of $|jm\rangle$ eigenkets are

$$\sum_{j=|j_1-j_2|}^{j_1+j_2} (2j+1) = \begin{cases} 2\sum\limits_{j_1-j_2}^{j_1+j_2} j + 2j_2 + 1 & \text{if } j_1 > j_2 \\ 2\sum\limits_{j_2-j_1}^{j_1+j_2} j + 2j_1 + 1 & \text{if } j_2 > j_1 \end{cases}$$

It may be noted that the first line corresponds to $j_1 > j_2$. While taking the summation, each term in it contributes 1 which occurs $(j_1 + j_2) - (j_1 - j_2) = 2j_2$ times. Since both $j_1 - j_2$ and $j_1 + j_2$ are included in the summation, an additional 1 is also added. Similar explanation holds for the $j_2 > j_1$ case. Taking $j_1 > j_2$, we get

$$\sum_{j|j_1-j_2|}^{|j_1+j_2|} (2j+1) = 2\frac{(j_1 + j_2)(j_1 + j_2 + 1)}{2} - 2\frac{(j_1 - j_2 - 1)(j_1 - j_2)}{2} + 2j_2 + 1$$

$$= 4j_1j_2 + 2j_1 + 2j_2 + 1 = 2j_1(2j_2 + 1) + (2j_2 + 1)$$

$$= (2j_1 + 1)\,(2j_2 + 1)$$

The number of simultaneous eigenstates of J^2 and $J_z = (2j_1 + 1)\,(2j_2 + 1)$.

7.30 If the eigenvalues of J^2 and J_z are given by $J^2|\lambda m\rangle = \lambda|\lambda m\rangle$ and $J_z|\lambda m\rangle = m|\lambda m\rangle$, show that $\lambda \geq m^2$.

Solution. Given $J^2|\lambda m\rangle = \lambda|\lambda m\rangle$. Find

$$(J_x^2 + J_y^2)|\lambda m\rangle + J_z^2|\lambda m\rangle = \lambda|\lambda m\rangle$$

$$\langle\lambda m|J_x^2|\lambda m\rangle + \langle\lambda m|J_y^2|\lambda m\rangle = \lambda\langle\lambda m|\lambda m\rangle - \langle\lambda m|J_z^2|\lambda m\rangle$$

$$\langle\lambda m|J_x^2|\lambda m\rangle + \langle\lambda m|J_y^2|\lambda m\rangle = \lambda - m^2$$

Since J_x and J_y are Hermitian, the LHS must be positive, i.e., $\lambda - m^2 \geq 0$.

7.31 The eigenfunctions of the Pauli spin operator σ_z are α and β. Show that $(\alpha + \beta)/\sqrt{2}$ and $(\alpha - \beta)/\sqrt{2}$ are the eigenfunctions of σ_x and $(\alpha + i\beta)/\sqrt{2}$ and $(\alpha - i\beta)/\sqrt{2}$ are the eigenfunctions of σ_y.

Solution. The Pauli operators are

$$\sigma_x = \begin{pmatrix} 0 & 1 \\ 1 & 0 \end{pmatrix}, \quad \sigma_y = \begin{pmatrix} 0 & -i \\ i & 0 \end{pmatrix}, \quad \sigma_z = \begin{pmatrix} 1 & 0 \\ 0 & -1 \end{pmatrix}$$

The eigenvalues of σ_x are +1 and –1. The eigenfunction corresponding to +1 eigenvalue is (refer Problem 7.24)

$$\frac{1}{\sqrt{2}}\begin{pmatrix} 1 \\ -1 \end{pmatrix} = \frac{1}{\sqrt{2}}\begin{pmatrix} 1+0 \\ 0+1 \end{pmatrix} = \frac{1}{\sqrt{2}}\left[\begin{pmatrix} 1 \\ 0 \end{pmatrix} + \begin{pmatrix} 0 \\ 1 \end{pmatrix}\right] = \frac{1}{\sqrt{2}}(\alpha + \beta)$$

The eigenfunction corresponding to the eigenvalue –1 is

$$\frac{1}{\sqrt{2}}\begin{pmatrix}1\\-1\end{pmatrix} = \frac{1}{\sqrt{2}}\left[\begin{pmatrix}1\\0\end{pmatrix} - \begin{pmatrix}0\\1\end{pmatrix}\right] = \frac{1}{\sqrt{2}}\begin{pmatrix}1\\0\end{pmatrix} - \frac{1}{\sqrt{2}}\begin{pmatrix}0\\1\end{pmatrix} = \frac{1}{\sqrt{2}}(\alpha - \beta)$$

Similarly, the eigenvectors of σ_y are $(\alpha + i\beta)/\sqrt{2}$ and $(\alpha - i\beta)/\sqrt{2}$.

7.32 An electron in a state is described by the wave function

$$\psi = \frac{1}{\sqrt{4\pi}}(e^{i\phi}\sin\theta + \cos\theta)\,R(r), \qquad \int_0^\infty |R(r)|^2 r^2\,dr = 1$$

where θ and ϕ are the polar and azimuth angles, respectively.

(i) Is the given wave function normalized?

(ii) What are the possible values expected in a measurement of the z-component L_z of the angular momentum of the electron in this state?

(iii) What is the probability of obtaining each of the possible values in (ii)?

Solution. The spherical harmonics

$$Y_{10} = \left(\frac{3}{4\pi}\right)^{1/2}\cos\theta, \qquad Y_{11} = -\left(\frac{3}{8\pi}\right)^{1/2}\sin\theta e^{i\phi}$$

Hence the wave function of the given state can be written as

$$\psi = \left(-\sqrt{\frac{2}{3}}Y_{11} + \sqrt{\frac{1}{3}}Y_{10}\right)R(r)$$

(i) $$\int\psi^*\psi\,d\tau = \int_0^\infty |R(r)|^2 r^2\,dr \int_0^\pi\int_0^{2\pi}\left|\left(-\sqrt{\frac{2}{3}}Y_{11} + \frac{1}{\sqrt{3}}Y_{10}\right)\right|^2 \sin\theta\,d\theta\,d\phi$$

$$\left|\left(-\sqrt{\frac{2}{3}}Y_{11} + \frac{1}{\sqrt{3}}Y_{10}\right)\right|^2 = \left(-\sqrt{\frac{2}{3}}Y_{11} + \frac{1}{\sqrt{3}}Y_{10}\right)\left(-\sqrt{\frac{2}{3}}Y_{11}^* + \frac{1}{\sqrt{3}}Y_{10}\right)$$

$$= \frac{2}{3}|Y_{11}|^2 + \frac{1}{3}Y_{10}^2 - \frac{\sqrt{2}}{3}Y_{11}Y_{10} - \frac{\sqrt{2}}{3}Y_{10}Y_{11}^*$$

$$= \frac{1}{4\pi}(\sin^2\theta + \cos^2\theta) + \frac{1}{4\pi}\sin\theta + \cos\theta\,(e^{i\phi} + e^{-i\phi})$$

$$= \frac{1}{4\pi}(1 + \sin 2\theta\cos\phi)$$

Hence,

$$\int\psi^*\psi\,d\tau = \frac{1}{4\pi}\int_0^\pi\int_0^{2\pi}(1 + \sin 2\theta\cos\phi)\sin\theta\,d\theta\,d\phi$$

$$= \frac{1}{4\pi}\int_0^\pi\int_0^{2\pi}\sin\theta\,d\theta\,d\phi + \frac{1}{4\pi}\int_0^\pi\int_0^{2\pi}\sin 2\theta\sin\theta\cos\phi\,d\theta\,d\phi$$

As the ϕ-part of second integral vanishes,

$$\int \psi^* \psi \, d\tau = \frac{12\pi}{4\pi} \int_0^{\pi} \sin\theta \, d\theta = 1$$

Therefore, the wavefunction ψ is normalized.

(ii) The m_l value in Y_{11} is 1 and in Y_{10} it is zero. Hence the possible values in a measurement of L_z are $\hbar$ and zero.

(iii) The probabilty density $P = |\psi|^2$. Since the wavefunction is normalized, the probability of

$$L_z = 1\hbar = \left(\sqrt{\frac{2}{3}}\right)^2 = \frac{2}{3}$$

and that of

$$L_z = 0 = \left(\frac{1}{\sqrt{3}}\right)^2 = \frac{1}{3}$$

7.33 The rotational part of the Hamiltonian of a diatomic molecule is

$$\frac{1}{2I}(L_x^2 + L_y^2) + \frac{1}{I}L_z^2, I$$

which is moment of inertia. Find the energy eigenvalues and eigenfunctions.

Solution.

$$\text{Hamiltonian } H = \frac{1}{2I}(L_x^2 + L_y^2) + \frac{1}{I}L_z^2$$

$$= \frac{1}{2I}(L_x^2 + L_y^2 + L_z^2) + \frac{1}{2I}L_z^2 = \frac{1}{2I}L^2 + \frac{1}{2I}L_z^2$$

The eigenkets are the spherical harmonics. Hence energy E is obtained as

$$E = \langle H \rangle = \frac{1}{2I}(L_x^2 + L_y^2) + \frac{1}{I}L_z^2$$

$$= \frac{1}{2I}l(l+1)\hbar^2 + \frac{1}{2I}m^2\hbar^2$$

$$= \frac{\hbar^2}{2I}[l(l+1) + m^2] \quad \left\} \begin{array}{l} l = 0, 1, 2, \ldots \\ m = 0, \pm 1, \pm 2, \ldots, \pm l \end{array} \right.$$

7.34 The spin functions for a free electron in a basis in which S^2 and S_z are diagonal are $\begin{pmatrix} 1 \\ 0 \end{pmatrix}$ and $\begin{pmatrix} 0 \\ 1 \end{pmatrix}$, with S_z eigenvalues $\frac{1}{2}\hbar$ and $-\frac{1}{2}\hbar$, respectively. Using this basis, find the eigenvalues and normalized eigenkets of S_x and S_y.

Solution. We have

$$S_x = \frac{\hbar}{2}\begin{pmatrix} 0 & 1 \\ 1 & 0 \end{pmatrix}, \qquad S_y = \frac{\hbar}{2}\begin{pmatrix} 0 & -i \\ i & 0 \end{pmatrix}$$

In the diagonal representation of S^2 and S_z, the eigenvalue eigenket equation for S_x is

$$\frac{\hbar}{2}\begin{pmatrix} 0 & 1 \\ 1 & 0 \end{pmatrix}\begin{pmatrix} a_1 \\ a_2 \end{pmatrix} = \lambda \begin{pmatrix} a_1 \\ a_2 \end{pmatrix}$$

where λ is the eigenvalue. Simplifying, we get

$$\frac{\hbar}{2}\begin{pmatrix} a_2 \\ a_1 \end{pmatrix} = \lambda \begin{pmatrix} a_1 \\ a_2 \end{pmatrix}$$

$$\lambda a_1 = \frac{\hbar}{2} a_2, \qquad \lambda a_2 = \frac{\hbar}{2} a_1$$

$$\lambda a_1 = \frac{\hbar}{2}\frac{\hbar}{2}\frac{a_1}{\lambda} \quad \text{or} \quad \lambda^2 = \frac{\hbar^2}{4}$$

$$\lambda = \pm \frac{\hbar}{2}$$

With $+\hbar/2$ eigenvalue, the above equations become

$$\begin{pmatrix} a_2 \\ a_1 \end{pmatrix} = \begin{pmatrix} a_1 \\ a_2 \end{pmatrix} \quad \text{or} \quad a_1 = a_2$$

The normalization condition gives

$$a_1^2 + a_2^2 = 1 \quad \text{or} \quad 2a_1^2 = 1 \quad \text{or} \quad a_1 = a_2 = \frac{1}{\sqrt{2}}$$

Hence, the normalized eigenket corresponding to the eigenvalue $(1/2)\hbar$ is

$$\frac{1}{\sqrt{2}}\begin{pmatrix} 1 \\ 1 \end{pmatrix}$$

Similarly, the normalized eigenket corresponding to $-(1/2)\hbar$ eigenvalue is

$$\frac{1}{\sqrt{2}}\begin{pmatrix} 1 \\ -1 \end{pmatrix}$$

Proceeding on similar lines, the eigenvalues of S_y are $(1/2)\hbar$ or $-(1/2)\hbar$ and the eigenkets are

$$\frac{1}{\sqrt{2}}\begin{pmatrix} 1 \\ i \end{pmatrix} \quad \text{and} \quad \frac{1}{\sqrt{2}}\begin{pmatrix} 1 \\ -i \end{pmatrix}$$

respectively.

7.35 Consider a spin (1/2) particle of mass m with charge $-e$ in an external magnetic field $\boldsymbol{B}$.

(i) What is the Hamiltonian of the system?

(ii) If $\boldsymbol{S}$ is the spin angular momentum vector, show that

$$\frac{d\boldsymbol{S}}{dt} = -\frac{e}{m}(\boldsymbol{S} \times \boldsymbol{B})$$

Solution.

(i) The magnetic moment of the particle is

$$\mu = -\frac{e}{m}\,\mathbf{S}$$

The interaction energy E of the moment μ in an external magnetic field $\mathbf{B}$ is given by

$$E = -\mu \cdot \mathbf{B} = \frac{e}{m}\,\mathbf{S} \cdot \mathbf{B}$$

$$\text{Hamiltonian } H = \frac{e}{m}\,\mathbf{S} \cdot \mathbf{B}$$

(ii) In the Heisenberg picture,

$$\begin{aligned}\frac{d\mathbf{S}}{dt} &= \frac{1}{i\hbar}[\mathbf{S}, \mathbf{H}] = \frac{e}{i\hbar m}[\mathbf{S}, \mathbf{S} \cdot \mathbf{B}] \\ &= \frac{e}{i\hbar m}[\mathbf{S}, S_x B_x + S_y B_y + S_z B_z]\end{aligned}$$

The x-component of the commutator on RHS is

$$[S_x, \mathbf{S} \cdot \mathbf{B}] = [S_x, S_x B_x] + [S_x, S_y B_y] + [S_x, S_z B_z]$$

Since B_x, B_y and B_z are constants,

$$\begin{aligned}[S_x, \mathbf{S} \cdot \mathbf{B}] &= [S_x, S_x]B_x + [S_x, S_y]B_y + [S_x, S_z]B_z \\ &= 0 + i\hbar S_z B_y - i\hbar S_y B_z \\ &= -i\hbar(S_y B_z - S_z B_y) = -i\hbar(\mathbf{S} \times \mathbf{B})_x\end{aligned}$$

Similarly,

$$[S_y, \mathbf{S} \cdot \mathbf{B}] = -i\hbar(\mathbf{S} \times \mathbf{B})_y$$

$$[S_z, \mathbf{S} \cdot \mathbf{B}] = -i\hbar(\mathbf{S} \times \mathbf{B})_z$$

Substituting these values, we get

$$[\mathbf{S}, \mathbf{S} \cdot \mathbf{B}] = -i\hbar(\mathbf{S} \times \mathbf{B})$$

$$\frac{d\mathbf{S}}{dt} = -\frac{e}{m}(\mathbf{S} \times \mathbf{B})$$

7.36 The sum of two noninteracting angular momenta $\mathbf{J}_1$ and $\mathbf{J}_2$ is given by $\mathbf{J} = \mathbf{J}_1 + \mathbf{J}_2$. Prove the following: (i) $[J_x, J_y] = i\hbar J_z$; (ii) $[J^2, J_1^2] = [J^2, J_2^2] = 0$.

Solution.

(i) $[J_x, J_y] = [J_{1x} + J_{2x}, J_{1y} + J_{2y}] = [J_{1x}, J_{1y}] + [J_{2x}, J_{2y}] + [J_{1x}, J_{2y}] + [J_{2x}, J_{1y}]$

Since the two angular momenta are noninteracting, the third and the fourth terms are zero. Hence,

$$\begin{aligned}[J_x, J_y] &= i\hbar J_{1z} + i\hbar J_{2z} = i\hbar(J_{1z} + J_{2z}) \\ &= i\hbar J_z\end{aligned}$$

(ii) $[J^2, J_1^2] = [(J_1 + J_2)^2, J_1^2] = [J_1^2, J_1^2] + [J_2^2, J_1^2] + [J_1J_2, J_1^2] + [J_2J_1, J_1^2]$

Since $\boldsymbol{J}_1$ and $\boldsymbol{J}_2$ are noninteracting, all term, except the first are zero. The first term is zero since both are J_1^2 in the commutator. Hence,

$$[J^2, J_1^2] = 0$$

Similarly, $[J^2, J_2^2] = 0$

7.37 Consider two noninteracting systems having angular momenta $\boldsymbol{J}_1$ and $\boldsymbol{J}_2$ with eigenkets $|\,j_1m_1\rangle$ and $|\,j_2m_2\rangle$, respectively. The total angular momentum vector $\boldsymbol{J} = \boldsymbol{J}_1 + \boldsymbol{J}_2$. For given values of j_1 and j_2, the simultaneous eigenket of J^2, J_z, J_1^2, J_2^2 is $|\,jm\rangle$. Show that (i) $m = m_1 + m_2$; (ii) the permitted values of j are $(j_1 + j_2)$, $(j_1 + j_2 - 1)$, $(j_1 + j_2 - 2)$ ..., $|j_1 - j_2|$.

Solution.

(i) From Eq. (7.25), we have

$$|\,jm\rangle = \sum_{m_1, m_2} |\,m_1m_2\rangle\langle m_1m_2\,|\,jm\rangle \tag{i}$$

where $\langle m_1m_2\,|\,jm\rangle$ are the Clebsh-Gordan coefficients. Operating Eq. (i) from left by J_z, we get

$$J_z\,|\,jm\rangle = \sum_{m_1, m_2} (J_{1z} + J_{2z})\,|\,m_1m_2\rangle\langle m_1m_2\,|\,jm\rangle$$

$$m\hbar|\,jm\rangle = \sum_{m_1, m_2} (m_1 + m_2)\,\hbar\,|\,m_1m_2\rangle\langle m_1m_2\,|\,jm\rangle$$

Replacing $|jm\rangle$ on the LHS by Eq. (i) and rearranging, we obtain

$$\sum_{m_1, m_2} (m - m_1 - m_2)|\,m_1m_2\rangle\langle m_1m_2\,|\,jm\rangle = 0 \tag{ii}$$

Equaton (ii) will be valid only if the coefficient of each term vanishes separately, i.e.,

$$(m - m_1 - m_2) = 0 \quad \text{or} \quad m = m_1 + m_2$$

which is one of the rules of the vector atom model.

(ii) m_1 can have values from j_1 to $-j_1$ and m_2 from j_2 to $-j_2$ in integral steps. Hence, the possible values of m are $(j_1 + j_2)$, $(j_1 + j_2 - 1)$, $(j_1 + j_2 - 2)$, ..., $-(j_1 + j_2)$. The largest value of $m = (j_1 + j_2)$ can occur only when $m_1 = j_1$ and $m_2 = j_2$. The value of j corresponding to this value of m is also $(j_1 + j_2)$.

The next largest value of m is $j_1 + j_2 - 1$ which can occur in two ways: $m_1 = j_1$, $m_2 = j_2 - 1$ or $m_1 = j_1 - 1$, $m_2 = j_2$. We can have $m = j_1 + j_2 - 1$ when $j = j_1 + j_2$ or $j = j_1 + j_2 - 1$ as can be seen from the following. When $j = (j_1 + j_2)$, m can have the values $(j_1 + j_2)$, $(j_1 + j_2 - 1)$, ..., $-(j_1 + j_2)$, and when $(j_1 + j_2 - 1)$, $m = (j_1 + j_2 - 1)$, $(j_1 + j_2 - 2)$, ..., $-(j_1 + j_2 - 1)$. That is, $m = (j_1 + j_2 - 1)$ can result from $j = (j_1 + j_2)$ and from $j = (j_1 + j_2 - 1)$. This process is continued and the results are summarized in Table 7.1.

Table 7.1 Values of j and m for Different Values of m_1 and m_2

m_1	m_2	m	j
j_1	j_2	$j_1 + j_2$	$j_1 + j_2$
j_1	$j_2 - 1$		$j_1 + j_2$
$j_1 - 1$	j_2	$j_1 + j_2 - 1$	$j_1 + j_2 - 1$
j_1	$j_2 - 2$		$j_1 + j_2$
$j_1 - 1$	$j_2 - 1$	$j_1 + j_2 - 2$	$j_1 + j_2 - 1$
$j_1 - 2$	j_2		$j_1 + j_2 - 2$
$\vdots$	$\vdots$	$\vdots$	$\vdots$
j_1	$j_2 - k$		$j_1 + j_2$
$j_1 - 1$	$j_2 - k + 1$		$j_1 + j_2 - 1$
$j_1 - 2$	$j_2 - k + 2$	$j_1 + j_2 - k$	$j_1 + j_2 - 2$
$\vdots$	$\vdots$	$\vdots$	$\vdots$
$j_1 - k$	j_2		$j_1 + j_2 - k$
$\vdots$	$\vdots$		$\vdots$

The smallest value of j occurs for $j_1 - k = -j_1$ or $j_2 - k = -j_2$, i.e., when $k = 2j_1$ or $2j_2$. The smallest value of j is then $j_1 + j_2 - k = j_1 + j_2 - 2j_1 = j_2 - j_1$ or $j_1 + j_2 - 2j_2 = j_1 - j_2$. In other words, the permitted values of j are

$$(j_1 + j_2),\ (j_1 + j_2 - 1),\ (j_1 + j_2 - 2),\ \ldots,\ |j_1 - j_2|$$

7.38 Consider a system of two spin-half particles, in a state with total spin quantum number $S = 0$. Find the eigenvalue of the spin Hamiltonian $H = A\ \boldsymbol{S}_1 \cdot \boldsymbol{S}_2$, where A is a positive constant in this state.

Solution. The total spin angular momentum S of the two-spin system is given by

$$\boldsymbol{S} = \boldsymbol{S}_1 + \boldsymbol{S}_2$$

$$S^2 = S_1^2 + S_2^2 + 2\boldsymbol{S}_1 \cdot \boldsymbol{S}_2$$

$$\boldsymbol{S}_1 \cdot \boldsymbol{S}_2 = \frac{S^2 - S_1^2 - S_2^2}{2}$$

$$\text{Eigenvalue of } S_1^2 = \frac{1}{2} \times \frac{3}{2}\hbar^2 = \frac{3}{4}\hbar^2$$

$$\text{Eigenvalue of } S_2^2 = \frac{3}{4}\hbar^2$$

$$\text{Eigenvalue of } S^2 = 0$$

$$\text{Eigenvalue of } A\boldsymbol{S}_1 \cdot \boldsymbol{S}_2 = A\left[\frac{0 - (3/4)\hbar^2 - (3/4)\hbar^2}{2}\right] = -\frac{3}{4}A\hbar^2$$

7.39 Consider two noninteracting angular momenta $\boldsymbol{J}_1$ and $\boldsymbol{J}_2$ and their eigenkets $|j_1 m_1\rangle$ and $|j_2 m_2\rangle$. Their sum $\boldsymbol{J} = \boldsymbol{J}_1 + \boldsymbol{J}_2$. Derive the expressions used for the computation of the Clebsh-Gordan coefficients with $j_1 = 1/2$, $j_2 = 1/2$.

Solution. We shall first derive the expressions needed for the evaluation of the coefficients. In Problem 7.17, we derived the relation

$$J_-|jm\rangle = [j(j+1) - m(m-1)]^{1/2}\,\hbar\,|j, m-1\rangle \tag{i}$$

The Clebsh-Gordan coefficients $\langle m_1 m_2 | jm\rangle$ are given by

$$|jm\rangle = \sum_{m_1, m_2} |m_1 m_2\rangle\langle m_1 m_2 | jm\rangle \tag{ii}$$

Operating from left by J_-, we get

$$J_- |jm\rangle = \sum_{m_1', m_2'} (J_{1-} + J_{2-})|m_1' m_2'\rangle\langle m_1' m_2' | jm\rangle$$

Using Eq. (i) and remembering that $|m_1 m_2\rangle$ stands for $|j_1 j_2 m_1 m_2\rangle$, we obtain

$$[j(j+1) - m(m-1)]^{1/2}|j, m-1\rangle = \sum_{m_1', m_2'} [j_1(j_1+1) - m_1'(m_1'-1)]^{1/2}\hbar|m_1'-1, m_2'\rangle\langle m_1' m_2' | jm\rangle$$

$$+ \sum_{m_1', m_2'} [j_2(j_2+1) - m_2'(m_2'-1)]^{1/2}\hbar|m_1', m_2'-1\rangle\langle m_1' m_2' | jm\rangle$$

Operating from left by bra $\langle m_1 m_2|$, we get

$$[j(j+1) - m(m-1)]^{1/2}\langle m_1 m_2 | j, m-1\rangle = [j_1(j_1+1) - m_1(m_1+1)]^{1/2}\langle m_1+1, m_2 | jm\rangle$$

$$+ [j_2(j_2+1) - m_2(m_2+1)]^{1/2}\langle m_1, m_2+1 | jm\rangle \tag{iii}$$

Repeating the procedure with J_+ instead of J_-, we have

$$[j(j+1) - m(m+1)]^{1/2}\langle m_1 m_2 | j, m+1\rangle = [j_1(j_1+1) - m_1(m_1-1)]^{1/2}\langle m_1-1, m_2 | jm\rangle$$

$$+ [j_2(j_2+1) - m_2(m_2-1)]^{1/2}\langle m_1, m_2-1 | jm\rangle \tag{iv}$$

The Clebsh-Gordan coefficient matrix has $(2j_1 + 1)$ $(2j_2 + 1)$ rows and columns. For the $j_1 = 1/2$, $j_2 = 1/2$ case, this will be a 4×4 matrix. It breaks up into smaller matrices depending on the value of m. The first such matrix will be a 1×1 submatrix for which $m = j_1 + j_2$ and $j = j_1 + j_2$. Then we have a 2×2 submatrix for which $m = j_1 + j_2 - 1$ and $j = j_1 + j_2$ or $j = j_1 + j_2 - 1$ (refer Table 7.1). Obviously, next we get a 1×1 submatrix. For convenience, the first 1×1 submatrix is selected as +1, i.e., the Clebsh-Gordan coefficient

$$\langle j_1, j_2 | j_1 + j_2, j_1 + j_2\rangle = 1 \tag{v}$$

To compute the 2×2 submatrix, set $m_1 = j_1$, $m_2 = j_2 - 1$, $j = j_1 + j_2$ and $m = j_1 + j_2$ in Eq. (iii). On simplification we get

$$(j_1 + j_2)^{1/2}\langle j_1, j_2 - 1 | j_1 + j_2, j_1 + j_2 - 1\rangle = j_2^{1/2}\langle j_1 j_2 | j_1 + j_2, j_1 + j_2\rangle$$

Using Eq. (v), we obtain

$$\langle j_1, j_2 - 1 | j_1 + j_2, j_1 + j_2 - 1\rangle = \left(\frac{j_1}{j_1 + j_2}\right)^{1/2} \tag{vi}$$

Proceeding on similar lines with $m_1 = j_1 - 1$, $m_2 = j_2$, $j = j_1 + j_2$ and $m = j_1 + j_2$, we get

$$\langle j_1 - 1, j_2 | j_1 + j_2, j_1 + j_2 - 1\rangle = \left(\frac{j_1}{j_1 + j_2}\right)^{1/2} \tag{vii}$$

Using the unitary character of the Clebsh-Gordan coefficient, the condition

$$\langle jm \,|\, m_1 m_2 \rangle = \langle m_1 m_2 \,|\, jm \rangle^*$$

and Eqs. (vi) and (vii), we can obtain

$$\langle j_1,\, j_2 - 1 \,|\, j_1 + j_2 - 1,\, j_1 + j_2 - 1 \rangle = \left(\frac{j_1}{j_1 + j_2} \right)^{1/2} \tag{viii}$$

$$\langle j_1 - 1,\, j_2 \,|\, j_1 + j_2 - 1,\, j_1 + j_2 - 1 \rangle = -\left(\frac{j_2}{j_1 + j_2} \right)^{1/2} \tag{ix}$$

The results are summarized in Table 7.2.

Table 7.2 Clebsh-Gordan Coefficients for $|m_1 m_1\rangle = |j_1, j_2 - 1\rangle$ and $|j_1 - 1, j_2\rangle$

m_1	m_2	$\lvert jm\rangle$	
		$\lvert j_1 + j_2, j_1 + j_2 - 1\rangle$	$\lvert j_1 + j_2 - 1, j_1 + j_2 - 1\rangle$
j_1	$j_2 - 1$	$\left(\frac{j_2}{j_1 + j_2} \right)^{1/2}$	$\left(\frac{j_1}{j_1 + j_2} \right)^{1/2}$
$j_1 - 1$	j_2	$\left(\frac{j_1}{j_1 + j_2} \right)^{1/2}$	$-\left(\frac{j_2}{j_1 + j_2} \right)^{1/2}$

7.40 Evaluate the Clebsh-Gordan coefficients for a system having $j_1 = 1/2$ and $j_2 = 1/2$.

Solution. The allowed values of j are 1, 0. For $j = 1$, $m = 1, 0, -1$ and for $j = 0$, $m = 0$. The number of eigenstates is 4. The 4×4 matrix reduces to two 1×1 and one 2×2 matrices, details of which are given in Table 7.2. The values of the elements $\langle 1/2, 1/2 \,|\, 1, 1\rangle$ and $\langle -1/2, -1/2 \,|\, 1, -1\rangle$ are unity. The elements $\langle 1/2, -1/2 \,|\, 1, 0\rangle$, $\langle 1/2, -1/2 \,|\, 0, 0\rangle$, $\langle -1/2, 1/2 \,|\, 1, 0\rangle$ and $\langle -1/2, 1/2 \,|\, 0, 0\rangle$ are easily evaluated with the help of Table 7.2. All the Clebsh-Gordan coefficients are listed in Table 7.3.

Table 7.3 Clebsh-Gordan Coefficients for $j_1 = 1/2$, $j_2 = 1/2$

j \ m		1	1	0	1
m_1	m_2	1	0	0	-1
1/2	1/2	1	0	0	0
1/2	$-1/2$	0	$\sqrt{1/2}$	$\sqrt{1/2}$	0
$-1/2$	1/2	0	$\sqrt{1/2}$	$-\sqrt{1/2}$	0
$-1/2$	$-1/2$	0	0	0	1

7.41 Obtain the Clebsh-Gordan coefficients for a system having $j_1 = 1$ and $j_2 = 1/2$.

Solution. The system has two angular momenta with $j_1 = 1$ and $j_2 = 1/2$. The allowed values of j are 3/2 and 1/2. For $j = 3/2$, $m = 3/2, 1/2, -1/2, -3/2$ and for $j = 1/2$, $m = 1/2$ and $-1/2$. The number of $|jm\rangle$ eigenstates is thus six, and the 6×6 matrix reduces to two 1×1 and two 2×2 matrices, details of which are given in Table 7.4. The elements $\langle 1, 1/2 \mid 3/2, 3/2\rangle$, $\langle 1, -1/2 \mid 3/2, 1/2\rangle$, $\langle 0, 1/2 \mid 3/2, 1/2\rangle$, $\langle 1, -1/2 \mid 1/2, 1/2\rangle$ and $\langle 0, 1/2 \mid 1/2, 1/2\rangle$ are easily evaluated (refer Problem 7.39) and are listed in Table 7.4. Evaluation of the remaining elements is done as detailed now.

Table 7.4 Clebsh-Gordan Coefficients for $j_1 = 1$ and $j_2 = 1/2$

m_1	m_2	$\left\|\frac{3}{2}, \frac{3}{2}\right\rangle$	$\left\|\frac{3}{2}, \frac{1}{2}\right\rangle$	$\left\|\frac{1}{2}, \frac{1}{2}\right\rangle$	$\left\|\frac{3}{2}, \frac{-1}{2}\right\rangle$	$\left\|\frac{1}{2}, \frac{-1}{2}\right\rangle$	$\left\|\frac{3}{2}, \frac{-3}{2}\right\rangle$
1	$\frac{1}{2}$	1					
1	$-\frac{1}{2}$		$\frac{1}{\sqrt{3}}$	$\sqrt{\frac{2}{3}}$			
0	$\frac{1}{2}$		$\sqrt{\frac{2}{3}}$	$-\frac{1}{\sqrt{3}}$			
0	$-\frac{1}{2}$				$\sqrt{\frac{2}{3}}$	$\frac{1}{\sqrt{3}}$	
−1	$\frac{1}{2}$				$\frac{1}{\sqrt{3}}$	$-\sqrt{\frac{2}{3}}$	
−1	$-\frac{1}{2}$						1

$\langle 0, -1/2 \mid 3/2, -1/2\rangle$:

Setting $j = 3/2$, $m = 1/2$, $m_1 = 0$ and $m_2 = -1/2$ in Eq. (iii) of Problem 7.39, we get

$$2\langle 0, -1/2 \mid 3/2, -1/2\rangle = 2^{1/2}\langle 1, -1/2 \mid 3/2, 1/2\rangle + \langle 0, 1/2 \mid 3/2, 1/2\rangle$$

Substituting the two coefficients on RHS from Table 7.4, we obtain

$$\langle 0, -1/2 \mid 3/2, -1/2\rangle = \sqrt{2/3}$$

$\langle -1, 1/2 \mid 3/2, -1/2\rangle$:

Setting $j = 3/2$, $m = 1/2$, $m_1 = -1$ and $m_2 = 1/2$ in Eq. (iii) of Problem 7.39 and proceeding as in the previous case, we get

$$2\langle -1, 1/2 \mid 3/2, -1/2\rangle = 2^{1/2}\langle 0, 1/2 \mid 3/2, 1/2\rangle$$

$$\langle -1, 1/2 \mid 3/2, -1/2\rangle = 1/\sqrt{3}.$$

$\langle 0, 1/2 \mid 1/2, -1/2\rangle$:

Setting $j = 1/2$, $m = 1/2$, $m_1 = 0$, $m_2 = -1/2$ in Eq. (iii) of Problem 7.39, we obtain the value as $1/\sqrt{3}$.

$\langle -1, 1/2 \mid 1/2, -1/2\rangle$:

Again, by setting $j = 1/2$, $m = 1/2$, $m_1 = -1$, $m_2 = 1/2$ in Eq. (iii) of Problem 7.39, we get the value as $-\sqrt{2/3}$.

Obviously, the last element $\langle -1, -1/2 \mid 3/2, -3/2\rangle = 1$.

7.42 Obtain the matrix of Clebsh-Gordan coefficients for $j_1 = 1$ and $j_2 = 1$.

Solution. The nonvanishing Clebsh-Gordan coefficients can be evaluated with the help of Tables 7.2 and 7.5. These coefficients are

$$\langle 1, 1 | 2, 2\rangle = \langle -1, -1 | 2, -2\rangle = 1$$

$$\langle 1, 0 | 2, 1\rangle = \langle 1, 0 | 1, 1\rangle = \langle 0, 1 | 2, 1\rangle = \langle 0, 1 | 2, -1\rangle = \langle 1, -1 | 1, -1\rangle$$

$$= \langle -1, 0 | 2-1\rangle = \langle 1, -1 | 1, 0\rangle = 1/\sqrt{2}$$

$$\langle 0, 1 | 1, 1\rangle = \langle -1, 1 | 1, 0\rangle = \langle -1, 0 | 1, -1\rangle = -1/\sqrt{2}$$

$$\langle 1, -1 | 2, 0\rangle = \langle -1, 1 | 2, 0\rangle = 1/\sqrt{6}$$

$$\langle 1, -1 | 0, 0\rangle = \langle -1, 1 | 0, 0\rangle = 1/\sqrt{3}$$

$$\langle 0, 0 | 2, 0\rangle = \sqrt{2/3}; \quad \langle 0, 0 | 0, 0\rangle = -1/\sqrt{3}\,; \quad \langle 0, 0 | 1, 0\rangle = 0$$

Table 7.5 Clebsh-Gordan Coefficients for $|m_1 m_2\rangle = |j_1, j_2 - 2\rangle$, $|j_1 - 1, j_2 - 1\rangle$ and $|j_1 - 2, j_2\rangle$

m_1	m_2	$\vert jm\rangle$		
		$\vert j_1 + j_2, j_1 + j_2 - 2\rangle$	$\vert j_1 + j_2 - 1, j_1 + j_2 - 2\rangle$	$\vert j_1 + j_2 - 2, j_1 + j_2 - 2\rangle$
j_1	$j_2 - 2$	$\left[\dfrac{j_2(2j_2 - 1)}{(j_1 + j_2)A}\right]^{1/2}$	$\left[\dfrac{j_1(2j_2 - 1)}{(j_1 + j_2)B}\right]^{1/2}$	$\left[\dfrac{j_1(2j_1 - 1)}{AB}\right]^{1/2}$
$j_1 - 2$	$j_2 - 1$	$\left[\dfrac{4j_1j_2}{(j_1 + j_2)A}\right]^{1/2}$	$\dfrac{j_1 - j_2}{[(j_1 + j_2)B]^{1/2}}$	$-\left[\dfrac{(2j_1 - 1)(2j_2 - 1)}{AB}\right]^{1/2}$
$j_1 - 2$	j_2	$\left[\dfrac{j_1(2j_1 - 1)}{(j_1 + j_2)A}\right]^{1/2}$	$-\left[\dfrac{j_2(2j_1 - 1)}{(j_1 + j_2)B}\right]^{1/2}$	$\left[\dfrac{j_2(2j_2 - 1)}{AB}\right]^{1/2}$

$A = 2j_1 + 2j_2 - 1, \qquad B = j_1 + j_2 - 1$

7.43 An electron is in a state described by the wave function

$$\psi = \frac{1}{\sqrt{4\pi}} (\cos\theta + e^{-i\phi}\sin\theta) R(r), \qquad \int_0^\infty |R(r)|^2 r^2\, dr = 1$$

where θ and ϕ are, respectively, the polar and azimuth angles: (i) What are the possible values of L_z? (ii) What is the probability of obtaining each of the possible values of L_z?

Solution.

(i) From Table 5.2 we have

$$Y_{10} = \left(\frac{3}{4\pi}\right)^{1/2} \cos\theta, \qquad Y_{1,-1} = \left(\frac{3}{8\pi}\right)^{1/2} \sin\theta\, e^{-i\phi}$$

Hence the given wave function can be written as

$$\psi = \frac{1}{\sqrt{3}} (Y_{10} + \sqrt{2} Y_{1,-1})\, R(r)$$

The possible values of L_z are 0 and $\hbar$.

(ii) $\int |\psi|^2\, d\tau = \frac{1}{3}\int |R(r)|^2\, |(Y_{10} + \sqrt{2}Y_{1,-1})|^2\, r^2 \sin\theta\, d\theta\, d\phi\, dr$

$$|(Y_{10} + \sqrt{2}Y_{1,-1})|^2 = (Y_{10} + \sqrt{2}Y_{1,-1})^* (Y_{10} + \sqrt{2}Y_{1,-1})$$

$$= Y_{10}^* Y_{10} + 2Y_{1,-1}^* Y_{1,-1} + \sqrt{2}(Y_{10}^* Y_{1,-1} + Y_{1,-1}^* Y_{10})$$

$$= \frac{3}{4\pi}(\cos^2\theta + \sin^2\theta) + \frac{3}{4\pi}\cos\theta\,\sin\theta\,(e^{-i\phi} + e^{i\phi})$$

$$= \frac{3}{4\pi}(1 + \sin 2\theta\,\cos\phi)$$

$$\int |\psi|^2\, d\tau = \frac{1}{4\pi}\int_0^\infty |R(r)|^2 r^2\, dr \int_0^\pi \sin\theta\, d\theta \int_0^{2\pi} (1 + \sin 2\theta\cos\phi)\, d\phi$$

$$= \frac{1}{2}\int_0^\pi \sin\theta\, d\theta = 1$$

i.e., the given wave function is normalized. The probability density is then $P = |\psi|^2$. Hence, the probability of obtaining $L_z = 0$ is $(1/\sqrt{3})^2 = 1/3$. The probability of obtaining $L_z = -1\hbar$ is $(\sqrt{2/3})^2 = 2/3$.

7.44 An operator P describing the interaction of two spin-half particles is $P = a + b\sigma_1 \cdot \sigma_2$, where a, b are constants, with σ_1 and σ_2 being the Pauli matrices of the two spins. The total spin angular momentum $\mathbf{S} = \mathbf{S}_1 + \mathbf{S}_2 = (1/2)\hbar\,(\sigma_1 + \sigma_2)$. Show that P, S^2 and S_z can be measured simultaneously.

Solution. P, S^2 and S_z can be measured simultaneously if

$$[P, S^2] = [P, S_z] = [S^2, S_z] = 0$$

We know that $[S^2, S_z] = 0$. From the definition

$$S^2 = \frac{\hbar^2}{4}(\sigma_1^2 + \sigma_1^2 + 2\sigma_1 \cdot \sigma_2)$$

we have

$$\sigma_1 \cdot \sigma_2 = \frac{2S^2}{\hbar^2} - \frac{1}{2}(\sigma_1^2 + \sigma_2^2)$$

Since for each particle,

$$\sigma^2 = \sigma_x^2 + \sigma_y^2 + \sigma_z^2 = 3I$$

where I is the unit matrix, we have

$$\frac{1}{2}(\sigma_1^2 + \sigma_2^2) = \frac{1}{2}(3I + 3I) = 3I$$

Hence,

$$\sigma_1 \cdot \sigma_1 = \frac{2S^2}{\hbar^2} - 3I$$

$$[S^2, P] = [S^2, a] + b[S^2, \boldsymbol{\sigma}_1 \cdot \boldsymbol{\sigma}_2] = b\left[S^2, \frac{2S^2}{\hbar^2} - 3I\right]$$

$$= b\left[S^2, \frac{2S^2}{\hbar^2}\right] - b[S^2, 3I] = 0$$

$$[S_z, P] = [S_z, a] + b\left[S_z, \frac{2S^2}{\hbar^2} - 3I\right] = 0$$

Since S^2 and S_z commute with P, all the three can be measured simultaneously.

7.45 Obtain the Hamiltonian operator for a free electron having magnetic moment $\boldsymbol{\mu}$ in an external magnetic field B_z in the z-direction in the electron's reference frame. If another constant magnetic field B_y is applied in the y-direction, obtain the time rate of change of $\boldsymbol{\mu}$ in the Heisenberg picture.

Solution. The magnetic moment of the electron is given by

$$\boldsymbol{\mu} = -\frac{e}{m}\boldsymbol{S} = -\frac{e\hbar}{2m}\boldsymbol{\sigma} = -\mu_B \boldsymbol{\sigma}$$

where $\boldsymbol{S} = 1/2\ \hbar\boldsymbol{\sigma}$ and μ_B is the Bohr magneton. The Hamiltonian

$$H' = -\boldsymbol{\mu} \cdot \boldsymbol{B} = -\mu_z B_z = \mu_B \sigma_z B_z$$

With the total magnetic field applied $B = B_y\hat{y} + B_z\hat{z}$, the total Hamiltonian

$$H = \mu_B\,(\sigma_z B_z + \sigma_y B_y)$$

From Eq. (3.30),

$$\frac{d\boldsymbol{\mu}}{dt} = \frac{1}{i\hbar}[\boldsymbol{\mu}, H] = \frac{1}{i\hbar}[-\mu_B\boldsymbol{\sigma}, \mu_B(\sigma_z B_z + \sigma_y B_y)]$$

$$= -\frac{\mu_B^2}{i\hbar}\,[\sigma_x\hat{x} + \sigma_y\hat{y} + \sigma_z\hat{z}, \sigma_z B_z + \sigma_y B_y]$$

$$= -\frac{\mu_B^2}{i\hbar}\,[\sigma_x, \sigma_z]\,B_z\hat{x} + [\sigma_x, \sigma_y]\,B_y\hat{x} + [\sigma_y, \sigma_z]\,B_z\hat{y}$$

$$+ [\sigma_y, \sigma_y]\,B_y\hat{y} + [\sigma_z, \sigma_z]\,B_z\hat{z} + [\sigma_z, \sigma_y]\,B_y\hat{z}$$

Using the commutation relations among σ_x, σ_y, σ_z, we get

$$\frac{d\boldsymbol{\mu}}{dt} = \frac{i}{\hbar}\mu_B^2\,[-2i\sigma_y B_z\hat{x} + 2i\sigma_z B_y\hat{x} + 2i\sigma_x B_z\hat{y} - 2i\sigma_x B_y\hat{z}]$$

$$= \frac{2}{\hbar}\mu_B^2\,[(\sigma_y B_z - \sigma_z B_y)\,\hat{x} - \sigma_x B_z\hat{y} + \sigma_x B_y\hat{z}]$$

$$= \frac{2}{\hbar}\mu_B^2\,[\boldsymbol{\sigma} \times \boldsymbol{B}] = -\frac{2}{\hbar}\mu_B^2\,[\boldsymbol{B} \times \boldsymbol{\sigma}]$$

$$= \frac{e}{m}\,[\boldsymbol{B} \times \boldsymbol{\mu}]$$

which is the time rate of change of the magnetic moment.

7.46 Obtain the energy levels of a symmetric top molecule with principal moments of inertia $I_1 = I_2 = I \neq I_3$.

Solution. Let (x, y, z) be the coordinates of a body-fixed coordinate system. The Hamiltonian

$$H = \frac{1}{2}\left(\frac{L_x^2}{I_1} + \frac{L_y^2}{I_2} + \frac{L_z^2}{I_3}\right) = \frac{1}{2I}(L_x^2 + L_y^2) + \frac{1}{2I_3}L_z^2$$

$$= \frac{1}{2I}L^2 + \frac{1}{2}\left(\frac{1}{I_3} - \frac{1}{I}\right)L_z^2$$

$|lm\rangle$ are the simultaneous eigenkets of L^2 and L_z. The Schrödinger equation is

$$\left[\frac{1}{2I}L^2 + \frac{1}{2}\left(\frac{1}{I_3} - \frac{1}{I}\right)L_z^2\right]|lm\rangle = E|lm\rangle$$

$$E_{lm} = \frac{\hbar^2}{2I}\,l(l+1) + \frac{\hbar^2}{2}\left(\frac{1}{I_3} - \frac{1}{I}\right)m^2$$

which is the energy equation for symmetric top. This energy equation can be expressed in the familiar form by writing

$$\frac{\hbar^2}{2I} = B, \qquad \frac{\hbar^2}{2I_3} = C$$

$$E_{lm} = Bl(l+1) + (C - B)\,m^2$$

The constants B and C are rotational constants.

$$l = 0, 1, 2, \ldots; \qquad m = 0, \pm 1, \pm 2, \ldots, \pm l$$

7.47 The kets $|j, m\rangle$ are the simultaneous eigenkets of J^2 and J_z. Show that $|j, m\rangle$ are also eigenkets of $[J_x, J_+]$ and of $[J_y, J_+]$. Find the eigenvalues of each of these commutators.

Solution. Operating $[J_x, J_+]$ on the eigenkets $|jm\rangle$, we obtain

$$[J_x, J_+]\,|jm\rangle = J_x J_+|jm\rangle - J_+ J_x|jm\rangle$$

$$= \frac{1}{2}(J_+ + J_-)J_+|jm\rangle - J_+\frac{1}{2}(J_+ + J_-)|jm\rangle$$

$$= \frac{1}{2}J_+J_+|jm\rangle - \frac{1}{2}J_-J_+|jm\rangle - \frac{1}{2}J_+J_+|jm\rangle - \frac{1}{2}J_+J_-|jm\rangle$$

$$= \frac{1}{2}J_-J_+|jm\rangle - \frac{1}{2}J_+J_-|jm\rangle$$

From Problem 7.14,

$$J_-J_+ = J^2 - J_z^2 - \hbar J_z, \qquad J_+J_- = J^2 - J_z^2 + \hbar J_z$$

Hence,

$$[J_x, J_+]|jm\rangle = \frac{1}{2}(J^2 - J_z^2 - \hbar J_z)|jm\rangle - \frac{1}{2}(J^2 - J_z^2 + \hbar J_z)|jm\rangle$$

$$= -\hbar J_z\,|jm\rangle = -m\hbar^2\,|jm\rangle$$

i.e., $|jm\rangle$ are eigenkets of $[J_x, J_+]$ with eigenvalues $-m\hbar^2$. Now,

$$[J_y, J_+]|jm\rangle = (J_yJ_+ - J_+J_y)|jm\rangle$$

$$= \frac{1}{2i}(J_+ - J_-)J_+|jm\rangle - \frac{1}{2i}J_+(J_+ - J_-)|jm\rangle$$

$$= -\frac{1}{2i}J_-J_+|jm\rangle + \frac{1}{2i}J_+J_-|jm\rangle$$

$$= -\frac{1}{2i}(J^2 - J_z^2 - \hbar J_z)|jm\rangle + \frac{1}{2i}(J^2 - J_z^2 + \hbar J_z)|jm\rangle$$

$$= \frac{1}{i}\hbar J_z)|jm\rangle = \frac{1}{i}m\hbar^2|jm\rangle$$

$$= -im\hbar^2|jm\rangle$$

That is, $|jm\rangle$ are eigenkets of the commutator $[J_y, J_+]$ with the eigenvalue $-im\hbar^2$.

7.48 The state of the hydrogen atom is 2p state. Find the energy levels of the spin-orbit interaction Hamiltonian $A\boldsymbol{L}\cdot\boldsymbol{S}$, where A is a constant.

Solution. The 2p state means $s = 1/2$, $l = 1$ and j = 1 + (1/2) = (3/2) or 1 – (1/2) = (1/2). The total angular momentum

$$\boldsymbol{J} = \boldsymbol{L} + \boldsymbol{S} \tag{i}$$

$$J^2 = L^2 + S^2 + 2\boldsymbol{L}\cdot\boldsymbol{S}$$

$$H_{so} = A\boldsymbol{L}\cdot\boldsymbol{S} = \frac{A}{2}(J^2 - L^2 - S^2) \tag{ii}$$

The eigenvector associated with the variable J^2, J_z, L^2, S_2 be $|jmls\rangle$. In this space,

$$J^2|jmls\rangle = j(j+1)\hbar^2|jmls\rangle \tag{iii}$$

$$S^2|jmls\rangle = s(s+1)\hbar^2|jmls\rangle \tag{iv}$$

$$L^2|jmls\rangle = l(l+1)\hbar^2|jmls\rangle \tag{v}$$

Using Eqs. (ii)–(v), the energy eigenvalue of H_{so} is given by

$$j = \frac{3}{2}: E_{so} = \frac{A}{2}\left[\frac{15}{4}\hbar^2 - 2\hbar^2 - \frac{3}{4}\hbar^2\right]$$

$$= \frac{A}{2}\hbar^2$$

$$j = \frac{1}{2}: E_{so} = \frac{A}{2}\left[\frac{3}{4}\hbar^2 - 2\hbar^2 - \frac{3}{4}\hbar^2\right]$$

$$= -A\hbar^2$$

7.49 The Hamiltonian of a system of 3 nonidentical spin-half particles is

$$H = A\mathbf{S}_1 \cdot \mathbf{S}_2 - B(\mathbf{S}_1 + \mathbf{S}_2) \cdot \mathbf{S}_3$$

where A and B are constants are $\mathbf{S}_1$, $\mathbf{S}_2$ and $\mathbf{S}_3$ are the spin angular momentum operators. Find their energy levels and their degeneracies.

Solution. Writing $\mathbf{S} = \mathbf{S}_1 + \mathbf{S}_2 + \mathbf{S}_3$ and $\mathbf{S}_{12} = \mathbf{S}_1 + \mathbf{S}_2$, we have

$$\mathbf{S}^2 = \mathbf{S}_{12}^2 + \mathbf{S}_3^2 + 2\mathbf{S}_{12} \cdot \mathbf{S}_3$$

$$\mathbf{S}_{12} \cdot \mathbf{S}_3 = \frac{1}{2}(\mathbf{S}^2 - \mathbf{S}_{12}^2 - \mathbf{S}_3^2)$$

Similarly,

$$\mathbf{S}_1 \cdot \mathbf{S}_2 = \frac{1}{2}(\mathbf{S}_{12}^2 - \mathbf{S}_1^2 - \mathbf{S}_2^2)$$

since $S_1 = 1/2$ and $S_2 = 1/2$, the possible values of the quantum number $S_{12} = 0$ and 1. When $S_{12} = 0$, the possible values of $S = 1/2$ and 1/3. The Hamiltonian

$$H = A\mathbf{S}_1 \cdot \mathbf{S}_2 - B(\mathbf{S}_1 + \mathbf{S}_2) \cdot \mathbf{S}_3$$

$$= \frac{A}{2}(\mathbf{S}_{12}^2 - \mathbf{S}_1^2 - \mathbf{S}_2^2) + \frac{B}{2}(\mathbf{S}^2 - \mathbf{S}_{12}^2 - \mathbf{S}_3^2)$$

In the basis $|SM_sS_{12}S_3\rangle$,

$$H|SM_sS_{12}S_3\rangle = \frac{A}{2}(\mathbf{S}_{12}^2 - \mathbf{S}_1^2 - \mathbf{S}_2^2)|SM_sS_{12}S_3\rangle + \frac{B}{2}(\mathbf{S}^2 - \mathbf{S}_{12}^2 - \mathbf{S}_3^2)|SM_sS_{12}S_3\rangle$$

The energy is then,

$$E = \frac{A}{2}\hbar^2[S_{12}(S_{12} + 1) - S_1(S_1 + 1) - S_2(S_2 + 1)]$$

$$+ \frac{B}{2}\hbar^2[S(S + 1) - S_{12}(S_{12} + 1) - S_3(S_3 + 1)]$$

since $S_1 = S_2 = S_3 = 1/2$. Now,

$$E_{S_{12},S} = \frac{A}{2}\hbar^2\left[S_{12}(S_{12} + 1) - \frac{3}{2}\right] + \frac{B}{2}\hbar^2\left[S(S + 1) - S_{12}(S_{12} + 1) - \frac{3}{4}\right]$$

As $S = 1/2$ when $S_{12} = 0$,

$$E_{0,1/2} = -\frac{3}{4}\hbar^2$$

which is $2S + 1 = 2$-fold degenerate. As $S = 1/2$ and $3/2$, when $S_{12} = 1$,

$$E_{1,1/2} = \frac{A}{4}\hbar^2\left(2 - \frac{3}{2}\right) + \frac{B}{4}\hbar^2\left(\frac{3}{4} - 2 - \frac{3}{4}\right)$$

$$= \frac{A}{4}\hbar^2 - B\hbar^2 = \left(\frac{A}{4} - B\right)\hbar^2$$

which is $2S + 1 = 2$-fold degenerate. We also have

$$E_{1,3/2} = \frac{A}{2}\hbar^2\left(2 - \frac{3}{2}\right) + \frac{B}{2}\hbar^2\left(\frac{15}{4} - 2 - \frac{3}{4}\right)$$

$$= \left(\frac{A}{4} + \frac{B}{2}\right)\hbar^2$$

which is four-fold degenerate.

7.50 Two electrons having spin angular momentum vectors $\boldsymbol{S}_1$ and $\boldsymbol{S}_2$ have an interaction of the type

$$H = A(\boldsymbol{S}_1 \cdot \boldsymbol{S}_2 - 3S_{1z}S_{2z}), \quad A \text{ being constant}$$

Express it in terms of $\boldsymbol{S} = \boldsymbol{S}_1 + \boldsymbol{S}_2$ and obtain its eigenvalues.

Solution. The sum of the angular momenta $\boldsymbol{S}_1$ and $\boldsymbol{S}_2$ is

$$\boldsymbol{S} = \boldsymbol{S}_1 + \boldsymbol{S}_2 \tag{i}$$

$$\boldsymbol{S}^2 = \boldsymbol{S}_1^2 + \boldsymbol{S}_2^2 + 2\boldsymbol{S}_1\boldsymbol{S}_2$$

$$\boldsymbol{S}_1 \cdot \boldsymbol{S}_2 = \frac{1}{2}(S^2 - S_1^2 - S_2^2) \tag{ii}$$

From Eq. (i),

$$S_z = S_{1z} + S_{2z}$$

$$S_z^2 = (S_{1z} + S_{2z})^2 = S_{1z}^2 + S_{2z}^2 + 2S_{1z}S_{2z}$$

$$S_{1z}S_{2z} = \frac{1}{2}(S_z^2 - S_{1z}^2 - S_{2z}^2) \tag{iii}$$

Hence,

$$\boldsymbol{S}_1 \cdot \boldsymbol{S}_2 - 3S_{1z}S_{2z} = \frac{1}{2}(S^2 - S_1^2 - S_2^2) - \frac{3}{2}(S_z^2 - S_{1z}^2 - S_{2z}^2) \tag{iv}$$

In the simultaneous eigenkets $|SM\rangle$ of S^2 and S_z,

$A(\boldsymbol{S}_1 \cdot \boldsymbol{S}_2 - 3S_{1z}S_{2z})\,|SM\rangle$

$$= \frac{A}{2}(S^2 - S_1^2 - S_2^2)\,|SM\rangle - \frac{3A}{2}(S_z^2 - S_{1z}^2 - S_{2z}^2)\,|SM\rangle$$

$$= \frac{A}{2}\left[S(S+1) - \frac{1}{2}\times\frac{3}{2} - \frac{1}{2}\times\frac{3}{2}\right]\hbar^2\,|SM\rangle - \frac{3A}{2}\left(M^2 - \frac{1}{4} - \frac{1}{4}\right)\hbar^2\,|SM\rangle$$

$$= \frac{A}{2}[S(S+1) - 3M^2]\,\hbar^2\,|SM\rangle \tag{v}$$

Since $\boldsymbol{S} = \boldsymbol{S}_1 + \boldsymbol{S}_2$, the quantum number S can have the values $\frac{1}{2} + \frac{1}{2} = 1$ or $\frac{1}{2} - \frac{1}{2} = 0$. When $S = 0$, $M = 0$ and when $S = 1$, $M = 1, 0, -1$. The eigenkets and the corresponding eigenvalues, see Eq. (v), are as follows:

$\|SM\rangle$	Eigenvalues
$\|0\ 0\rangle$	0
$\|1\ 1\rangle$	$-\frac{1}{2}A\hbar^2$
$\|1\ 0\rangle$	$1\ A\hbar^2$
$\|1, -1\rangle$	$-\frac{1}{2}A\hbar^2$

7.51 The wave function $\psi = c_1\psi_{n_1l_1m_1} + c_2\psi_{n_2l_2m_2}$ is a combination of the normalized stationary state wave functions ψ_{nlm}. For ψ to be normalized, show that c_1 and c_2 must satisfy $|c_1|^2 + |c_2|^2 = 1$. Calculate the expectation values of L^2 and L_z.

Solution. Let us evaluate the value of

$$\begin{aligned}\langle\psi|\psi\rangle &= \langle(c_1\psi_{n_1l_1m_1} + c_2\psi_{n_2l_2m_2})|(c_1\psi_{n_1l_1m_1} + c_2\psi_{n_2l_2m_2})\rangle \\ &= |c_1|^2\langle\psi_{n_1l_1m_1}|\psi_{n_1l_1m_1}\rangle + |c_2|^2\langle\psi_{n_2l_2m_2}|\psi_{n_2l_2m_2}\rangle \\ &= |c_1|^2 + |c_2|^2\end{aligned}$$

For ψ to be normalized, it is necessary that

$$\langle\psi|\psi\rangle = |c_1|^2 + |c_2|^2 = 1$$

The expectation value of L^2 is

$$\begin{aligned}\langle\psi|L^2|\psi\rangle &= \langle(c_1\psi_{n_1l_1m_1} + c_2\psi_{n_2l_2m_2})|L^2|(c_1\psi_{n_1l_1m_1} + c_2\psi_{n_2l_2m_2})\rangle \\ &= |c_1|^2\langle\psi_{n_1l_1m_1}|L^2|\psi_{n_1l_1m_1}\rangle + |c_2|^2\langle\psi_{n_2l_2m_2}|L^2|\psi_{n_2l_2m_2}\rangle \\ &= |c_1|^2 l_1(l_1+1)\hbar^2 + |c_2|^2 l_2(l_2+1)\hbar^2\end{aligned}$$

The expectation value of L_z is

$$\begin{aligned}\langle\psi|L_z|\psi\rangle &= \langle(c_1\psi_{n_1l_1m_1} + c_2\psi_{n_2l_2m_2})|L_z|(c_1\psi_{n_1l_1m_1} + c_2\psi_{n_2l_2m_2})\rangle \\ &= |c_1|^2\langle\psi_{n_1l_1m_1}|L_z|\psi_{n_1l_1m_1}\rangle + |c_2|^2\langle\psi_{n_2l_2m_2}|L_z|\psi_{n_2l_2m_2}\rangle \\ &= |c_1|^2 m_1\hbar + |c_2|^2 m_2\hbar\end{aligned}$$

7.52 Verify that $\psi = A\sin\theta\exp(i\phi)$, where A is a constant, is an eigenfunction of L^2 and L_z. Find the eigenvalues.

Solution. The operators for L^2 and L_z are

$$L^2 = -\hbar^2\left[\frac{1}{\sin\theta}\frac{\partial}{\partial\theta}\left(\sin\theta\frac{\partial}{\partial\theta}\right) + \frac{1}{\sin^2\theta}\frac{\partial^2}{\partial\phi^2}\right]$$

$$L_z = -i\hbar\frac{\partial}{\partial\phi}$$

$$L^2\psi = -\hbar^2\left[\frac{1}{\sin\theta}\frac{\partial}{\partial\theta}\left(\sin\theta\frac{\partial}{\partial\theta}\right)+\frac{1}{\sin^2\theta}\frac{\partial^2}{\partial\phi^2}\right]A\sin\theta e^{i\phi}$$

$$= -A\hbar^2\left[\frac{1}{\sin\theta}\frac{\partial}{\partial\theta}(\sin\theta\cos\theta)-\frac{1}{\sin^2\theta}\sin\theta\right]e^{i\phi}$$

$$= -A\hbar^2\left[-\sin\theta+\frac{\cos^2\theta}{\sin\theta}-\frac{1}{\sin\theta}\right]e^{i\phi}$$

$$= -A\hbar^2\left[-\sin\theta+\frac{1}{\sin\theta}(\cos^2\theta-1)\right]e^{i\phi}$$

$$= -A\hbar^2\left[-\sin\theta+\frac{1}{\sin\theta}(-\sin^2\theta)\right]e^{i\phi}$$

$$= 2A\hbar^2\sin\theta\, e^{i\phi} = 2\hbar^2\psi$$

That is, ψ is an eigenfunction of L^2 with the eigenvalue $2\hbar^2$, and hence

$$L_z\psi = -i\hbar\frac{\partial}{\partial\phi}(A\sin\theta\, e^{i\phi}) = \hbar A\sin\theta\, e^{i\phi} = \hbar\psi$$

The function ψ is an eigenfunction of L_z also with an eigenvalue $\hbar$.

7.53 State Pauli's spin matrices and their eigenvectors. For Pauli's spin matrices, prove the following relations:

(i) $\sigma_x^2 = \sigma_y^2 = \sigma_z^2 = 1.$

(ii) $\sigma_x\sigma_y = i\sigma_z;\ \sigma_y\sigma_z = i\sigma_x;\ \sigma_z\sigma_x = i\sigma_y.$

(iii) $\sigma_x\sigma_y + \sigma_y\sigma_x = \sigma_y\sigma_z + \sigma_z\sigma_y = \sigma_z\sigma_x + \sigma_x\sigma_z = 0.$

Solution. The Pauli spin matrix $\boldsymbol{\sigma}$ is defined by

$$\mathbf{S} = \frac{1}{2}\hbar\boldsymbol{\sigma}$$

$$\sigma_x = \begin{pmatrix}0 & 1\\ 1 & 0\end{pmatrix},\qquad \sigma_y = \begin{pmatrix}0 & -i\\ i & 0\end{pmatrix},\qquad \sigma_z = \begin{pmatrix}1 & 0\\ 0 & -1\end{pmatrix}$$

σ_x, σ_y, σ_z are the Pauli spin matrices. From the definition it is evident that their eigenvalues are ± 1. Their eigenvectors are (refer Problem 7.21).

Matrix σ_x: eigenvector for +1 eigenvalue $\dfrac{1}{\sqrt{2}}\begin{pmatrix}1\\ 1\end{pmatrix}$

eigenvector for –1 eigenvalue $\dfrac{1}{\sqrt{2}}\begin{pmatrix}1\\ -1\end{pmatrix}$

Matrix σ_y: eigenvector for +1 eigenvalue $\frac{1}{\sqrt{2}}\begin{pmatrix}1\\ i\end{pmatrix}$

eigenvector for –1 eigenvalue $\frac{1}{\sqrt{2}}\begin{pmatrix}1\\ -i\end{pmatrix}$

Matrix σ_z: eigenvector for +1 eigenvalue $\frac{1}{\sqrt{2}}\begin{pmatrix}1\\ 0\end{pmatrix}$

eigenvector for –1 eigenvalue $\frac{1}{\sqrt{2}}\begin{pmatrix}0\\ 1\end{pmatrix}$

(i) $\sigma_x^2 = \begin{pmatrix}0 & 1\\ 1 & 0\end{pmatrix}\begin{pmatrix}0 & 1\\ 1 & 0\end{pmatrix} = \begin{pmatrix}1 & 0\\ 0 & 1\end{pmatrix} = I$

Similarly, $\sigma_y^2 = \sigma_z^2 = 1$.

(ii) $\sigma_x\sigma_y = \begin{pmatrix}0 & 1\\ 1 & 0\end{pmatrix}\begin{pmatrix}0 & -i\\ i & 0\end{pmatrix} = \begin{pmatrix}i & 0\\ 0 & -i\end{pmatrix} = i\begin{pmatrix}1 & 0\\ 0 & -1\end{pmatrix} = i\sigma_z$

The same procedure gives the other relations.

(iii) $\sigma_x\sigma_y + \sigma_y\sigma_x = \begin{pmatrix}0 & 1\\ 1 & 0\end{pmatrix}\begin{pmatrix}0 & -i\\ i & 0\end{pmatrix} + \begin{pmatrix}0 & -i\\ i & 0\end{pmatrix}\begin{pmatrix}0 & 1\\ 1 & 0\end{pmatrix}$

$$= \begin{pmatrix}i & 0\\ 0 & -i\end{pmatrix}\begin{pmatrix}-i & 0\\ 0 & i\end{pmatrix} = 0$$

The same procedure proves the other relations too.

7.54 The kets $|jm\rangle$ are the simultaneous eigenkets of J^2 and J_z with eigenvalues $j(j+1)\hbar^2$ and $m\hbar$, respectively. Show that:

(i) $J_+|jm\rangle$ and $J_-|jm\rangle$ are also eigenkets of J^2 with the same eigenvalue.
(ii) $J_+|jm\rangle$ is an eigenket of J_z with the eigenvalue $(m+1)\hbar$.
(iii) $J_-|jm\rangle$ is an eigenket of J_z with the eigenvalue $(m-1)\hbar$.
(iv) Comment on the results.

Solution. Given

$$J^2|jm\rangle = j(j+1)\hbar^2|jm\rangle \tag{i}$$

$$J_z|jm\rangle = m\hbar|jm\rangle \tag{ii}$$

(i) Operating Eq. (i) from left by J_+ and using the result $[J_2, J_+] = 0$, we have

$$J_+J^2|jm\rangle = j(j+1)\hbar^2 J_+|jm\rangle$$

$$J^2J_+|jm\rangle = j(j+1)\hbar^2 J_+|jm\rangle$$

Similarly,

$$J^2J_-|jm\rangle = j(j+1)\hbar^2 J_-|jm\rangle$$

(ii) Operating Eq. (ii) from left by J_+, we get

$$J_+J_z|jm\rangle = m\hbar\, J_+|jm\rangle$$

Since $[J_z, J_+] = \hbar J_+$, $J_+J_z = J_zJ_+ - \hbar J_+$.
we have

$$(J_z, J_+ - \hbar J_+)|jm\rangle = m\hbar J_+|jm\rangle$$

$$J_zJ_+|jm\rangle = (m + 1)\hbar J_+|jm\rangle$$

(iii) Operating Eq. (ii) from left by J_- and using the result $[J_z, J_-] = -\hbar J_-$, we get

$$J_zJ_-|jm\rangle = (m - 1)\hbar J_-|jm\rangle$$

(iv) $J_+|jm\rangle$ is an eigenket of J_z with the eigenvalue $(m + 1)\hbar$ and of J^2 with the same eigenvalue $j(j + 1)\hbar^2$. Since operation by J_+ generates a state with the same magnitude of angular momentum but with a z-component higher by $\hbar$, J_+ is called a raising operator. Similarly, J_- is called a lowering operator.

7.55 The two spin – half particles are described by the Hamiltonian

$$H = A(S_{1z} + S_{2z}) + B(\mathbf{S}_1 \cdot \mathbf{S}_2)$$

where A and B are constants and $\mathbf{S}_1$ and $\mathbf{S}_2$ are the spin angular momenta of the two spins. Find the energy levels of the system.

Solution. Let the total angular momentum

$$\mathbf{S} = \mathbf{S}_1 + \mathbf{S}_2, \qquad S_z = S_{1z} + S_{2z}$$

$$\mathbf{S}_1 \cdot \mathbf{S}_2 = \frac{1}{2}(S^2 - S_1^2 - S_2^2)$$

Let the spin quantum number associated with $\mathbf{S}_1$ be s_1 and that with $\mathbf{S}_2$ be S_2. Since $S_1 = 1/2$ and $S_2 = 1/2$, the possible values of S are 0 and 1. When $S = 0$, the possible values of $M_s = 0$. When $S = 1$, the possible values of $M_s = 1, 0, -1$. The Hamiltonian

$$H = A(S_{1z} + S_{2z}) + B(\mathbf{S}_1 \cdot \mathbf{S}_2)$$

$$AS_z + \frac{B}{2}(S^2 - S_1^2 - S_2^2)$$

Selecting $|SM_sS_1S_2\rangle$ as the eigenkets, we get

$$H|SM_sS_1S_2\rangle = AS_z|SM_sS_1S_2\rangle + \frac{B}{2}(S^2 - S_1^2 - S_2^2)|SM_sS_1S_2\rangle$$

The energy

$$E_{s,M_s} = AM_s\hbar + \frac{B}{2}\hbar^2\left[S(S+1) - \frac{3}{4} - \frac{3}{4}\right]$$

$$E_{0,0} = -\frac{3}{4}B\hbar^2$$

$$E_{1,1} = A\hbar + \frac{B}{4}\hbar^2$$

$$E_{1,0} = \frac{B}{4}\hbar^2$$

$$E_{1,-1} = -A\hbar + \frac{B}{4}\hbar^2$$

E_{00} is a singlet whereas the other three form a triplet.

CHAPTER 8

Time-Independent Perturbation

The potential energy of most of the real systems are different from those considered, and an exact solution is not possible. Different approximate methods have therefore been developed to obtain approximate solutions of systems. One such method is the time-independent perturbation.

8.1 Correction of Nondegenerate Energy Levels

In the time independent perturbation approach, the Hamiltonian operator H of the system is written as

$$H = H^0 + H' \tag{8.1}$$

where H^0 is the unperturbed Hamiltonian, whose nondegenerate eigenvalues E_n^0, $n = 1, 2, 3 \ldots$, and eigenfunctions ψ_n^0 are assumed to be known. The functions ψ_n^0, $n = 1, 2, 3 \ldots$, form a complete orthonormal basis. The time-independent operator H' is the perturbation. The first-order correction to the energy and wave function of the nth state are given by

$$E_n^{(1)} = \langle \psi_n^0 | H' | \psi_n^0 \rangle = \langle n | H' | n \rangle \tag{8.2}$$

$$\psi_n^{(1)} = \sum_m{}' \frac{\langle m | H' | n \rangle}{E_n^0 - E_m^0} | \psi_m^0 \rangle \tag{8.3}$$

where the prime on the sum means that the state $m = n$ should be excluded. The second order correction to the energy

$$E_n^{(2)} = \sum_m{}' \frac{|\langle m | H' | n \rangle|^2}{E_n^0 - E_m^0} \tag{8.4}$$

8.2 Correction to Degenerate Energy Levels

When a degeneracy exists, a linear combination of the degenerate wave functions can be taken as the unperturbed wave function. As an example, consider the case in which E_n^0 is two-fold degenerate. Let ψ_n^0 and ψ_l^0 be eigenfunctions corresponding to the eigenvalues $E_n^0 = E_l^0$ and let the linear combination be

$$\phi = C_n \psi_n^0 + C_l \psi_l^0 \tag{8.5}$$

where C_n and C_l constants. The first order correction to the energies are the solutions of the determinant

$$\begin{vmatrix} H'_{nn} - E_n^{(1)} & H'_{nl} \\ H'_{nl} & H'_{ll} - E_n^{(1)} \end{vmatrix} = 0 \tag{8.6}$$

The corrected energies are

$$E_n = E_n^0 + E_{n+}^{(1)}, \qquad E_l = E_n^0 + E_{n-}^{(1)}$$

PROBLEMS

8.1 Calculate the first order correction to the ground state energy of an anharmonic oscillator of mass m and angular frequency ω subjected to a potential $V(x) = 1/2\ m\omega^2x^2 + bx^4$, where b is a parameter independent of x. The ground state wave function is

$$\psi_0^0 = \left(\frac{m\omega}{\pi\hbar}\right)^{1/4} \exp\left(-\frac{m\omega x^2}{2\hbar}\right)$$

Solution. The first order correction to the ground state energy

$$E_0^{(1)} = \langle\psi_0^0|\,H'\,|\psi_0^0\rangle = \left(\frac{m\omega}{\pi\hbar}\right)^{1/2} b \int_{-\infty}^{\infty} x^4 \exp\left(-\frac{m\omega x^2}{\hbar}\right)dx$$

Using the result given in the Appendix, we get

$$E_0^{(1)} = b\left(\frac{m\omega}{\pi\hbar}\right)^{1/2} 2\cdot\frac{3\sqrt{\pi}}{8}\left(\frac{\hbar}{m\omega}\right)^{5/2} = \frac{3b\hbar^2}{4m^2\omega^2}$$

8.2 A simple harmonic oscillator of mass m_0 and angular frequency ω is perturbed by an additional potential bx^3. Evaluate the second order correction to the ground state energy of the oscillator.

Solution. The second order correction to the ground state energy is given by

$$E_0^{(2)} = \sum_m{}' \frac{|\langle 0|H'|m\rangle|^2}{E_0^0 - E_m^0}, \qquad H' = bx^3$$

In terms of $a^\dagger$ and a,

$$x = \left(\frac{\hbar}{2m_0\omega}\right)^{1/2}(a + a^\dagger)$$

$$\langle 0|x^3|m\rangle = \left[\frac{\hbar}{2m_0\omega}\right]^{3/2} \langle 0|(a + a^\dagger)(a + a^\dagger)(a + a^\dagger)|m\rangle, \qquad m = 1, 2, 3, \ldots$$

$$= \left(\frac{\hbar}{2m_0\omega}\right)^{3/2} [\langle 0|aaa|3\rangle + \langle 0|aaa^\dagger + aa^\dagger a|1\rangle]$$

The other contributions vanish. For the nonvanishing contributions, we have

$$\langle 0|aaa|3\rangle = \sqrt{6}, \qquad \langle 0|aaa^\dagger + aa^\dagger a|1\rangle = 2 + 1 = 3$$

$$E_0^{(2)} = b^2\left(\frac{\hbar}{2m_0\omega}\right)^3\left(\frac{6}{-3\hbar\omega} + \frac{9}{-\hbar\omega}\right) = -\frac{11b^2\hbar^2}{8m_0^3\omega^4}$$

8.3 Work out the splitting of the ^{1}P $\rightarrow$ ^{1}S transition of an atom placed in a magnetic field B along the z-axis.

Solution. For ^{1}P level, $S = 0$ and, therefore, the magnetic moment of the atom is purely orbital. The interaction energy between magnetic moment and the field is

$$H' = -\mu_z B = \frac{e}{2m_0}L_Z B$$

m_0 is the mass of electron and L_z is the z-component of the orbital angular momentum. The first order correction to energy of the ^1P state is

$$E^{(1)} = \left\langle lm \left| \left(\frac{e}{2m_0} \right) L_z B \right| lm \right\rangle = \frac{e\hbar}{2m_0} B m_l, \qquad m_l = 1, 0, -1$$

The ^1P level thus splits into three levels as shown in Fig. 8.1. The ^1S level has neither orbital nor spin magnetic moment. Hence it is not affected by the field and the $^1P \to {}^1S$ transition splits into three lines.

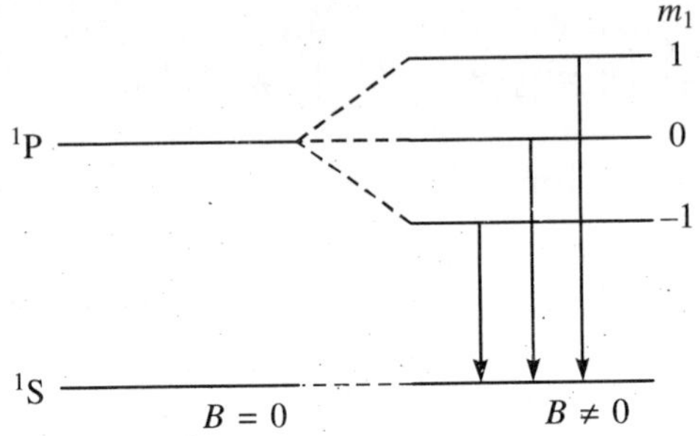

Fig. 8.1 Splitting of ^{1}P → ^{1}S transition of an atom in a magnetic field.

Note: (i) If the system has more than one electron, $l_z = (l_{1z} + l_{2z} + \cdots)$.
(ii) Splitting of a spectral line into three components in the presence of a magnetic field is an example of **normal Zeeman effect.**

8.4 The unperturbed wave functions of a particle trapped in an infinite square well of bottom a are $\psi_n^0 = (2/a)^{1/2} \sin (n\pi x/a)$. If the system is perturbed by raising the floor of the well by a constant amount V_0, evaluate the first and second order corrections to the energy of the nth state.

Solution. The first order correction to the energy of the nth state is

$$\langle \psi_n^0 | H' | \psi_n^0 \rangle = \langle \psi_n^0 | V_0 | \psi_n^0 \rangle = V_0 \langle \psi_n^0 | \psi_n^0 \rangle = V_0$$

Hence, the corrected energy levels are lifted by the amount V_0. The second order correction to the energy is

$$E_n^{(2)} = \sum_m{}' \frac{|\langle \psi_m^0 | H' | \psi_n^0 \rangle|^2}{E_n^0 - E_m^0} = \sum_m{}' \frac{V_0^2 |\langle \psi_m^0 | \psi_n^0 \rangle|^2}{E_n^0 - E_m^0} = 0$$

The second order correction to the energy is zero.

8.5 A particle of mass m_0 and charge e oscillates along the x-axis in a one-dimensional harmonic potential with an angular frequency ω. If an electric field ε is applied along the x-axis, evaluate the first and second order corrections to the energy of the nth state.

Solution. The potential energy due to the field $\varepsilon = -e\varepsilon x$.
The perturbation $H' = -e\varepsilon x$.

$$\text{First order correction } E_n^{(1)} = -e\varepsilon \langle n | x | n \rangle$$

In terms of a and $a^\dagger$,

$$x = \left(\frac{\hbar}{2m_0\omega} \right)^{1/2} (a + a^\dagger)$$

$$E_n^{(1)} = -e\varepsilon\left(\frac{\hbar}{2m_0\omega}\right)^{1/2} \langle n|(a + a^\dagger)|n\rangle = 0$$

$$E_n^{(2)} = \sum_m{}' \frac{|\langle n|H'|m\rangle|^2}{E_n^0 - E_m^0}$$

$$\langle n|H'|m\rangle = -e\varepsilon\left(\frac{\hbar}{2m_0\omega}\right)^{1/2} \langle n|a + a^\dagger|m\rangle$$

Here, m can take all integral values except n. The nonvanishing elements correspond to $m = (n + 1)$ and $(n - 1)$. Hence,

$$E_n^{(2)} = e^2\varepsilon^2 \frac{\hbar}{2m_0\omega}\left[\frac{(\sqrt{n+1})^2}{-\hbar\omega} + \frac{(\sqrt{n})^2}{\hbar\omega}\right] = -\frac{e^2\varepsilon^2}{2m_0\omega^2}$$

8.6 Evaluate the first and second order correction to the energy of the $n = 1$ state of an oscillator of mass m and angular frequency w subjected to a potential

$$V(x) = \frac{1}{2}m\omega^2x^2 + bx, \qquad bx \ll \frac{1}{2}m\omega^2x^2$$

Solution. The first order correction to energy for the $n = 1$ state is given by

$$E_1^{(1)} = \langle 1|bx|1\rangle = b\left(\frac{\hbar}{2m\omega}\right)^{1/2} \langle 1|(a + a^\dagger)|1\rangle$$

$$= b\left(\frac{\hbar}{2m\omega}\right)^{1/2} [\langle 1|a|1\rangle + \langle 1|a^\dagger|1\rangle] = 0$$

Since $a|n\rangle = \sqrt{n}\,|(n-1)\rangle$ and $a^\dagger|n\rangle = \sqrt{n+1}\,|(n+1)\rangle$,

$$E_1^{(2)} = b^2\left(\frac{\hbar}{2m\omega}\right)\sum{}' \frac{|\langle 1|(a + a^\dagger)|k\rangle|^2}{E_1^0 - E_k^0} = b^2\left(\frac{\hbar}{2m\omega}\right)\left[\frac{1}{E_1^0 - E_0^0} + \frac{2}{E_1^0 - E_2^0}\right]$$

$$= b^2\left(\frac{\hbar}{2m\omega}\right)\left(\frac{1}{\hbar\omega} - \frac{2}{\hbar\omega}\right) = -\frac{b^2}{2m\omega^2}$$

8.7 Calculate the ground state energy up to first order of the anharmonic oscillator having a potential energy $V = 1/2\ m\omega^2x^2 + ax^3$; $ax^3 \ll 1/2\ m\omega^2x^2$, where a is independent of x.

Solution. $E_0^{(1)} = \langle 0|ax^3|0\rangle$. The integrand of this integral is an odd function of x and, therefore, the first order correction to the ground state energy is zero.

8.8 Evaluate the first order correction to the energy of the nth state of the anharmonic oscillator having the potential energy

$$V = \frac{1}{2}m\omega^2x^2 + bx^4, \qquad bx^4 \ll \frac{1}{2}m\omega^2x^2$$

Solution.

$$E_n^{(1)} = \langle n|H'|n\rangle = b\langle n|x^4|n\rangle$$

$$= b\left(\frac{\hbar}{2m\omega}\right)^2 \langle n|(a + a^\dagger)(a + a^\dagger)(a + a^\dagger)(a + a^\dagger)|n\rangle$$

The six nonvanishing matrix elements are

1. $\langle n|(aaa^\dagger a^\dagger|n\rangle = (n + 1)(n + 2)$
2. $\langle n|(aa^\dagger aa^\dagger|n\rangle = (n + 1)^2$
3. $\langle n|(aa^\dagger a^\dagger a\,|n\rangle = n(n + 1)$
4. $\langle n|(a^\dagger aaa^\dagger\,|n\rangle = n(n + 1)$
5. $\langle n|(a^\dagger aa^\dagger a|n\rangle = n^2$
6. $\langle n|(a^\dagger a^\dagger aa|n\rangle = n(n - 1)$

Now,

$$E_n^{(1)} = b\left(\frac{\hbar}{2m\omega}\right)^2 [(n + 1)(n + 2) + (n + 1)^2 + 2n(n + 1) + n^2 + n(n - 1)]$$

$$= 3b\left(\frac{\hbar}{2m\omega}\right)^2 (2n^2 + 2n + 1)$$

8.9 A simple harmonic oscillator of mass m and angular frequency ω is perturbed by an additional potential 1/2 bx^2. Obtain the first and second order corrections to the ground state energy.

Solution.

$$E_0^{(1)} = \frac{1}{2}b\langle 0\,|x^2|\,0\rangle = \frac{1}{2}b\left(\frac{\hbar}{2m\omega}\right)\langle 0|(a + a^\dagger)(a + a^\dagger)\,|\,0\rangle$$

$$= \frac{1}{2}b\left(\frac{\hbar}{2m\omega}\right)\langle 0|(aa^\dagger|\,0\rangle = \frac{b\hbar}{4m\omega}$$

$$E_0^{(2)} = \sum_n{}' \frac{|\langle 0|H'|n\rangle|^2}{E_0^0 - E_n^0}$$

$$\langle 0|H'|n\rangle = \frac{1}{2}b\left(\frac{\hbar}{2m\omega}\right)\langle 0|aa + aa^\dagger + a^\dagger a + a^\dagger a^\dagger|n\rangle, \qquad n \neq 0$$

$$= \frac{b\hbar}{4m\omega}\langle 0|aa|n\rangle, \qquad n = 2$$

$$= \frac{\sqrt{2}b\hbar}{4m\omega}$$

$$E_0^{(2)} = -\frac{2b^2\hbar^2}{16\,m^2\omega^2}\frac{1}{2\hbar\omega} = -\frac{b^2\hbar}{16\,m^2\omega^3} \qquad \text{since } E_0 - E_2 = -2\hbar\omega$$

8.10 A rotator having a moment of inertia I and an electric dipole moment μ executes rotational motion in a plane. Estimate the first and second order corrections to the energy levels when the rotator is acted on by an electric field ε in the plane of rotation.

Solution. The energy eigenvalues and eigenfunctions of a plane rotator (Problem 5.3) are

$$E_m = \frac{\hbar^2 m^2}{2I}, \qquad \psi(\phi) = \frac{1}{\sqrt{2\pi}} \exp(im\phi), \qquad m = 0, \pm 1, \pm 2, \ldots$$

The perturbation $H' = -\mu\varepsilon \cos\phi = -\frac{\mu\varepsilon}{2}(e^{i\phi} + e^{-i\phi})$

$$E_n^{(1)} = \langle n|H'|n\rangle = -\frac{\mu\varepsilon}{2\pi}\int_0^{2\pi} \cos\phi\, d\phi = 0$$

$$E_n^{(2)} = \sum_m{}' \frac{|\langle n|H'|m\rangle|^2}{E_n^0 - E_m^0}$$

$$\langle n|H'|m\rangle = -\frac{\mu\varepsilon}{4\pi}\int_0^{2\pi} e^{-in\phi}(e^{i\phi} + e^{-i\phi})e^{im\phi}\, d\phi$$

$$= -\frac{\mu\varepsilon}{4\pi}\left[\int_0^{2\pi} e^{i(m+1-n)\phi}\, d\phi + \int_0^{2\pi} e^{i(m-1-n)\phi}\, d\phi\right]$$

The integrals are finite when $m = n - 1$ (first one) and $m = n + 1$ (second one). Therefore,

$$E_n^{(2)} = \left(-\frac{\mu\varepsilon}{4\pi}\right)^2 \left[\frac{4\pi^2}{E_n^0 - E_{n-1}} + \frac{4\pi^2}{E_n^0 - E_{n+1}}\right]$$

$$= \left(-\frac{\mu\varepsilon}{4\pi}\right)^2 \frac{4\pi^2 2I}{\hbar^2}\left(\frac{1}{2n-1} - \frac{1}{2n+1}\right) = \frac{\mu^2\varepsilon^2 I}{\hbar^2(4n^2-1)}$$

8.11 The Hamiltonian matrix of a system is

$$H = \begin{pmatrix} 1 & \varepsilon & 0 \\ \varepsilon & 1 & 0 \\ 0 & 0 & 2 \end{pmatrix}, \qquad \varepsilon \ll 1$$

Find the energy eigenvalues corrected to first order in the perturbation. Also, find the eigenkets if the unperturbed eigenkets are $|\phi_1\rangle$, $|\phi_2\rangle$ and $|\phi_3\rangle$.

Solution. The Hamiltonian matrix can be written as

$$H = \begin{pmatrix} 1 & 0 & 0 \\ 0 & 1 & 0 \\ 0 & 0 & 2 \end{pmatrix} + \begin{pmatrix} 0 & \varepsilon & 0 \\ \varepsilon & 0 & 0 \\ 0 & 0 & 0 \end{pmatrix} \tag{i}$$

In this form, we can identify the unperturbed part H^0 and the perturbation H' as

$$H^0 = \begin{pmatrix} 1 & 0 & 0 \\ 0 & 1 & 0 \\ 0 & 0 & 2 \end{pmatrix} \qquad H' = \begin{pmatrix} 0 & \varepsilon & 0 \\ \varepsilon & 0 & 0 \\ 0 & 0 & 0 \end{pmatrix} \tag{ii}$$

The unperturbed energies are 1, 1, 2 units. The energy 1 units are two-fold degenerate. The secular determinant corresponding to H' is

$$\begin{vmatrix} -E^{(1)} & \varepsilon & 0 \\ \varepsilon & -E^{(1)} & 0 \\ 0 & 0 & -E^{(1)} \end{vmatrix} = 0 \quad \text{or} \quad E^{(1)^2} - \varepsilon^2 = 0 \text{ and } E^{(1)} = 0$$

where $E^{(1)}$ is the first order correction. The solution gives

$$E^{(1)} = \varepsilon, -\varepsilon, 0 \tag{iii}$$

Hence, the state $|\phi_3\rangle$ is not affected by the perturbation. The eigenkets corresponding to states 1 and 2 can easily be obtained. Let these states be

$$\phi_n' = c_1|\phi_1\rangle + c_2|\phi_2\rangle, \qquad n = 1, 2 \tag{iv}$$

The coefficients must obey the condition

$$-E^{(1)}c_1 + \varepsilon c_2 = 0 \tag{v}$$

For the eigenvalue $E^{(1)} = \varepsilon$, this equation reduces to

$$-\varepsilon c_1 + \varepsilon c_2 = 0 \quad \text{or} \quad c_1 = c_2$$

Normalization gives $c_1 = c_2 = 1/\sqrt{2}$. Hence,

$$\phi_1' = \frac{1}{\sqrt{2}}[|\phi_1\rangle + |\phi_2\rangle] \tag{vi}$$

With the value $E^{(1)} = -\varepsilon$, Eq. (v) reduces to

$$\varepsilon c_1 + \varepsilon c_2 = 0 \quad \text{or} \quad c_1 = -c_2$$

Normalization gives $c_1 = -c_2 = 1/\sqrt{2}$. This leads to

$$\phi_2' = \frac{1}{\sqrt{2}}[|\phi_1\rangle - |\phi_2\rangle] \tag{vii}$$

Thus, the corrected energies and eigenkets are

$$1 + \varepsilon \qquad \frac{1}{\sqrt{2}}[|\phi_1\rangle + |\phi_2\rangle]$$

$$1 - \varepsilon \qquad \frac{1}{\sqrt{2}}[|\phi_1\rangle - |\phi_2\rangle]$$

$$2 \qquad |\phi_3\rangle$$

8.12 A rigid rotator in a plane is acted on by a perturbation represented by

$$H' = \frac{V_0}{2}(3\cos^2\phi - 1), \qquad V_0 = \text{constant}$$

Calculate the ground state energy up to the second order in the perturbation.

Solution. The energy eigenvalues and eigenfunctions of a plane rotator (refer Problem 5.3) are given by

$$E_m = \frac{m^2\hbar^2}{2I}, \qquad m = 0, \pm 1, \pm 2, \ldots$$

$$\psi_m(\phi) = \frac{1}{\sqrt{2\pi}} \exp(im\phi)$$

Except the ground state, all levels are doubly degenerate. The first order correction to the ground state energy is

$$E_0^{(1)} = \langle\psi|H'|\psi\rangle = \left\langle\psi\left|\frac{V_0}{2}(3\cos^2\phi - 1)\right|\psi\right\rangle$$

$$= \left\langle\psi\left|\frac{3V_0}{2}\cos^2\phi\right|\psi\right\rangle - \left\langle\psi\left|\frac{V_0}{2}\right|\psi\right\rangle$$

$$= \frac{3}{4}V_0 - \frac{V_0}{2} = \frac{V_0}{4}$$

The second order energy correction

$$E_0^{(2)} = \sum_m{}' \frac{|\langle 0|H'|m\rangle|^2}{E_0^0 - E_m^0}$$

$$\langle 0|H'|m\rangle = \frac{V_0}{2}\int_0^{2\pi} \frac{1}{\sqrt{2\pi}}(3\cos^2\phi - 1)\frac{1}{\sqrt{2\pi}} e^{im\phi}\, d\phi$$

$$= \frac{3V_0}{4\pi}\int_0^{2\pi} \cos^2\phi\, e^{im\phi}\, d\phi - \frac{V_0}{4\pi}\int_0^{2\pi} e^{im\phi}\, d\phi$$

We can write $\cos^2\phi = (1 + \cos 2\phi)/2$. Also, the second integral vanishes. Hence,

$$\langle 0|H'|m\rangle = \frac{3V_0}{8\pi}\int_0^{2\pi} (1 + \cos 2\phi)\, e^{im\phi}\, d\phi = \frac{3V_0}{8\pi}\int_0^{2\pi} \cos 2\phi\, e^{im\phi}\, d\phi$$

since the other integral vanishes. Putting $\cos 2\phi$ in the exponential, we get

$$\langle 0|H'|m\rangle = \frac{3V_0}{16\pi}\int_0^{2\pi} (e^{i2\phi} + e^{-i2\phi})\, e^{im\phi}\, d\phi$$

$$= \frac{3V_0}{16\pi}\int_0^{2\pi} e^{i(m+2)\phi}\, d\phi + \frac{3V_0}{16\pi}\int_0^{2\pi} e^{i(m-2)\phi}\, d\phi$$

The first integral is finite when $m = -2$, the second integral is finite when $m = +2$ and their values are equal to $3V_0/8$. $E_{\pm 2} = 2\hbar^2/I$, $E_0 = 0$. Hence,

$$E_0^0 - E_2^0 = E_0^0 - E_{-2}^0 = -\frac{2\hbar^2}{I}$$

Thus,

$$E_0^{(2)} = \frac{(3V_0 / 8)^2}{-2\hbar^2/I} + \frac{(3V_0 / 8)^2}{-2\hbar^2/I} = -\frac{9}{64}\frac{V_0^2 I}{\hbar^2}$$

8.13 A plane rigid rotator in the first excited state is subjected to the interaction

$$H' = \frac{V_0}{2}(3\cos^2\phi - 1)$$

where V_0 is constant. Calculate the energies to first order in H'.

Solution. For a plane rotator,

$$E_m = \frac{\hbar^2 m^2}{2I}, \qquad \psi(\phi) = \frac{1}{\sqrt{2\pi}}e^{im\phi}, \qquad m = 0, \pm 1, \pm 2, \ldots$$

Except the $m = 0$ state, all states are doubly degenerate. The energy and wave function of the first excited state are

$$E_{\pm 1} = \frac{\hbar^2}{2I}, \qquad \psi(\phi) = \frac{1}{\sqrt{2\pi}}e^{\pm i\phi}$$

The first order energy corrections are given by the roots of Eq. (8.6):

$$\begin{vmatrix} H'_{11} - E_1^{(1)} & H'_{12} \\ H'_{21} & H'_{22} - E_1^{(1)} \end{vmatrix} = 0$$

$$H'_{11} = H'_{22} = \frac{1}{2\pi}\int_0^{2\pi}\frac{V_0}{2}(3\cos^2\phi - 1)\,d\phi$$

$$= \frac{V_0}{2\pi}\left[3\int_0^{2\pi}\cos^2\phi\,d\phi - \int_0^{2\pi}d\phi\right] = \frac{V_0}{2\pi}(3\pi - 2\pi) = \frac{V_0}{4}$$

$$H'_{12} = H'_{21} = \frac{1}{2\pi}\int_0^{2\pi}e^{-i\phi}\frac{V_0}{2}(3\cos^2\phi - 1)\,e^{-i\phi}\,d\phi = \frac{3V_0}{8}$$

The secular determinant takes the form

$$\begin{vmatrix} \dfrac{V_0}{4} - E_1^{(1)} & \dfrac{3V_0}{8} \\ \dfrac{3V_0}{8} & \dfrac{V_0}{4} - E_1^{(1)} \end{vmatrix} = 0$$

$$[E_1^{(1)}]^2 - \frac{V_0}{2}E_1^{(1)} - \frac{5V_0^2}{64} = 0$$

The roots of this equation are $-(V_0/8)$ and $-(5V_0/8)$. The corrected energies are

$$E = \frac{\hbar^2}{2I} + \frac{5V_0}{8} \quad \text{and} \quad \frac{\hbar^2}{2I} - \frac{V_0}{8}$$

8.14 A one-dimensional box of length a contains two particles each of mass m. The interaction between the particles is described by a potential of the type $V(x_1, x_2) = \lambda\delta(x_1 - x_2)$, which is the δ-Dirac delta function. Calculate the ground state energy to first order in λ.

Solution. The interaction between the particles can be treated as the perturbation. The Hamiltonian without that will be the unperturbed part. Without the δ-potential

$$V(x_1, x_2) = \begin{cases} 0, & 0 \le x_1, x_2 \le a \\ \infty, & \text{Otherwise} \end{cases}$$

$$H_0 = -\frac{\hbar^2}{2m}\frac{d^2}{dx_1^2} - \frac{\hbar^2}{2m}\frac{d^2}{dx_2^2} + V(x_1, x_2)$$

From the results of an infinitely deep potential well, the energy and wave functions are

$$E_{nk} = \frac{\pi^2\hbar^2}{2ma^2}(n^2 + k^2), \qquad n, k = 1, 2, 3, \ldots$$

$$\psi_{nk}(x_1, x_2) = \psi_n(x_1)\psi_k(x_2) = \frac{2}{a}\sin\left(\frac{n\pi x_1}{a}\right)\sin\left(\frac{k\pi x_2}{a}\right)$$

For the ground state, $n = k = 1$, we have

$$E_{11}^0 = \frac{\pi^2\hbar^2}{ma^2}, \qquad \psi_{11}^0(x_1, x_2) = \frac{2}{a}\sin\left(\frac{\pi x_1}{a}\right)\sin\left(\frac{\pi x_2}{a}\right)$$

$$H' = \lambda\delta(x_1 - x_2)$$

The first order correction to the ground state energy

$$\Delta E = \langle 11|H'|11\rangle$$

$$= \left(\frac{2}{a}\right)^2 \int_0^a\int_0^a \lambda\delta(x_1 - x_2)\sin^2\left(\frac{\pi x_1}{a}\right)\sin^2\left(\frac{\pi x_2}{a}\right)dx_1\,dx_2$$

$$= \left(\frac{2}{a}\right)^2 \lambda\int_0^a \sin^4\left(\frac{\pi x_1}{a}\right)dx_1 = \frac{4\lambda}{a^2}\frac{3}{8}a = \frac{3\lambda}{2a}$$

The corrected energy

$$E' = E_{11}^0 + \Delta E = \frac{\pi^2\hbar^2}{ma^2} + \frac{3\lambda}{2a}$$

8.15 Consider the infinite square well defined by

$$V(x) = 0 \quad \text{for} \quad 0 \le x < a$$

$V(x) = \infty$ otherwise

Using the first order perturbation theory, calculate the energy of the first two states of the potential well if a portion defined by $V(x) = V_0 x/a$, where V_0 is a small constant, with $0 \le x \le a$ being sliced off.

Solution. From Problem 4.1, the energy eigenvalues and eigenfunctions of the the unperturbed Hamiltonian are

$$E_n^0 = \frac{n^2\pi^2\hbar^2}{2ma^2}, \qquad \psi_n^0 = \sqrt{\frac{2}{a}}\sin\frac{n\pi x}{a}, \qquad n = 1, 2, 3, \ldots$$

The perturbation $H' = V_0 x/a$ which is depicted in Fig. 8.2.

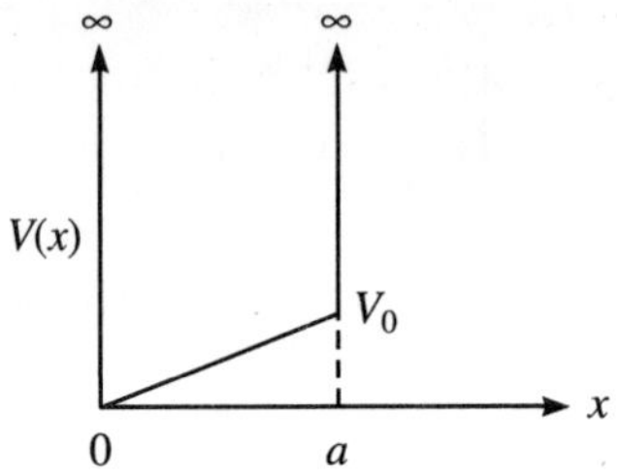

Fig. 8.2 Sliced infinite potential well.

The first order correction to the energy for the $n = 1$ state is

$$\left\langle \psi_1^0 \left| \frac{V_0 x}{a} \right| \psi_1^0 \right\rangle = \frac{V_0}{a}\frac{2}{a}\int_0^a x \sin^2\frac{\pi x}{a}\,dx$$

$$= \frac{2V_0}{a^2}\int_0^a \frac{x}{2}\left(1 - \cos\frac{2\pi x}{a}\right)dx$$

$$= \frac{2V_0}{a^2}\int_0^a \frac{x}{2}\,dx - \frac{2V_0}{a^2}\int_0^a \frac{x}{2}\cos\frac{2\pi x}{a}\,dx$$

$$= \frac{V_0}{2} + 0 = \frac{V_0}{2}$$

The first order correction to the $n = 2$ state is

$$\left\langle \psi_2^0 \left| \frac{V_0 x}{a} \right| \psi_2^0 \right\rangle = \frac{V_0}{a}\frac{2}{a}\int_0^a x \sin^2\frac{2\pi x}{a}\,dx = \frac{V_0}{2}$$

The corrected energies are

$$\frac{\pi^2\hbar^2}{2ma^2} + \frac{V_0}{2} \quad \text{and} \quad \frac{2\pi^2\hbar^2}{ma^2} + \frac{V_0}{2}$$

8.16 The energy levels of the one-electron atoms are doublets, except the s-states because of spin-orbit interaction. The spin-orbit Hamiltonian

$$H_{so} = \frac{1}{2m^2c^2}\frac{1}{r}\frac{dV}{dr}\mathbf{L}\cdot\mathbf{S}$$

Treating H_{so} as a perturbation, evaluate the spin-orbit interaction energy. For hydrogenic atoms, assume that the expectation value is

$$\left\langle \frac{1}{r_3} \right\rangle = \frac{2z^3}{n^3 a_0^3\, l(l+1)(2l+1)}$$

where a_0 is the Bohr radius.

Solution. For the valence electron in a hydrogen-like atom, the potential

$$V(r) = -\frac{Ze^2}{4\pi\varepsilon_0 r} \quad \text{or} \quad \frac{dV}{dr} = \frac{Ze^2}{4\pi\varepsilon_0 r^2} \tag{i}$$

Substituting the value of dV/dr, we get

$$H_{so} = \frac{Ze^2}{8\pi\varepsilon_0 m^2 c^2} \frac{\boldsymbol{L}\cdot\boldsymbol{S}}{r^3} \tag{ii}$$

Since $\boldsymbol{J} = \boldsymbol{L} + \boldsymbol{S}$,

$$J^2 = L^2 + S^2 + 2\boldsymbol{L}\cdot\boldsymbol{S} \quad \text{or} \quad \boldsymbol{L}\cdot\boldsymbol{S} = \frac{J^2 - L^2 - S^2}{2} \tag{iii}$$

Using the basis $|\,lsjm\rangle$, the expectation value of $J^2 - L^2 - S^2$ is given by

$$\langle (J^2 - L^2 - S^2)\rangle = [\,j(j+1) - l(l+1) - s(s+1)]\,\hbar^2 \tag{iv}$$

Since the first order correction to the energy constitutes the diagonal matrix elements, substituting the values of $\langle 1/r^3\rangle$ and $\langle (J^2 - L^2 - S^2)\rangle$, we get

$$E_{so} = \frac{z^4 e^2 \hbar^2}{8\pi\varepsilon_0 m^2 c^2 a_0^3} \frac{j(j+1) - l(l+1) - s(s+1)}{n^3 l(l+1)(2l+1)} \tag{v}$$

The Bohr radius a_0 and the fine structure constant α are defined as

$$a_0 = \frac{4\pi\varepsilon_0 \hbar^2}{me^2}, \qquad \alpha = \frac{e^2}{4\pi\varepsilon_0 c\hbar} \tag{vi}$$

Using Eq. (vi), we get

$$E_{so} = \frac{z^4 e^2 \hbar^2}{8\pi\varepsilon_0 m^2 c^2 a_0^3} \frac{j(j+1) - l(l+1) - s(s+1)}{n^3 l(l+1)(2l+1)} \tag{vii}$$

This makes the state $j = l - (1/2)$ to have a lower energy than that with $j = l + (1/2)$.

8.17 The spin -orbit interaction energy

$$E_{so} = \frac{z^4 \alpha^4 mc^2}{2n^3} \frac{j(j+1) - l(l+1) - s(s+1)}{l(l+1)(2l+1)}$$

Calculate the doublet separation ΔE_{so} of states with the same n and l. Apply the result to the 2p state of hydrogen and obtain the doublet separation in units of eV.

Solution. For a given value of l, j can have the values $j = l + (1/2)$ and $j = l - (1/2)$. The difference in energy between these two is the doublet separation ΔE_{so}. Hence,

$$\Delta E_{so} = \frac{z^4\alpha^4 mc^2}{2n^3 l(l+1)(2l+1)}\left[\left(l+\frac{1}{2}\right)\left(l+\frac{3}{2}\right)-\left(l-\frac{1}{2}\right)\left(l+\frac{1}{2}\right)\right]$$

$$= \frac{z^4\alpha^4 mc^2\,(2l+1)}{2n^3 l(l+1)(2l+1)} = \frac{z^4\alpha^4 mc^2}{2n^3 l(l+1)}$$

For the 2p state of hydrogen, $n = 2$, $l = 1$, $z = 1$. So,

$$\Delta E_{so} = \frac{(9.1\times 10^{-31}\text{ kg})\,(3\times 10^8\text{ ms}^{-1})}{(137)^4 \times 2\times 2^3\times 2} = 7.265\times 10^{-24}\text{ J}$$

$$= \frac{1.765\times 10^{-24}\text{ J}}{1.6\times 10^{-19}\text{ J/eV}} = 4.5\times 10^{-5}\text{ eV}$$

8.18 The matrices for the unperturbed (H^0) and perturbation (H') Hamiltonians in the orthonormal basis $|\phi_1\rangle$ and $|\phi_2\rangle$ are

$$H^0 = \begin{pmatrix} E_o+\varepsilon & 0 \\ 0 & E_o-\varepsilon \end{pmatrix}, \qquad H' = \begin{pmatrix} 0 & A \\ A & 0 \end{pmatrix}$$

Determine (i) the first order correction to energy, (ii) second order correction to energy, and (iii) the wave function corrected to first order.

Solution.

(i) The first order correction to the energy is zero since the perturbation matrix has no diagonal element.

(ii) $$E_n^{(2)} = \sum_m \frac{|\langle n|H'|m\rangle|^2}{E_n^0 - E_m^0}, \qquad E_1^{(2)} = \frac{|\langle 1|H'|2\rangle|^2}{E_1^0 - E_2^0} = \frac{|A|^2}{2\varepsilon} = \frac{A^2}{2\varepsilon}$$

$$E_2^{(2)} = \frac{|\langle 2|H'|1\rangle|^2}{E_2^0 - E_1^0} = \frac{A^2}{-2\varepsilon}$$

$$E_1 = E_0 + \varepsilon + \frac{A^2}{2\varepsilon}, \qquad E_2 = E_0 - \varepsilon - \frac{A^2}{2\varepsilon}$$

The wave function corrected to first order is given by

$$\psi_n = \psi_n^0 + \sum_m \frac{\langle m|H'|n\rangle}{E_n^0 - E_m^0}\left|\psi_m^0\right\rangle$$

$$\psi_1 = |\phi_1\rangle + \frac{A}{E_1^0 - E_2^0}\left|\phi_2\right\rangle = |\phi_1\rangle + \frac{A}{2\varepsilon}\left|\phi_2\right\rangle$$

$$\psi_2 = |\phi_2\rangle - \frac{A}{2\varepsilon}\left|\phi_1\right\rangle$$

8.19 Given the matrix for H^0 and H':

$$H^0 = \begin{pmatrix} E_0 & 0 \\ 0 & E_0 \end{pmatrix}, \qquad H' = \begin{pmatrix} 0 & -A \\ -A & 0 \end{pmatrix}$$

In the orthonormal basis $|1\rangle$ and $|2\rangle$, determine (i) the energy eigenvalues, and (ii) energy eigenfunctions.

Solution. This is a case of degenerate states $|1\rangle$ and $|2\rangle$ with energy eigenvalue E_0. The secular determinant is, then,

$$\begin{vmatrix} -E^{(1)} & -A \\ -A & -E^{(1)} \end{vmatrix} = 0 \quad \text{or} \quad E^{(1)} = \pm A$$

The eigenfunctions corresponding to these eigenvalues are obtained by a linear combination of $|1\rangle$ and $|2\rangle$. Let the combination be $c_1|1\rangle + c_2|2\rangle$. For +A eigenvalue, the equation $(H'_{11} - E_1^{(1)})c_1 + H'_{12}c_2 = 0$ reduces to

$$-Ac_1 - Ac_2 = 0 \quad \text{or} \quad \frac{c_1}{c_2} = -1$$

Normalization gives $c_1 = 1/\sqrt{2}$, $c_2 = 1/\sqrt{2}$. Hence, the combination is $(|1\rangle - |2\rangle)/\sqrt{2}$. The other combination is $(|1\rangle + |2\rangle)/\sqrt{2}$. The energy eigenvalues and eigenfunctions are

$$E_0 + A \text{ and } (|1\rangle - |2\rangle)/\sqrt{2}$$

$$E_0 - A \text{ and } (|1\rangle + |2\rangle)/\sqrt{2}$$

8.20 Prove the Lande interval rule which states that in a given L-S term, the energy difference between two adjacent J-levels is proportional to the larger of the two values of J.

Solution. For a given L-S term the total orbital angular momentum J can have the values $J = L + S,\ L + S - 1,\ \ldots\ |L - S|$. The spin-orbit coupling energy E_{so}, Problem 8.16 for a given L-S term is

$$E_{so} = \text{constant } [J(J + 1) - L(L + 1) - S(S + 1)]$$

The energy difference between $J - 1$ and J levels is ΔE_{so} given by

$$\begin{aligned} \Delta E_{so} &= \text{constant } [J(J + 1) - L(L + 1) - S(S + 1) - J(J - 1) + L(L + S) + S(S + 1)] \\ &= \text{constant} \times 2J \end{aligned}$$

That is, the energy difference between two adjacent J-levels is proportional to the larger of the two values of J.

8.21 An interaction of the nuclear angular momentum of an atom (I) with electronic angular momentum (J) causes a coupling of the $\boldsymbol{I}$ and $\boldsymbol{J}$ vectors: $\boldsymbol{F} = \boldsymbol{I} + \boldsymbol{J}$. The interaction Hamiltonian is of the type H_{int} = constant $\boldsymbol{I} \cdot \boldsymbol{J}$. Treating this as a perturbation, evaluate the first order correction to the energy.

Solution. Though the unperturbed Hamiltonian has degenerate eigenvalues, one can avoid working with degenerate perturbation theory (refer Problem 8.16). The perturbing Hamiltonian

$$H' = \text{costant } \boldsymbol{I} \cdot \boldsymbol{J}$$

The first order correction to energy is the diagonal matrix element of $H' = \langle H' \rangle$ which can be obtained as

$$F^2 = (\boldsymbol{I} + \boldsymbol{J})^2 = I^2 + J^2 + 2\boldsymbol{I} \cdot \boldsymbol{J}$$

$$\boldsymbol{I} \cdot \boldsymbol{J} = \frac{F^2 - I^2 - J^2}{2}$$

$$\langle H' \rangle = \text{constant } [F(F+1) - I(I+1) - J(J+1)] \frac{\hbar^2}{2}$$

Hence, the first order correction

$$E^{(1)} = a\,[F(F+1) - I(I+1) - J(J+1)]$$

where a is a constant.

8.22 A particle in a central potential has an orbital angular momentum quantum number $l = 3$. If its spin $s = 1$, find the energy levels and degeneracies associated with the spin-orbit interaction.

Solution. The spin-orbit interaction

$$H_{so} = \xi(r)\, \boldsymbol{L} \cdot \boldsymbol{S}$$

where $\xi(r)$ is a constant. The total angular momentum

$$\boldsymbol{J} = \boldsymbol{L} + \boldsymbol{S} \quad \text{or} \quad \boldsymbol{L} \cdot \boldsymbol{S} = \frac{1}{2}(J^2 - L^2 - S^2)$$

Hence,

$$H_{so} = \frac{1}{2}\xi(r)\,(J^2 - L^2 - S^2)$$

In the $|jm_j ls\rangle$ basis, the first order correction

$$E_{so} = \left\langle jm_j ls \left| \frac{1}{2}\xi(r)\,(J^2 - L^2 - S^2) \right| jm_j ls \right\rangle$$

$$= \frac{1}{2}\xi(r)\,[j(j+1) - l(l+1) - s(s+1)]\,\hbar^2$$

Since $l = 3$ and $s = 1$, the possible values of j are 4, 3, 2. Hence

$$E_{so} = \begin{cases} 3\xi(r)\hbar^2, & j = 4 \\ -\xi(r)\hbar^2, & j = 3 \\ -4\xi(r)\hbar^2, & j = 2 \end{cases}$$

The degeneracy d is given by the $(2j + 1)$ value

$$d = \begin{cases} 9, & j = 4 \\ 7, & j = 3 \\ 5, & j = 2 \end{cases}$$

8.23 Consider the infinite square well

$$V(x) = 0 \quad \text{for } -a \le x \le a$$

$V(x) = \infty$ for $|x| > a$

with the bottom defined by $V(x) = V_0 x/a$, where V_0 constant, being sliced off. Treating the sliced-off part as a perturbation to the regular infinite square well, evaluate the first order correction to the energy of the ground and first excited states.

Solution. For the regular infinite square well, the energy and eigenfunctions are given by Eqs. (4.2) and (4.3).

$$E_1^0 = \frac{\pi^2\hbar^2}{8ma^2}, \qquad \psi_1^0 = \frac{1}{\sqrt{a}}\cos\frac{\pi x}{2a}$$

$$E_2^0 = \frac{\pi^2\hbar^2}{2ma^2}, \qquad \psi_2^0 = \frac{1}{\sqrt{a}}\sin\frac{\pi x}{a}$$

The portion sliced off is illustrated in Fig. 8.3.

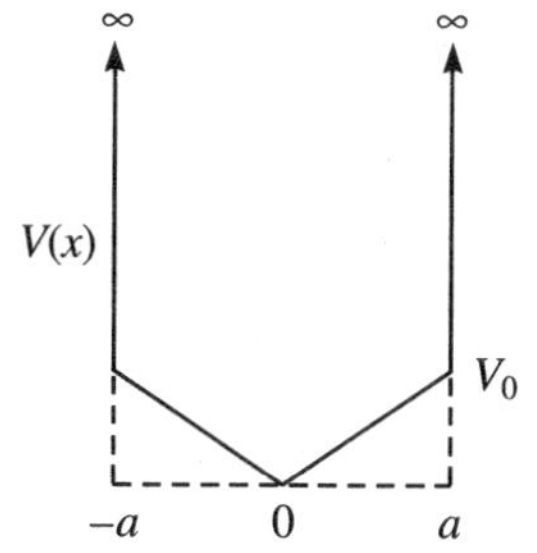

Fig. 8.3 Infinite square well with the bottom sliced off.

$$\text{Perturbation } H' = \frac{V_0 x}{a}$$

The first order correction to the ground state energy is

$$E_1^{(1)} = \langle 1|H'|1\rangle = \frac{V_0}{a^2}\int_{-a}^{a} x\cos^2\frac{\pi x}{2a}\,dx = 0$$

since the integrand is odd. The first order correction to the first excited state is

$$E_2^{(1)} = \left\langle \psi_2^0 \left| \frac{V_0 x}{a} \right| \psi_2^0 \right\rangle = \frac{V_0}{a^2}\int_{-a}^{a} x\sin^2\frac{\pi x}{a}\,dx = 0$$

since the integrand is odd.

8.24 Draw the energy levels, including the spin-orbit interaction for $n = 3$ and $n = 2$ states of hydrogen atom and calculate the spin-orbit doublet separation of the 2p, 3p and 3d states. The Rydberg constant of hydrogen is $1.097 \times 10^7\ \text{m}^{-1}$.

Solution. Figure 8.4 represents the energy level for n = 3 and n = 2 states of hydrogen (Z = 1), including the spin-orbit interaction.

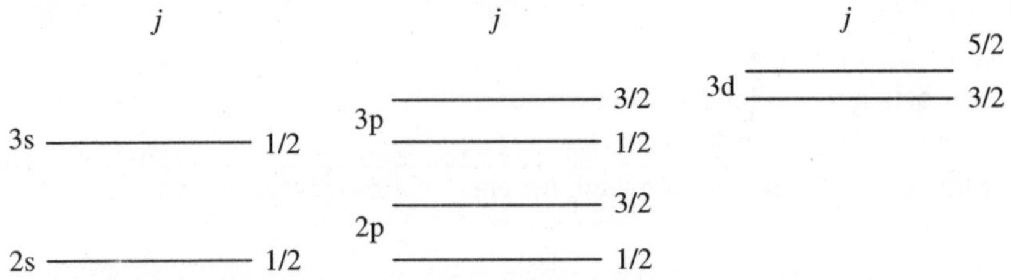

Fig. 8.4 Energy levels for n = 3 and n = 2 states of hydrogen.

The doublet separation

$$\Delta E = \frac{Z^4 \alpha^2 R}{n^3 l(l+1)}$$

For the 2p state, n = 2, l = 1, and hence

$$(\Delta E)_{2p} = \frac{(1/137)^2 (1.097 \times 10^7 \text{ m}^{-1})}{8 \times 2} = 36.53 \text{ m}^{-1}$$

For the 3p state, n = 3, l = 1, and so

$$(\Delta E)_{3p} = \frac{(1/137)^2 (1.097 \times 10^7 \text{ m}^{-1})}{27 \times 2} = 10.82 \text{ m}^{-1}$$

For the 3d state n = 3, l = 2 and, therefore,

$$(\Delta E)_{3d} = \frac{(1/137)^2 (1.097 \times 10^7 \text{ m}^{-1})}{27 \times 2 \times 3} = 3.61 \text{ m}^{-1}$$

Note: The doublet separation decreases as l increases. The 2p doublet separation is greater than the 3p doublet which will be greater than the 4p separation (if evaluated), and so on. The d-electron doublet splitting are also similar.

8.25 A hydrogen atom in the ground state is placed in an electric field ε along the z-axis. Evaluate the first order correction to the energy.

Solution. Consider an atom situated at the origin. If $\boldsymbol{r}$ is the position vector of the electron, the dipole moment

$$\boldsymbol{\mu} = -e\boldsymbol{r}$$

The additional potential energy in the electric field ε is $-\boldsymbol{\mu} \cdot \varepsilon$, where θ is the angle between vectors $\boldsymbol{r}$ and ε. This energy can be treated as the perturbation

$$H' = er\, \varepsilon \cos \theta$$

The unperturbed Hamiltonian

$$H^0 = -\frac{\hbar^2}{2\mu}\nabla^2 - \frac{e^2}{4\pi\varepsilon_0 r}$$

The unperturbed wave function

$$\psi_{100} = -\frac{1}{\pi^{1/2} a_0^{3/2}} e^{-r/a_0}$$

The first order correction to the energy

$$E_l^{(1)} = \langle 100 \,|\, er\varepsilon \cos\theta \,|\, 100 \rangle$$

The angular part of this equation is

$$\int_0^{\pi} \cos\theta \sin\theta \, d\theta = 0$$

i.e., the first order correction to the energy is zero.

8.26 A particle of mass m moves in an infinite one-dimensional box of bottom a with a potential dip as defined by

$$V(x) = \infty \quad \text{for } x < 0 \text{ and } x > a$$

$$V(x) = -V_0 \quad \text{for } 0 < x < \frac{a}{3}$$

$$V(x) = 0 \quad \text{for } \frac{a}{3} < x < a$$

Find the first order energy of the ground state.

Solution. For a particle in the infinite potential well (Fig. 8.5) defined by $V(x) = 0$ for $0 < x < a$ and $V(x) = \infty$ otherwise, the energy eigenvalues and eignfunctions are

$$E_n = \frac{n^2\pi^2\hbar^2}{2ma^2}, \qquad \psi_n = \sqrt{\frac{2}{a}} \sin\frac{n\pi x}{a}, \qquad n = 1, 2. 3, \ldots$$

The perturbation $H' = -V_0$, $0 < x < (a/3)$. Hence, the first order energy correction to the ground state is

$$E^{(1)} = -\frac{2}{a} V_0 \int_0^{a/3} \sin^2 \frac{\pi x}{a} \, dx$$

$$= -\frac{2}{a} V_0 \int_0^{a/3} \frac{1}{2}\left(1 - \cos\frac{2\pi x}{a}\right) dx$$

$$-\frac{a}{3} = -\frac{V_0}{a}[x]_0^{a/3} + \frac{V_0}{a}\frac{a}{2\pi}\left[\sin\frac{2\pi x}{a}\right]_0^{a/3}$$

$$= -\frac{V_0}{3} + \frac{V_0}{4\pi} \times 0.866 = -\,0.264\, V_0$$

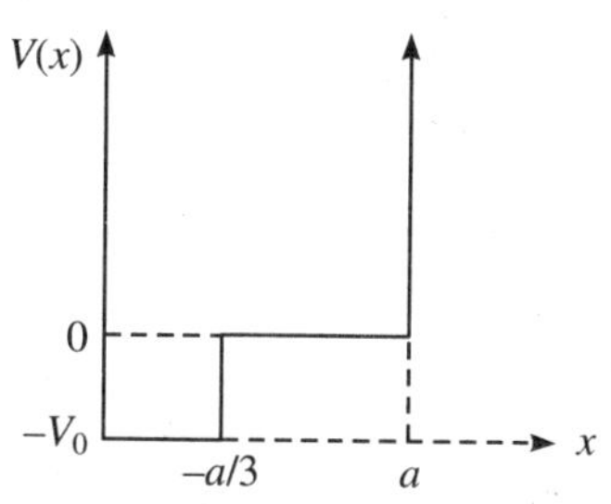

Fig. 8.5 Infinite square well with potential dip.

The energy of the ground state corrected to first order is

$$E = \frac{\pi^2\hbar^2}{2ma^2} - 0.264 V_0$$

8.27 A particle of mass m moves in a one-dimensional potential well defined by

$$V(x) = \begin{cases} 0 & \text{for } -2a < x < -a \text{ and } a < x < 2a \\ \infty & \text{for } x > 2a \text{ and } x < -2a \\ V_0 & \text{for } -a < x < a \end{cases}$$

Treating V_0 for $-a < x < a$ as perturbation on the flat bottom box $V(x) = 0$ for $-2a < x < 2a$ and $V(x) = \infty$ otherwise, calculate the energy of the ground state corrected up to first order.

Solution. The unperturbed energy and wave function of the ground state is

$$E_1^0 = \frac{\pi^2\hbar^2}{32ma^2}$$

$$\psi_1^0 = \frac{1}{\sqrt{2a}} \cos\frac{\pi x}{4a}$$

The first order correction to the energy

$$E^{(1)} = \frac{V_0}{2a}\int_{-a}^{a} \cos^2\frac{\pi x}{4a}\,dx = \frac{V_0}{2a}\int_{-a}^{a} \frac{1}{2}\left(1 + \cos\frac{\pi x}{2a}\right)dx$$

$$= \frac{V_0}{4a}(x)_{-a}^{a} + \frac{V_0}{4a}\frac{2a}{\pi}\left(\sin\frac{\pi x}{2a}\right)_{-a}^{a}$$

$$= \frac{V_0}{2} + \frac{V_0}{\pi} = V_0\left(\frac{1}{2} + \frac{1}{\pi}\right)$$

The corrected ground state energy

$$E_1 = \frac{\pi^2\hbar^2}{32ma^2} + V_0\left(\frac{1}{2} + \frac{1}{\pi}\right)$$

8.28 A particle of mass m moves in an infinite one dimensional box of bottom $2a$ with a potential dip as defined by

$$V(x) = \infty \qquad \text{for } x < -a \text{ and } x > a$$

$$V(x) = -V_0 \quad \text{for } -a < x < -\frac{a}{3}$$

$$V(x) = 0 \qquad \text{for } -\frac{a}{3} < x < a$$

Find the energy of the ground state corrected to first order.

Solution. The unperturbed part of the Hamiltonian is that due to a particle in an infinite potential defined by $V(x)$ for $-a < x < a$ and $V(x) = \infty$ otherwise. The unperturbed ground state energy and eignfunctions are

$$E_1 = \frac{\pi^2\hbar^2}{8ma^2}, \qquad \psi_1 = \frac{1}{\sqrt{a}} \cos\frac{\pi x}{2a}$$

The perturbation $H' = -V_0$, $-a < x < -(a/3)$. The first order correction is

$$E^{(1)} = -\frac{V_0}{a}\int_{-a}^{-a/3} \cos^2 \frac{\pi x}{2a}\, dx = -\frac{V_0}{2a}\int_{-a}^{-a/3}\left(1 + \cos\frac{\pi x}{a}\right)dx$$

$$= -\frac{V_0}{a}(x)_{-a}^{-a/3} - \frac{V_0}{2a}\frac{a}{\pi}\left(\sin\frac{\pi x}{a}\right)_{-a}^{-a/3}$$

$$= -\frac{V_0}{3} + \frac{V_0}{2\pi}\sin 60° = -\frac{V_0}{3} + \frac{V_0}{2\pi}\times 0.866$$

$$E_1^{(1)} = 0.195V_0$$

The ground state energy corrected to first order is

$$E = \frac{\pi^2\hbar^2}{8ma^2} - 0.195V_0$$

8.29 A hydrogen atom in the first excited state is placed in a uniform electric field ε along the positive z-axis. Evaluate the second order correction to the energy. Draw an energy level diagram illustrating the different states in the presence of the field. Given

$$\psi_{200} = \frac{1}{\pi^{1/2}}\left(\frac{1}{2a_0}\right)^{3/2}\left(1 - \frac{r}{2a_0}\right)e^{-r/2a_0}$$

$$\psi_{210} = \frac{1}{\pi^{1/2}}\left(\frac{1}{2a_0}\right)^{5/2} re^{-r/2a_0}\cos\theta$$

$$\int_0^\infty x^n e^{-ax}\, dx = \frac{n!}{a^{n+1}}$$

Solution. The first excited state ($n = 2$) is four-fold degenerate. The possible (l, m) values are (0,0), (1,0), (1,1) and (1,–1). The four degerate states are $|nlm\rangle$: $|200\rangle$, $|210\rangle$, $|211\rangle$, and $|21, -1\rangle$. The additional potential energy in the field can be taken as the perturbation, i.e.,

$$H' = er\varepsilon\cos\theta \tag{i}$$

The energy of the $n = 2$ state, E_2^0 is the unperturbed energy. Out of the 12 off-diagonal elements, in 10 we have the factor

$$\int_0^{2\pi} e^{i(m'-m)\phi}\, d\phi$$

which is equal to zero if $m' \neq m$. Only two off-diagonal elements will be nonvanishing; these are

$$\langle 200|er\varepsilon\cos\theta|210\rangle = \frac{e\varepsilon}{16\pi a_0^4}\int_0^{2\pi}\int_0^{\pi}\int_0^{\infty}\left(1 - \frac{r}{2a_0}\right)r^4 e^{-r/a_0}\cos^2\theta\sin\theta\, dr\, d\theta\, d\phi$$

$$= \frac{e\varepsilon 2\pi}{16\pi a_0^4}\int_0^{\pi}\cos^2\theta\sin\theta\, d\theta\int_0^{\infty}\left(r^4 - \frac{r^5}{2a_0}\right)e^{-r/a_0}\, dr \tag{ii}$$

The integral in θ is very straightforward. The integral in the variable r can be evaluated with the data given. Then,

$$\int_0^{\pi} \cos^2\theta \sin\theta \, d\theta = \frac{2}{3} \tag{iii}$$

$$\int_0^{\infty} \left(r^4 - \frac{r^5}{2a_0} \right) e^{-r/a_0} \, dr = -36a_0^5 \tag{iv}$$

Substituting these integrals in Eq. (ii), we get

$$\langle 200|H'|210\rangle = \frac{e\varepsilon}{8a_0^4} \times \frac{2}{3}(36a_0^5) = -3ea_0\varepsilon \tag{v}$$

Then the perturbation matrix is

$$\begin{array}{c} (nlm) \to \\ \downarrow \\ (200) \\ (210) \\ (211) \\ (21,-1) \end{array} \begin{array}{c} (200) \quad (210) \quad (211) \quad (21,-1) \\ \\ \begin{bmatrix} 0 & -3ea_0\varepsilon & 0 & 0 \\ -3ea_0\varepsilon & 0 & 0 & 0 \\ 0 & 0 & 0 & 0 \\ 0 & 0 & 0 & 0 \end{bmatrix} \end{array} \tag{vi}$$

and the secular determinant is

$$\begin{vmatrix} -E_2^{(1)} & -3ea_0\varepsilon & 0 & 0 \\ -3ea_0\varepsilon & -E_2^{(1)} & 0 & 0 \\ 0 & 0 & -E_2^{(1)} & 0 \\ 0 & 0 & 0 & -E_2^{(1)} \end{vmatrix} = 0 \tag{vii}$$

The four roots of this determinant are $3ea_0\varepsilon$, $-3ea_0\varepsilon$, 0 and 0. The states $|200\rangle$ and $|210\rangle$ are affected by the electric field, whereas the states $|211\rangle$ and $|21, -1\rangle$ are not. Including the correction, the energy of the states are

$$E_2^0 - 3ea_0\varepsilon, \qquad E_2^0 \text{ and } E_2^0 + 3ea_0\varepsilon$$

This is illustrated below (The eigenstates are also noted these).

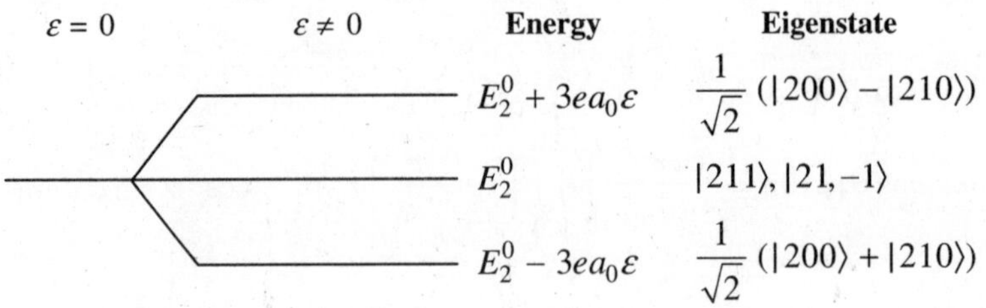

Note: The electric field has affected the energy means that the atom has a permanent magnetic moment. The states $|211\rangle$ and $|21, -1\rangle$ do not possess dipole moment and therefore do not have first order interaction.

8.30 The ground state of the Hydrogen atom is split by the hyperfine interaction. Work out the interaction energy using first order perturbation theory and indicate the level diagram.

Solution. Hyperfine interaction is one that takes place between the electronic angular momentum and the nuclear spin angular momentum. Hydrogen atom in the ground state has no orbital angular momentum. Hence the electronic angular momentum is only due to electron spin and the interaction is simply between the intrinsic angular momenta of the electron ($\mathbf{S}_e$) and proton ($\mathbf{S}_p$); both are spin-half particles. The resultant angular momentum

$$\boldsymbol{I} = \boldsymbol{S}_e + \boldsymbol{S}_p$$

$$\boldsymbol{S}_e \cdot \boldsymbol{S}_p = \frac{1}{2}(I^2 - S_e^2 - S_p^2)$$

Since both are spin half particles, the possible values of I are 0 and 1. $I = 0$ corresponds to a singlet state and $I = 1$ to a triplet state.

$$\langle \boldsymbol{S}_e \cdot \boldsymbol{S}_p \rangle = \frac{1}{2}\left[I(I+1) - \frac{1}{2} \times \frac{3}{2} - \frac{1}{2} \times \frac{3}{2} \right] \hbar^2$$

$$= \begin{cases} -\dfrac{3}{4}\hbar^2, & I = 0 \text{ (singlet state)} \\ \dfrac{1}{4}\hbar^2, & I = 1 \text{ (triplet state)} \end{cases}$$

The hyperfine interaction causes the ground state to split into two, a singlet ($I = 0$) and a triplet ($I = 1$), see Fig. 8.6.

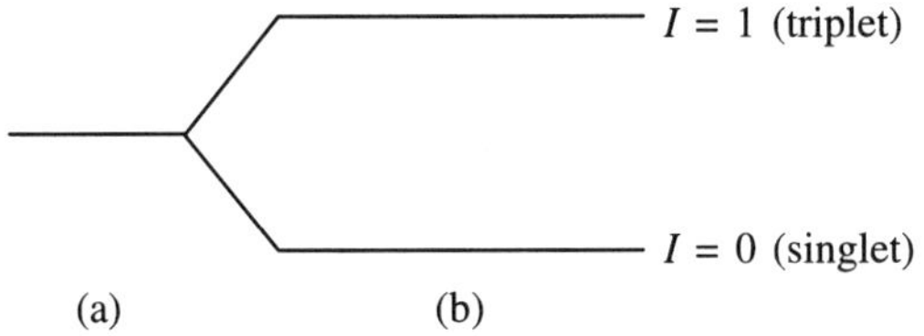

Fig. 8.6 Energy level: (a) without hyperfine interaction; (b) with hyperfine interaction.

8.31 Consider an atomic electron with angular momentum quantum number $l = 3$, placed in a magnetic field of 2 T along the z-direction. Into how many components does the energy level of the atom split. Find the separation between the energy levels.

Solution. For $l = 3$, m can have the values 3, 2, 1, 0, –1, –2, –3. The interaction Hamiltonian $H' = -\boldsymbol{\mu} \cdot \boldsymbol{B}$, where $\boldsymbol{\mu}$ is the magnetic moment of the electron which is given by

$$\boldsymbol{\mu} = -\frac{e}{2m_0}\boldsymbol{L}$$

Here, $\boldsymbol{L}$ is the orbital angular momentum of the electron and m_0 is its rest mass.

$$H' = \frac{e}{2m_0}\,\boldsymbol{L}\cdot\boldsymbol{B} = \frac{eB}{2m_0}\,L_z$$

In the $|lm\rangle$ basis, the energy

$$E = \frac{eB}{2m_0}m\hbar = \frac{e\hbar}{2m_0}Bm = \mu_B Bm$$

where μ_B is the Bohr magneton which has a value of 9.27×10^{-24} J/T. Since m can have seven values, the energy level splits into seven. The energies of these seven levels are

$$3\mu_B B,\quad 2\mu_B B,\quad 1\mu_B B,\quad 0,\quad -1\mu_B B,\quad -2\mu_B B,\quad -3\mu_B B$$

The lines are equally spaced and the separation between any two is

$$\mu_B B = (9.27 \times 10^{-24}\ \text{J/T}) \times 2T$$
$$= 18.54 \times 10^{-24}\ \text{J}$$

8.32 A system described by the Hamiltonian $H = \alpha L^2$, where L^2 is the square of the angular momentum and α is a constant, exhibits a line spectrum where the line A represents transition from the second excited state to the first excited state. The system is now placed in an external magnetic field and the Hamiltonian changes to $H = \alpha L^2 + \beta L_z$, where L_z is the z-component of the angular momentum. How many distinct lines will the original line A split into?

Solution. The Hamiltonian $H = \alpha L^2$. The eigenkets are $|lm\rangle$, $l = 0, 1, 2, \ldots$, $m = 0, \pm1, \pm2, \ldots$ The first excited state is $l = 1$, $m = 0, \pm1$. The second excited state is $l = 2$, $m = 0, \pm1, \pm2$. In the presence of magnetic field, $H = \alpha L^2 + \beta L_z$. The perturbation $H' = \beta L_z$.

$$\text{First order correction} = \langle lm|\beta L_z|lm\rangle$$
$$= \beta m\hbar \text{ for a given value of } l$$

For the first excited state,

$$\beta m\hbar = \beta\hbar,\ 0,\ -\beta\hbar$$

For the second excited state

$$\beta m\hbar = 2\beta\hbar,\ \beta\hbar,\ 0,\ -\beta\hbar,\ -2\beta\hbar$$

Figure 8.7 illustrates the splitting of the two energy levels. The allowed transitions

$$\Delta l = \pm1, \qquad \Delta m = 0, \pm1$$

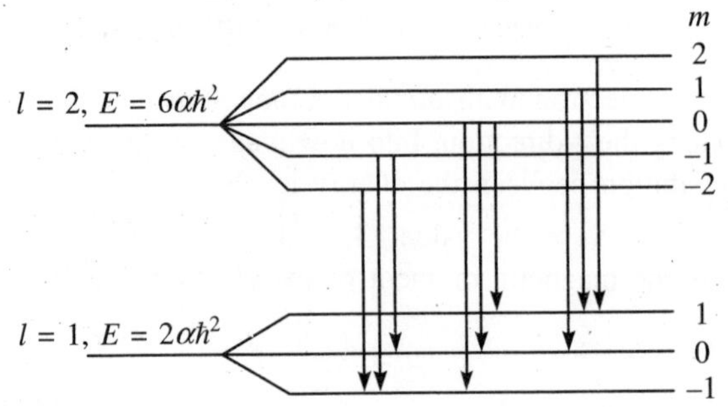

Fig. 8.7 Transitions in the presence of magnetic field.

Transitions are also shown in Figure 8.7. The energies of the levels are also given, from which the transition energies can be evaluated. The original line will split into eight lines.

8.33 The Hamiltonian of a two-electron syatem is perturbed by an interaction $\alpha \mathbf{S}_1 \cdot \mathbf{S}_2$, where α is a constant and $\mathbf{S}_1$ and $\mathbf{S}_2$ are the spin angular momenta of the electrons. Calculate the splitting between the $S = 0$ and $S = 1$ states by first order perturbation, where S is the magnitude of the total spin.

Solution. We have $\mathbf{S} = \mathbf{S}_1 + \mathbf{S}_2$. Then,

$$S^2 = S_1^2 + S_2^2 + 2\mathbf{S}_1 \cdot \mathbf{S}_2$$

$$\mathbf{S}_1 \cdot \mathbf{S}_2 = \frac{S^2 - S_1^2 - S_2^2}{2}$$

Since the spin of electron is 1/2 when the two electrons combine, the total spin $S = 0$ or 1. The state, for which $S = 0$, is called a *singlet state* with $m_s = 0$. The state, for which $S = 1$, is called a *triplet state* with $m_s = 1, 0, -1$. The first order correction to $S = 0$ state in the $|sm_s\rangle$ basis

$$E_0^{(1)} = \left\langle sm_s \left| \frac{(S^2 - S_1^2 - S_2^2)\,\alpha}{2} \right| sm_s \right\rangle$$

$$= \frac{\alpha}{2}\,[s(s+1) - s_1(s_1+1) - s_2(s_2+1)]\,\hbar^2$$

$$= \frac{\alpha}{2}\left(0 - \frac{3}{4} - \frac{3}{4}\right)\hbar^2 = -\frac{3}{4}c\hbar^2$$

The first order correction to the $S = 1$ state is

$$E_1^{(1)} = \frac{\alpha}{2}\left[1 \times 2 - \frac{1}{2} \times \frac{3}{2} - \frac{1}{2} \times \frac{3}{2}\right]\hbar^2$$

$$= \frac{\alpha}{4}\hbar^2$$

$$\text{Splitting between the two states} = \frac{\alpha}{4}\hbar^2 - \left(-\frac{3}{4}\,a\hbar^2\right)$$

$$= \alpha\hbar^2$$

8.34 The unperturbed Hamiltonian of a system is

$$H_0 = \frac{p^2}{2m} + \frac{1}{2}m\omega^2 x^2$$

If a small perturbation

$$V' = \begin{cases} \lambda x & \text{for } x > 0 \\ 0 & \text{for } x \le 0 \end{cases}$$

acts on the system, evaluate the first order correction to the ground state energy.

Solution. The given H_0 is the one for a simple harmonic oscillator. Hence the unperturbed ground state energy is

$$\psi_0(x) = \left(\frac{m\omega}{\hbar\pi}\right)^{1/4} \exp\left(\frac{m\omega x^2}{2\hbar}\right)$$

The first order correction to the energy is

$$\begin{aligned} E_0^{(1)} &= \langle \psi_0(x) | \lambda x | \psi_0(x) \rangle \\ &= \left(\frac{m\omega}{\hbar\pi}\right)^{1/2} \lambda \int_0^\infty x \exp\left(-\frac{m\omega x^2}{\hbar}\right) dx \\ &= \left(\frac{m\omega}{\hbar\pi}\right)^{1/2} \lambda \left(\frac{\hbar}{2m\omega}\right) = \frac{\lambda}{2}\sqrt{\frac{\hbar}{\pi m\omega}} \end{aligned}$$

8.35 Consider an atomic state specified by angular momenta $\boldsymbol{L}$, $\boldsymbol{S}$ and $\boldsymbol{J} = \boldsymbol{L} + \boldsymbol{S}$ placed in a magnetic field $\boldsymbol{B}$. Treating the interaction representing the magnetic moment of the electron in the magnetic field as the perturbing Hamiltonian and writing $\boldsymbol{L} + 2\boldsymbol{S} = g_J \boldsymbol{J}$, obtain an expression for (i) the g factor of the Jth state are (ii) the corrected energy.

Solution. When placed in the magnetic field $\boldsymbol{B}$, the interaction Hamiltonian

$$H' = -\boldsymbol{\mu} \cdot \boldsymbol{B} = -(\boldsymbol{\mu}_L + \boldsymbol{\mu}_S) \cdot \boldsymbol{B} \qquad \text{(i)}$$

where $\boldsymbol{\mu}_L$ and $\boldsymbol{\mu}_S$ are the orbital and spin magnetic moments of the electron. We have

$$\boldsymbol{\mu}_L = -\frac{e}{2m}\boldsymbol{L}, \qquad \boldsymbol{\mu}_S = -\frac{e}{2m}\boldsymbol{S} \qquad \text{(ii)}$$

$\boldsymbol{L}$ is the orbital angular momentum and $\boldsymbol{S}$ is the spin angular momentum. Substituting these values of $\boldsymbol{\mu}_L$ and $\boldsymbol{\mu}_S$, we get

$$H' = \frac{e}{m}(\boldsymbol{L} + 2\boldsymbol{S}) \cdot \boldsymbol{B}$$

Given

$$g_J \boldsymbol{J} = \boldsymbol{L} + 2\boldsymbol{S}$$

where g_J is a constant. Taking the dot product with $\boldsymbol{J}$, we obtain

$$\begin{aligned} g_J J^2 &= \boldsymbol{J} \cdot (\boldsymbol{L} + 2\boldsymbol{S}) = \boldsymbol{J} \cdot (\boldsymbol{L} + \boldsymbol{S} + \boldsymbol{S}) \\ &= \boldsymbol{J} \cdot (\boldsymbol{J} + \boldsymbol{S}) = \boldsymbol{J} \cdot \boldsymbol{J} + \boldsymbol{J} \cdot \boldsymbol{S} \\ &= J^2 + \boldsymbol{J} \cdot \boldsymbol{S} \end{aligned}$$

Since $\boldsymbol{L} = \boldsymbol{J} - \boldsymbol{S}$,

$$L^2 = J^2 + S^2 - 2\boldsymbol{J} \cdot \boldsymbol{S}$$

$$\boldsymbol{J} \cdot \boldsymbol{S} = \frac{J^2 + S^2 - L^2}{2}$$

$$g_J J^2 = J^2 + \frac{J^2 + S^2 - L^2}{2}$$

In the simultaneous eigenkets of J^2, J_z, L^2, S^2,

$$g_J \langle J^2 \rangle = \langle J^2 \rangle + \frac{1}{2} \langle J^2 + S^2 - L^2 \rangle$$

$$g_J J(J+1)\hbar^2 = J(J+1)\hbar^2 + \frac{1}{2}[J(J+1) + S(S+1) - L(L+1)]\hbar^2$$

$$g_J = 1 + \frac{J(J+1) + S(S+1) - L(L+1)}{2J(J+1)}$$

where J, L and S are the quantum numbers associated with the angular momenta $\boldsymbol{J}$, $\boldsymbol{L}$ and $\boldsymbol{S}$, respectively.

(ii) The interaction Hamiltonian

$$H' = \frac{e}{2m} g\boldsymbol{J} \cdot \boldsymbol{B} = \frac{e}{2m} gJB \cos\theta$$

$$= \frac{e}{2m} gJB \frac{J_z}{J} = \frac{e}{2m} gBJ_z$$

The first order correction to the energy is the diagonal matrix element

$$E^{(1)} = \frac{e}{2m} gBM_J\hbar = \frac{e\hbar}{2m} Bg_J M_J$$

The corrected energy

$$E = E^0 + \frac{e\hbar}{2m} Bg_J M_J$$

Since M_J can have $(2J + 1)$-fold degenerate, each energy level is split into $2J + 1$ equally spaced levels.

8.36 The nuclear spin of bismuth atom is 9/2. Find the number of levels into which a $^2D_{5/2}$ term of bismuth splits due to nuclear spin-electron angular momentum interaction. If the separation of $^2_7D_{5/2}$ term from $^2_6D_{5/2}$ is 70 cm^{-1}, what is the separation between the other adjacent levels?

Solution. $^2D_{5/2}$ term means $2S + 1 = 2$, $S = (1/2)$, $L = 2$ and $J = (5/2)$. Given $I = (9/2)$. The total angular momentum is $\boldsymbol{F} = \boldsymbol{I} + \boldsymbol{J}$. The possible values of the quantum number F are 7, 6, 5, 4, 3, 2. Hence, the $^2D_{5/2}$ level splits into six sublevels corresponding to the F values, 7, 6, 5, 4, 3, and 2. From Problem 8.21, we have the correction to energy as

$$E^{(1)} = a\,[F(F+1) - I(I+1) - J(J+1)]$$

Hence, the energy difference ΔE between successive levels $(F + 1)$ and F is given by

$$\Delta E = a\,[(F+1)(F+2) - I(I+1) - J(J+1)] - a\,[F(F+1) - I(I+1) - J(J+1)]$$

Given the separation between $J = 7$ and $J = 6$ is 70 cm^{-1}, i.e.,

$$70 \text{ cm}^{-1} = 2a \times 7 \quad \text{or} \quad a = 5 \text{ cm}^{-1}$$

Hence,

$$^2_6D_{5/2} - {}^2_5D_{5/2} = 60 \text{ cm}^{-1}$$

$$^2_5D_{5/2} - {}^2_4D_{5/2} = 50 \text{ cm}^{-1}$$

$$^2_4D_{5/2} - ^2_3D_{5/2} = 40 \text{ cm}^{-1}$$

$$^2_3D_{5/2} - ^2_2D_{5/2} = 30 \text{ cm}^{-1}$$

8.37 Discuss the splitting of atomic energy levels in a weak magnetic fièld and show that an energy level of the atom splits into $(2J + 1)$ levels. Use L-S coupling and $\boldsymbol{L} + 2\boldsymbol{S} = g\boldsymbol{J}$, where g is the Lande g-factor, $\boldsymbol{L}$, $\boldsymbol{S}$ and $\boldsymbol{J}$ are respectively the orbital, spin and total angular momenta of the atom.

Solution. Let μ be the magnetic moment of the atom. Its orbital magnetic moment be μ_L and spin magnetic moment be μ_S. The Hamiltonian representing the interactionof the magnetic field $\boldsymbol{B}$ with μ is

$$H' = -\mu \cdot \mathrm{B} = -(\mu_L + \mu_S) \cdot B$$

Since

$$\mu_L = -\frac{e}{2m}\boldsymbol{L}, \qquad \mu_S = -\frac{e}{m}\boldsymbol{S} = -\frac{e}{2m}2\boldsymbol{S}$$

$$H' = \frac{e}{2m}(\boldsymbol{L} + 2\boldsymbol{S}) \cdot \boldsymbol{B} = \frac{e}{2m}g\boldsymbol{J} \cdot \boldsymbol{B} = \frac{e}{2m}gJB \cos(\boldsymbol{J}, \boldsymbol{B})$$

Since $(\boldsymbol{J}, \boldsymbol{B}) = (J_z/J)$,

$$H' = \frac{e}{2m}gJB\frac{J_z}{J} = \frac{e}{2m}gBJ_z$$

The first order correction to energy in the common state of J^2 and J_z is

$$E^{(1)} = \left\langle Jm_J \left| \frac{e}{2m}gBJ_z \right| Jm_J \right\rangle$$

$$= \frac{e}{2m}gBm_J\hbar = \frac{e\hbar}{2m}gBm_J$$

$$= \mu_B gBm_J$$

where $\mu_B = e\hbar/2m$ is the Bohr magneton. As m_j can have $(2J + 1)$ values, each level splits into $2(J + 1)$ equally spaced levels. Hence the energy of the system

$$E = E_{nl} + m_B gBM_J$$

8.38 Discuss the splitting of atomic energy levels in a strong magnetic field. (the Paschen-Back effect).

Solution. In a strong magnetic field, the magnetic field interaction energy is stronger than the spin-orbit interaction energy. Hence the L–S coupling breaks. The Hamiltonian representing the interaction of the magnetic field with μ is

$$H' = -\mu \cdot \boldsymbol{B} = -(\mu_L + \mu_S) \cdot \boldsymbol{B}$$

$$= \frac{e}{2m}\boldsymbol{L} \cdot \boldsymbol{B} + \frac{e}{2m}2\boldsymbol{S} \cdot \boldsymbol{B}$$

$$= \frac{e}{2m}LB\cos(\boldsymbol{L}, B) + \frac{e}{2m}2SB\cos(\boldsymbol{S}, \boldsymbol{B})$$

$$= \frac{e}{2m} LB\frac{L_z}{L} + \frac{e}{2m} 2SB\frac{S_z}{S}$$

$$= \frac{e}{2m} BL_z + \frac{e}{2m} 2BS_z$$

The first order correction in the common eigenstate of L^2, L_z, S^2 and S_z is

$$E^{(1)} = \frac{e}{2m} Bm_L\hbar + \frac{e}{2m} 2SBm_s\hbar$$

$$= \mu_B B(m_L + 2m_s)$$

The energy of the level becomes

$$E = E_{nl} + \mu_B B(m_L + 2m_s)$$

8.39 A simple pendulum of length l swings in a vertical plane under the influence of gravity. In the small angle approximation, find the energy levels of the system. Also evaluate the first order correction to the ground state energy, taking one more term in the small angle approximation.

Solution. The first part of the problem is discussed in Problem 4.58. The energy eigenvalues and eigenfunctions are the same as those of a linear harmonic oscillator with angular frequency $\omega = \sqrt{g/l}$, where l is the length of the pendulum. While evaluating the energy eigenvalues, we assumed the angle θ (Fig. 4 .5) to be small and retained only two terms in the expansion of cos θ. Retaining one more term, we get

$$\cos\theta = 1 - \frac{\theta^2}{2} + \frac{\theta^4}{24}$$

The potential is, then,

$$V = mgl\,(1 - \cos\theta) = mgl\left(\frac{\theta^2}{2} - \frac{\theta^4}{24}\right)$$

$$= \frac{mgl\theta^2}{2} - \frac{mgl\theta^4}{24}$$

Since $\theta = x/l$,

$$\text{Perturbation } H' = -\frac{mgl\theta^4}{24} = -\frac{mgx^4}{24l^3}$$

The first order correction to the ground state energy is

$$E_0^{(1)} = \left\langle 0 \left| -\frac{mgx^4}{24l^3} \right| 0 \right\rangle$$

In terms of the raising and lowering operators, we have

$$x = \sqrt{\frac{\hbar}{2m\omega}}(a + a^\dagger)$$

With this value of x,

$$E_0^{(1)} = \left(-\frac{mg}{24l^3}\right)\left(\frac{\hbar}{2m\omega}\right)^2 \langle 0|(a + a^\dagger)(a + a^\dagger)(a + a^\dagger)(a + a^\dagger)|0\rangle$$

In all, there will be 16 terms on the RHS. However, only two will be nonvanishing. They are $\langle 0|aaa^\dagger a^\dagger|0\rangle$ and $\langle 0|a\ a^\dagger\ a\ a^\dagger|0\rangle$. Consequently,

$$\langle 0|aa^\dagger aa^\dagger|0\rangle = 1, \qquad \langle 0|aaa^\dagger a^\dagger|0\rangle = 2$$

Hence,

$$E_0^{(1)} = -\frac{g\hbar^2}{8ml^3\omega^2}$$

8.40 Obtain the hyperfine splitting in the ground state of the hydrogen atom to first order in perturbation theory, for the perturbation

$$H' = A\mathbf{S}_\mathrm{p} \cdot \mathbf{S}_\mathrm{e}\delta^3(\boldsymbol{r}), \qquad A \text{ being constant}$$

where S_p and S_e denote the spins of the proton and electron, respectively.

Solution. The hydrogen ground state wave function is

$$\psi_{100} = \left(\frac{1}{\pi a_0^3}\right)^{1/2} e^{-r/a_0}$$

The perturbation $H' = A\mathbf{S}_\mathrm{p} \cdot \mathbf{S}_\mathrm{e}\delta^3(\boldsymbol{r})$. Denoting the spin function by χ_s, the total wave function of the ground state is

$$\psi = \psi_{100}\,\chi_s$$

The first order correction to energy

$$\begin{aligned}
E_0^{(1)} &= \langle \psi_{100}\chi_s|A\mathbf{S}_\mathrm{p} \cdot \mathbf{S}_\mathrm{e}\delta^3(\boldsymbol{r})|\psi_{100}\chi_s\rangle \\
&= \langle \psi_{100}|A\delta^3(\boldsymbol{r})|\psi_{100}\rangle\langle \chi_s|\mathbf{S}_\mathrm{p} \cdot \mathbf{S}_\mathrm{e}|\chi_s\rangle \\
&= \frac{A}{\pi a_0^3}\langle \chi_s|\mathbf{S}_\mathrm{p} \cdot \mathbf{S}_\mathrm{e}|\chi_s\rangle
\end{aligned}$$

Writing

$$\mathbf{F} = \mathbf{S}_\mathrm{p} + \mathbf{S}_\mathrm{e} \quad \text{or} \quad \mathbf{S}_\mathrm{p} \cdot \mathbf{S}_\mathrm{e} = \frac{F^2 - S_\mathrm{p}^2 - S_\mathrm{e}^2}{2}$$

$$\begin{aligned}
E_0^{(1)} &= \frac{A}{\pi a_0^3}\left\langle \chi_s \left| \frac{F^2 - S_\mathrm{p}^2 - S_\mathrm{e}^2}{2} \right| \chi_s \right\rangle \\
&= \frac{A}{2\pi a_0^3}[F(F+1) - S_\mathrm{p}(S_\mathrm{p}+1) - S_\mathrm{e}(S_\mathrm{e}+1)]\hbar^2
\end{aligned}$$

As $S_\mathrm{p} = (1/2)$ and $S_\mathrm{e} = (1/2)$, the possible values of F are 1, 0. The separation between the two F states is the hyperfine splitting ΔE. Thus,

$$\Delta E = \frac{A}{2\pi a_0^3}\left[\left(1\times 2-\frac{1}{2}\times\frac{3}{2}-\frac{1}{2}\times\frac{3}{2}\right)\left(0-\frac{1}{2}\times\frac{3}{2}-\frac{1}{2}\times\frac{3}{2}\right)\right]$$

$$= \frac{A}{\pi a_0^3}$$

8.41 In the nonrelativistic limit, the kinetic energy of a particle moving in a potential $V(x) = 1/2m\omega^2$ is $p^2/2m$. Obtain the relativistic correction to the kinetic energy. Treating the correction as a perturbation, compute the first order correction to the ground state energy.

Solution. The relativistic expression for kinetic energy is

$$T = \sqrt{m_0^2c^4 + c^2p^2} - m_0c^2$$

$$= m_0c^2\left(1+\frac{p^2}{m_0^2c^2}\right)^{1/2} - m_0c^2$$

$$= m_0c^2\left(1+\frac{p^2}{2m_0^2c^2}-\frac{p^4}{4m_0^4c^4}\right) - m_0c^2$$

$$= \frac{p^2}{2m_0} - \frac{p^4}{8m_0^3c^2}$$

Perturbation

$$H' = \frac{p^4}{8m_0^3c^2}$$

The operators a and $a^\dagger$ are defined by

$$a = \sqrt{\frac{m\omega}{2\hbar}}x + \frac{i}{\sqrt{2m\hbar\omega}}p$$

$$a^\dagger = \sqrt{\frac{m\omega}{2\hbar}}x - \frac{i}{\sqrt{2m\hbar\omega}}p$$

where

$$p = \frac{\sqrt{2m\hbar\omega}}{2i}(a - a^\dagger)$$

The first order correction to the ground state energy is

$$E_0^{(1)} = \left\langle 0\left|\left(-\frac{p^4}{8m_0^3c^2}\right)\right|0\right\rangle = -\frac{1}{8m_0^3c^2}$$

$$\times\left\langle 0\left|\left(\frac{2m\hbar\omega}{4}\right)^2 (a-a^\dagger)(a-a^\dagger)(a-a^\dagger)(a-a^\dagger)\right|0\right\rangle$$

$$E_0^{(1)} = -\frac{1}{8m_0^3c^2}\left(\frac{2m\hbar\omega}{4}\right)^2 \langle 0|(a-a^\dagger)(a-a^\dagger)(a-a^\dagger)(a-a^\dagger)|0\rangle$$

When expanded, the expression will have 16 terms. Only two terms will be nonvanishing; these terms are

$$\langle 0 | a a a^\dagger a^\dagger | 0 \rangle \quad \text{and} \quad \langle 0 | a a^\dagger a a^\dagger | 0 \rangle$$

Since

$$a^\dagger | n \rangle = \sqrt{n+1} \, | n+1 \rangle, \qquad a | n \rangle = \sqrt{n} \, | n-1 \rangle$$

we have

$$\langle 0 | a a a^\dagger a^\dagger | 0 \rangle = 2, \qquad \langle 0 | a a^\dagger a a^\dagger | 0 \rangle = 1$$

Hence,

$$E_0^{(1)} = -\frac{3}{32} \frac{(\hbar\omega)^2}{m_0 c^2}$$

8.42 The Hamiltonian matrix of a system in the orthonormal basis

$$\begin{pmatrix} 1 \\ 0 \\ 0 \end{pmatrix}, \quad \begin{pmatrix} 0 \\ 1 \\ 0 \end{pmatrix}, \quad \begin{pmatrix} 0 \\ 0 \\ 1 \end{pmatrix}$$

is given by

$$H = \begin{pmatrix} 1 & 2\varepsilon & 0 \\ 2\varepsilon & 2+\varepsilon & 3\varepsilon \\ 0 & 3\varepsilon & 3+\varepsilon \end{pmatrix}$$

Find the energy levels corrected up to second order in the small parameter ε.

Solution. The matrix H can be written as

$$H = \begin{pmatrix} 1 & 0 & 0 \\ 0 & 2 & 0 \\ 0 & 0 & 3 \end{pmatrix} + \begin{pmatrix} 1 & 2\varepsilon & 0 \\ 2\varepsilon & 2+\varepsilon & 3\varepsilon \\ 0 & 3\varepsilon & 3+\varepsilon \end{pmatrix}$$

$$= H^0 + H'$$

Identifying H^0 and H' as the unperturbed and perturbation part, the eigenvalues of the unperturbed Hamiltonian H^0 are 1, 2 and 3. The first order correction to the energy is given by the diagonal matrix element of H'. Then,

$$H'_{11} = (1 \;\; 0 \;\; 0) \begin{pmatrix} 0 & 2 & 0 \\ 2 & 1 & 3 \\ 0 & 3 & 1 \end{pmatrix} \begin{pmatrix} 1 \\ 0 \\ 0 \end{pmatrix} \varepsilon = 0$$

$$H'_{22} = (0 \;\; 1 \;\; 0) \begin{pmatrix} 0 & 2 & 0 \\ 2 & 1 & 3 \\ 0 & 3 & 1 \end{pmatrix} \begin{pmatrix} 0 \\ 1 \\ 0 \end{pmatrix} \varepsilon = 1\varepsilon$$

$$H'_{33} = (0 \;\; 0 \;\; 1) \begin{pmatrix} 0 & 2 & 0 \\ 2 & 1 & 3 \\ 0 & 3 & 1 \end{pmatrix} \begin{pmatrix} 0 \\ 0 \\ 1 \end{pmatrix} \varepsilon = 1\varepsilon$$

The first order correction to the energies are 0, 1ε, 1ε, respectively. The second order correction is given by

$$E_n^{(2)} = \sum_m{}' \frac{|\langle m|H'|n\rangle|^2}{E_n^0 - E_m^0}$$

$$H'_{12} = (1 \;\; 0 \;\; 0)\begin{pmatrix} 0 & 2 & 0 \\ 2 & 1 & 3 \\ 0 & 3 & 1 \end{pmatrix}\begin{pmatrix} 0 \\ 1 \\ 0 \end{pmatrix}\varepsilon = (1 \;\; 0 \;\; 0)\begin{pmatrix} 2 \\ 1 \\ 3 \end{pmatrix}\varepsilon = 2\varepsilon$$

$$H'_{13} = (1 \;\; 0 \;\; 0)\begin{pmatrix} 0 & 2 & 0 \\ 2 & 1 & 3 \\ 0 & 3 & 1 \end{pmatrix}\begin{pmatrix} 0 \\ 0 \\ 1 \end{pmatrix}\varepsilon = (1 \;\; 0 \;\; 0)\begin{pmatrix} 0 \\ 3 \\ 1 \end{pmatrix}\varepsilon = 0$$

$$H'_{23} = (0 \;\; 1 \;\; 0)\begin{pmatrix} 0 & 2 & 0 \\ 2 & 1 & 3 \\ 0 & 3 & 1 \end{pmatrix}\begin{pmatrix} 0 \\ 0 \\ 1 \end{pmatrix}\varepsilon = (0 \;\; 1 \;\; 0)\begin{pmatrix} 0 \\ 3 \\ 1 \end{pmatrix}\varepsilon = 3\varepsilon$$

$$E_1^{(2)} = \frac{|H'_{21}|^2}{1-2} + \frac{|H'_{31}|^2}{1-3} = -4\varepsilon^2 + 0 = -4\varepsilon^2$$

$$E_2^{(2)} = \frac{|H'_{12}|^2}{2-1} + \frac{|H'_{32}|^2}{2-3} = 4\varepsilon^2 - 9\varepsilon^2 = -5\varepsilon^2$$

$$E_3^{(2)} = \frac{|H'_{13}|^2}{3-1} + \frac{|H'_{23}|^2}{3-2} = 0 + 9\varepsilon^2 = 9\varepsilon^2$$

The energies of the three levels corrected to second order are

$$E_1 = 1 + 0 - 4\varepsilon^2 = 1 - 4\varepsilon^2$$

$$E_2 = 2 + \varepsilon - 5\varepsilon^2$$

$$E_3 = 3 + 1\varepsilon + 9\varepsilon^2$$

CHAPTER 9

Variation and WKB Methods

The variation method is usually applied to obtain the ground state energy and wave functions of quantum mechanical systems. Extension to excited states is also possible. The WKB method is based on the expansion of the wave function of a one-dimensional system in powers of $\hbar$.

9.1 Variation Method

The essential idea of the method is to evaluate the expectation value $\langle H \rangle$ of the Hamiltonian operator H of the system with respect to a trial wave function ϕ. The variational principle states that the ground state energy

$$E_1 \leq \langle H \rangle = \langle \phi \,|H|\, \phi \rangle \tag{9.1}$$

In practice, the trial function is selected in terms of one or more variable parameters and the value of $\langle H \rangle$ is evaluated. The value of $\langle H \rangle$ is then minimized with respect to each of the parameters. The resulting value is the closest estimate possible with the selected trial function. If the trial wave function is not a normalized one, then

$$\langle H \rangle = \frac{\langle \phi | H | \phi \rangle}{\langle \phi | \phi \rangle} \tag{9.2}$$

9.2 WKB Method

The WKB method is based on the expansion of the wave function in powers of $\hbar$. This method is applicable when the potential $V(x)$ is slowly varying. When $E > V(x)$, the Schrodinger equation for a one-dimensional system is given by

$$\frac{d^2\psi}{dx^2} + k^2\psi = 0, \qquad k^2 = \frac{2m}{\hbar^2}[E - V(x)] \tag{9.3}$$

The solution is given by

$$\psi = \frac{A}{\sqrt{k}} \exp\left(\pm i \int k\, dx\right) \tag{9.4}$$

where A is a constant. The general solution will be a linear combination of the two. When $E < V(x)$, the basic equation becomes

$$\frac{d^2\psi}{dx^2} - \gamma^2\psi = 0, \qquad \gamma^2 = \frac{2m[V(x) - E]}{\hbar^2} \tag{9.5}$$

Then the solution of Eq. (9.5) is

$$\psi = \frac{B}{\sqrt{\gamma}} \exp\left(\pm\int \gamma\, dx\right) \tag{9.6}$$

where B is a constant.

9.3 The Connection Formulas

When $E \cong V(x)$, both the quantities k and $\gamma \to 0$. Hence, ψ goes to infinity. The point at which $E = V(x)$ is called the **turning point**. On one side the solution is exponential and on the other side, it is oscillatory. The solutions for the regions $E > V(x)$ and $E < V(x)$ must be connected. The connection formulas are as follows:

Barrier to the right of the turning point at x_1:

$$\frac{2}{\sqrt{k}} \cos\left(\int_x^{x_1} k\, dx - \frac{\pi}{4}\right) \leftarrow \frac{1}{\sqrt{\gamma}} \exp\left(-\int_{x_1}^{x} \gamma\, dx\right) \tag{9.7}$$

$$\frac{1}{\sqrt{k}} \sin\left(\int_x^{x_1} k\, dx - \frac{\pi}{4}\right) \rightarrow -\frac{1}{\sqrt{\gamma}} \exp\left(\int_{x_1}^{x} \gamma\, dx\right)$$

Barrier to the left of the turning point at x_2:

$$\frac{1}{\sqrt{\gamma}} \exp\left(-\int_x^{x_2} \gamma\, dx\right) \rightarrow \frac{2}{\sqrt{k}} \cos\left(\int_{x_2}^{x} k\, dx - \frac{\pi}{4}\right) \tag{9.8}$$

$$-\frac{1}{\sqrt{\gamma}} \exp\left(\int_x^{x_2} \gamma\, dx\right) \leftarrow \frac{1}{\sqrt{k}} \sin\left(\int_{x_2}^{x} k\, dx - \frac{\pi}{4}\right)$$

The approximation breaks down if the turning points are close to the top of the barrier. *Barrier penetration*: For a broad high barrier, the transmission coefficient

$$T = \exp\left(-2\int_{x_1}^{x_2} \gamma\, dx\right) \tag{9.9}$$

PROBLEMS

9.1 Optimize the trial function exp $(-\alpha r)$ and evaluate the ground state energy of the hydrogen atom.

Solution. The trial function $\phi = \exp(-\alpha r)$.

$$\text{Hamiltonian of the atom } H = -\frac{\hbar^2}{2\mu}\nabla^2 - \frac{ke^2}{r}$$

The trial function depends only on r. Hence, ∇^2 in the spherical polar coordinates contains only the radial derivatives. So,

$$\nabla^2 = \frac{1}{r^2}\frac{d}{dr}\left(r^2\frac{d}{dr}\right) = \frac{d^2}{dr^2} + \frac{2}{r}\frac{d}{dr}$$

From Eq. (9.2),

$$\langle H\rangle\langle\phi|\phi\rangle = -\frac{\hbar^2}{2\mu}\left[\left\langle\phi\left|\frac{d^2}{dr^2}\right|\phi\right\rangle + \left\langle\phi\left|\frac{2}{r}\frac{d}{dr}\right|\phi\right\rangle\right] - \left\langle\phi\left|\frac{ke^2}{r}\right|\phi\right\rangle$$

The angular part of $d\tau$ contributes a factor 4π to the integrals in the above equation. Hence,

$$\left\langle\phi\left|\frac{d^2}{dr^2}\right|\phi\right\rangle = 4\pi\alpha^2\int_0^\infty r^2\exp(-2\alpha r)\,dr = \frac{\pi}{\alpha}$$

$$\left\langle\phi\left|\frac{2}{r}\frac{d}{dr}\right|\phi\right\rangle = -8\pi\alpha\int_0^\infty r\exp(-2\alpha r)\,dr = -\frac{2\pi}{\alpha}$$

$$\left\langle\phi\left|\frac{ke^2}{r}\right|\phi\right\rangle = 4\pi e^2\int_0^\infty r\exp(-2\alpha r)\,dr = \frac{\pi ke^2}{\alpha^2}$$

$$\langle\phi|\phi\rangle = 4\pi\int_0^\infty r^2\exp(-2\alpha r)\,dr = \frac{\pi}{\alpha^3}$$

Substituting these integrals, we get

$$\langle H\rangle\frac{\pi}{\alpha^3} = -\frac{\hbar^2}{2\mu}\left[\frac{\pi}{\alpha} + \frac{2\pi}{\alpha}\right] - \frac{\pi ke^2}{\alpha^2}$$

$$\langle H\rangle = \frac{\hbar^2\alpha^2}{2\mu} - \alpha ke^2$$

Minimizing with respect to α, we obtain

$$0 = \frac{\hbar^2\alpha}{\mu} - ke^2 \quad \text{or} \quad \alpha = \frac{k\mu e^2}{\hbar^2}$$

With this value of α,

$$E_{\min} = \langle H\rangle_{\min} = -\frac{\mu k^2e^4}{2\hbar^2}$$

and the optimum wave function is

$$\phi = \left(\frac{1}{\pi a_0^3}\right)^{1/2} \exp\left(\frac{-r}{a_0}\right)$$

where a_0 is the Bohr radius.

9.2 Estimate the ground state energy of a one-dimensional harmonic oscillator of mass m and angular frequency ω using a Gaussian trial function.

Solution. The Hamiltonian of the system $H = \dfrac{-\hbar^2}{2m}\dfrac{d^2}{dx^2} + \dfrac{1}{2}m\omega^2 x^2$

$$\text{Gaussian trial function } \phi(x) = A \exp(-\alpha x^2)$$

where A and α are constants. The normalization condition gives

$$1 = |A|^2 \int_{-\infty}^{\infty} \exp(-2\alpha x^2)\, dx = |A|^2 \left(\frac{\pi}{2\alpha}\right)^{1/2}$$

$$\text{Normalized trial function } \phi(x) = \left(\frac{2\alpha}{\pi}\right)^{1/4} \exp(-\alpha x^2)$$

$$\langle H \rangle = -\frac{\hbar^2}{2m}\left\langle \phi \left| \frac{d^2}{dx^2} \right| \phi \right\rangle + \frac{1}{2}m\omega^2 \langle \phi | x^2 | \phi \rangle$$

$$\left\langle \phi \left| \frac{d^2}{dx^2} \right| \phi \right\rangle = -\left(\frac{2\alpha}{\pi}\right)^{1/2} 2\alpha \int_{-\infty}^{\infty} \exp(-2\alpha x^2)\, dx + \left(\frac{2\alpha}{\pi}\right)^{1/2} 4\alpha^2 \int_{-\infty}^{\infty} x^2 \exp(-2\alpha x^2)\, dx$$

$$= -\left(\frac{2\alpha}{\pi}\right)^{1/2} 2\alpha \left(\frac{\pi}{2\alpha}\right)^{1/2} + \left(\frac{2\alpha}{\pi}\right)^{1/2} 4\alpha^2 \frac{1}{4\alpha}\left(\frac{\pi}{2\alpha}\right)^{1/2} = -\alpha$$

$$\langle \phi | x^2 | \phi \rangle = \left(\frac{2\alpha}{\pi}\right)^{1/2} \int_{-\infty}^{\infty} x^2 \exp(-2\alpha x^2)\, dx = \frac{1}{4\alpha}$$

$$\langle H \rangle = \frac{\hbar^2 \alpha}{2m} + \frac{1}{2}m\omega^2 \frac{1}{4\alpha} = \frac{\hbar^2 \alpha}{2m} + \frac{m\omega^2}{8\alpha}$$

Minimizing with respect to α, we get

$$0 = \frac{d\langle H \rangle}{d\alpha} = \frac{\hbar^2}{2m} - \frac{m\omega^2}{8\alpha^2} \quad \text{or} \quad \alpha = \frac{m\omega}{\hbar}$$

With this value of α,

$$\langle H \rangle_{\min} = \frac{1}{2}\hbar\omega$$

which is the same as the value we obtained in Chapter 4. Thus, the trial wave function is the exact eigenfunction.

9.3 The Schrödinger equation of a particle confined to the positive x-axis is

$$\frac{-\hbar^2}{2m}\frac{d^2\psi}{dx^2} + mgx\psi = E\psi$$

with $\psi(0) = 0$, $\psi(x) \to 0$ as $x \to \infty$ and E is the energy eigenvalue. Use the trial function $x \exp(-ax)$ and obtain the best value of the parameter a.

Solution.

$$\text{Hamiltonian } H = \frac{-\hbar^2}{2m}\frac{d^2}{dx^2} + mgx$$

$$\text{Trial function } \phi(x) = x \exp(-ax)$$

$$\langle\phi|\phi\rangle = \int_0^\infty x^2 \exp(-2ax)\, dx = \frac{1}{4a^3}$$

$$\left\langle \phi \left| \frac{-\hbar^2}{2m}\frac{d^2}{dx^2} \right| \phi \right\rangle = \frac{\hbar^2}{m}\alpha \int_0^\infty x \exp(-2\alpha x)\, dx - \frac{\hbar^2 a^2}{2m}\int_0^\infty x^2 \exp(-2\alpha x)\, dx$$

$$= \frac{\hbar^2}{4ma} - \frac{\hbar^2}{8ma} = \frac{\hbar^2}{8ma}$$

$$\langle\phi|mgx|\phi\rangle = mg\int_0^\infty x^3 \exp(-2\alpha x)\, dx = \frac{3mg}{8a^4}$$

$$\langle H\rangle = \frac{\langle\phi|H|\phi\rangle}{\langle\phi|\phi\rangle} = \frac{[\hbar^2/(8ma)] + (3mg/8a^4)}{1/4a^3} = \frac{\hbar^2 a^2}{2m} + \frac{3}{2}\frac{mg}{a}$$

Minimizing $\langle H\rangle$ with respect to a, we get

$$0 = \frac{\hbar^2}{2m}2a - \frac{3}{2}\frac{mg}{a^2} \quad \text{or} \quad a = \left(\frac{3}{2}\frac{m^2 g}{\hbar^2}\right)^{1/3}$$

which is the best value of the parameter a so that $\langle H\rangle$ is minimum.

9.4 A particle of mass m moves in the attractive central potential $V(r) = -g^2/r^{3/2}$, where g is a constant. Using the normalized function $(k^3/8\pi)^{1/2}\, e^{-kr/2}$ as the trial function, estimate an upper bound to the energy of the lowest state. Given

$$\int_0^\infty x^n e^{-ax}\, dx = \frac{n!}{a^{n+1}} \quad \text{if } n \text{ is positive and a} > 0$$

we have

$$\int_0^\infty \sqrt{x} e^{-ax}\, dx = \frac{1}{2a}\sqrt{\frac{x}{a}}$$

Solution. The expectation value of the Hamiltonian

$$\langle H\rangle = \langle\phi|H|\phi\rangle = \frac{k^3}{8\pi}4\pi\int_0^\infty r^2 e^{-kr/2} \times \left[-\frac{\hbar^2}{2m}\frac{1}{r^2}\frac{d}{dr}\left(r^2\frac{d}{dr}\right) - \frac{g^2}{r^{3/2}}\right] e^{-kr/2}\, dr$$

The factor 4π outside the integral comes from the integration of the angular part, and r^2 inside the integral comes from the volume element $d\tau$. Then,

$$\frac{1}{r^2}\frac{d}{dr}\left(r^2\frac{d}{dr}\right)e^{-kr/2} = \left(\frac{d^2}{dr^2}+\frac{2}{r}\frac{d}{dr}\right)e^{-kr/2} = \left(\frac{k^2}{4}-\frac{k}{r}\right)e^{-kr/2}$$

Hence,

$$\langle H\rangle = \frac{k^3}{2}\left(-\frac{\hbar^2}{2m}\right)\int_0^\infty r^2 e^{-kr/2}\left(\frac{k^2}{4}-\frac{k}{r}\right)e^{-kr/2}\,dr - \frac{k^3}{2}g^2\int_0^\infty r^{1/2}e^{-kr}\,dr$$

$$= -\frac{\hbar^2k^5}{16m}\int_0^\infty r^2e^{-kr}dr + \frac{\hbar^2k^4}{4m}\int_0^\infty re^{-kr}dr - -\frac{k^3g^2}{2}\int_0^\infty r^{1/2}e^{-kr}dr$$

$$= -\frac{\hbar^2k^5}{16m}\frac{2}{k^3} + \frac{\hbar^2k^4}{4m}\frac{1}{k^2} - \frac{k^3g^2}{2}\frac{1}{2k}\sqrt{\frac{\pi}{k}}$$

$$= \frac{\hbar^2k^2}{8m} - \frac{\sqrt{\pi}g^2k^{3/2}}{4}$$

For $\langle H\rangle$ to be minimum, $\partial\langle H\rangle/\partial k = 0$, i.e.,

$$\frac{\hbar^2k}{4m} - \frac{3\sqrt{\pi}}{8}g^2k^{1/2} = 0$$

This leads to two values for k, and so

$$k = 0, \qquad k^{1/2} = \frac{3\sqrt{\pi}g^2m}{2\hbar^2}$$

The first value can be discarded as it leads to $\psi = 0$. Hence the upper bound to the energy of the lowest state is

$$\langle H\rangle_{\min} = \frac{81\pi^2g^8m^3}{128\hbar^6} - \frac{27\pi^2g^8m^3}{32\hbar^6} = -\frac{27\pi^2g^8m^3}{128\hbar^2}$$

9.5 A trial function ϕ differs from an eigenfunction ψ_E so that $\phi = \psi_E + \alpha\phi_1$, where ψ_E and ϕ_1 are orthonormal and normalized and $\alpha << 1$. Show that $\langle H\rangle$ differs from E only by a term of order α^2 and find this term.

Solution. Given $H\psi_E = E\psi_E$. We have

$$\langle H\rangle = \frac{\langle\phi|H|\phi\rangle}{\langle\phi|\phi\rangle} = \frac{\langle(\psi_E+\alpha\phi_1)|H|(\psi_E+\alpha\phi_1)\rangle}{\langle(\psi_E+\alpha\phi_1)|(\psi_E+\alpha\phi_1)\rangle}$$

$$= \frac{\langle\psi_E|H|\psi_E\rangle + \alpha\langle\psi_E|H|\phi_1\rangle + \alpha\langle\phi_1|H|\psi_E\rangle + \alpha^2\langle\phi_1|H|\phi_1\rangle}{\langle\psi_E|\psi_E\rangle + \alpha\langle\psi_E|\phi_1\rangle + \alpha\langle\phi_1|\psi_E\rangle + \alpha^2\langle\phi_1|\phi_1\rangle}$$

Since H is Hermitian,

$$\langle\psi_E|H|\phi_1\rangle = E\langle\psi_E|\phi_1\rangle = 0$$

$$\langle H\rangle = \frac{E + \alpha^2\langle\phi_1|H|\phi_1\rangle}{1+\alpha^2} = E + \alpha^2\langle\phi_1|H|\phi_1\rangle$$

as $1 + \alpha^2 \cong 1$. Hence the result. $\langle H\rangle$ differs from E by the term $\alpha^2\langle\phi_1|H|\phi_1\rangle$.

9.6 Evaluate the ground state energy of a harmonic oscillator of mass m and angular frequency ω using the trial function

$$\phi(x) = \begin{cases} \cos\left(\dfrac{\pi x}{2a}\right), & -a \le x \le a \\ 0, & |x| > a \end{cases}$$

Solution.

$$\langle H\rangle = \frac{\langle\phi|H|\phi\rangle}{\langle\phi|\phi\rangle} = \frac{\left(\dfrac{-\hbar^2}{2m}\right)\left\langle\phi\left|\dfrac{d^2}{dx^2}\right|\phi\right\rangle + \dfrac{1}{2}m\omega^2\langle\phi|x^2|\phi\rangle}{\langle\phi|\phi\rangle}$$

$$\langle\phi|\phi\rangle = \int_{-a}^{a}\cos^2\frac{\pi x}{2a}\,dx = a$$

$$\left(\frac{-\hbar^2}{2m}\right)\left\langle\phi\left|\frac{d^2}{dx^2}\right|\phi\right\rangle = \frac{\hbar^2\pi^2}{8ma^2}\int_{-a}^{a}\cos^2\frac{\pi x}{2a}\,dx = \frac{\hbar^2\pi^2}{8ma}$$

$$\langle\phi|x^2|\phi\rangle = \int_{-a}^{a}x^2\cos^2\frac{\pi x}{2a}\,dx = \int_{-a}^{a}\frac{x^2}{2}\,dx + \frac{1}{2}\int_{-a}^{a}x^2\cos\frac{\pi x}{a}\,dx$$

$$= \frac{a^3}{3} - \frac{2a^3}{\pi^2} = 2a^3\left(\frac{1}{6} - \frac{1}{\pi^2}\right)$$

$$\langle H\rangle = \frac{\hbar^2\pi^2}{8ma^2} + m\omega^2a^2\left(\frac{1}{6} - \frac{1}{\pi^2}\right)$$

For $\langle H\rangle$ to be minimum, $\partial\langle H\rangle/\partial a = 0$. Minimizing $a^4 = \dfrac{6\hbar^2\pi^4}{8m^2\omega^2(\pi^2-6)}$, we get

$$\langle H\rangle_{\min} = \frac{1}{2}\hbar\omega\left(\frac{\pi^2-6}{3}\right)^{1/2} = 0.568\,\hbar\omega$$

9.7 For a particle of mass m moving in the potential,

$$V(x) = \begin{cases} kx, & x > 0 \\ \infty, & x < 0 \end{cases}$$

where k is a constant. Optimize the trial wavefunction $\phi = x\exp(-ax)$, where a is the variable parameter, and estimate the groundstate energy of the system.

Solution. In the region $x < 0$, the wave function is zero since $V(x) = \infty$. The Hamiltonian of the system

$$H = -\frac{\hbar^2}{2m}\frac{d^2}{dx^2} + kx, \quad x > 0, \qquad \langle H \rangle = \frac{\langle \phi | H | \phi \rangle}{\langle \phi | \phi \rangle}$$

$$\langle \phi | \phi \rangle = \int_0^\infty x^2 e^{-2ax} dx = \frac{1}{4a^3}$$

$$\frac{d^2}{dx^2}(xe^{-ax}) = a^2 xe^{-ax} - 2ae^{-ax}$$

$$\int_0^\infty xe^{-ax} \frac{d^2}{dx^2}(xe^{-ax})\, dx = \int_0^\infty a^2 x^2 e^{-2ax} dx - 2a \int_0^\infty xe^{-2ax} dx$$

$$= \frac{1}{4a} - \frac{1}{2a} = -\frac{1}{4a}$$

$$\int_0^\infty xe^{-ax}\,(kx)\, xe^{-ax} dx = k \int_0^\infty x^3 e^{-2ax}\, dx = \frac{3k}{8a^4}$$

$$\langle H \rangle = \left(\frac{\hbar^2}{8ma} + \frac{3k}{8a^4} \right) 4a^3 = \frac{\hbar^2 a^2}{2m} + \frac{3k}{2a}$$

Minimizing with respect to a, we get

$$a = \left(\frac{3km}{2\hbar^2} \right)^{1/3}, \qquad \langle H \rangle_{\min} = \frac{9}{4}\left(\frac{2k^2\hbar^2}{3m} \right)^{1/3}$$

9.8 The Hamiltonian of a particle of mass m is

$$H = -\frac{\hbar^2}{2m}\frac{d^2}{dx^2} + bx^4$$

where b is a constant. Use the trial function $\phi(x) = Ae^{-\alpha^2 x^2}$, where α is the variable parameter, to evaluate the energy of the ground state. Given

$$\int_0^\infty \exp(-\alpha x^2)\, dx = \frac{1}{2}\left(\frac{\pi}{\alpha} \right)^{1/2}$$

$$\int_0^\infty x^2 \exp(-\alpha x^2)\, dx = \frac{\sqrt{\pi}}{4}\frac{1}{\alpha^{3/2}}$$

$$\int_0^\infty x^4 \exp(-\alpha x^2)\, dx = \frac{3\sqrt{\pi}}{8}\frac{1}{\alpha^{5/2}}$$

Solution. The Hamiltonian H and the trial function $\phi(x)$ are

$$H = -\frac{\hbar^2}{2m}\frac{d^2}{dx^2} + bx^4 \qquad \phi(x) = Aee^{-\alpha^2 x^2}$$

The normalization condition gives

$$1 = |A|^2 \int_{-\infty}^{\infty} e^{-2\alpha^2 x^2}\, dx$$

$$1 = |A|^2 \left(\frac{\pi}{2\alpha^2}\right)^{1/2} \quad \text{or} \quad |A|^2 \left(\frac{2\alpha^2}{\pi}\right)^{1/2}$$

$$\langle H\rangle = \langle \phi|H|\phi\rangle = \left\langle \phi \left| -\frac{\hbar^2}{2m}\frac{d^2}{dx^2} + bx^4 \right| \phi \right\rangle$$

$$= \frac{\hbar^2}{2m}|A|^2\, 2\alpha^2 \int_{-\infty}^{\infty} e^{-2\alpha^2 x^2} dx - \frac{\hbar^2}{2m}|A|^2\, 4\alpha^4 \int_{-\infty}^{\infty} x^2 e^{-2\alpha^2 x^2} dx + b|A|^2 \int_{-\infty}^{\infty} x^4 e^{-2\alpha^2 x^2} dx$$

$$= \frac{\hbar^2\alpha^2}{m} - \frac{\hbar^2\alpha^2}{2m} + \frac{3b}{16}\frac{1}{\alpha^4}$$

$$= \frac{\hbar^2\alpha^2}{2m} + \frac{3}{16}\frac{b}{\alpha^4}$$

Minimizing $\langle H\rangle$ with respect to α, we have

$$\frac{\partial\langle H\rangle}{\partial\alpha} = 0 = \frac{\hbar^2\alpha}{m} - \frac{3}{4}\frac{b}{\alpha^5}$$

$$\alpha^2 = \left(\frac{3}{4}\frac{bm}{\hbar^2}\right)^{1/3}$$

Substituting this value of α, we get

$$\langle H\rangle_{\min} = \frac{3}{4}\left(\frac{3}{4}\right)^{1/3}\left(\frac{b\hbar^4}{m^2}\right)^{1/3} = \left(\frac{3}{4}\right)^{4/3}\left(\frac{b\hbar^4}{m^2}\right)^{1/3}$$

9.9 An anharmonic oscillator is described by the Hamiltonian

$$H = -\frac{\hbar^2}{2m}\frac{d^2}{dx^2} + Ax^4$$

Determine its ground state energy by selecting

$$\psi = \frac{\lambda^{1/2}}{\pi^{1/4}} \exp\left(\frac{-\lambda^2 x^2}{2}\right)$$

λ being a variable parameter as the variational trial wave function.

Solution. With the trial function ψ, the expectation value of H is

$$\langle H\rangle = \lambda\pi^{-1/2}\int_{-\infty}^{\infty} e^{-\lambda^2x^2/2}\left(-\frac{\hbar^2}{2m}\frac{d^2}{dx^2} + Ax^4\right)e^{-\lambda^2x^2/2}\,dx$$

Using the values of the first three integrals from the Appendix, we obtain

$$\langle H\rangle = \frac{\hbar^2\lambda^2}{4m} + \frac{3A}{4\lambda^4}$$

Minimizing $\langle H\rangle$ with respect the variable parameter λ, we get

$$0 = \frac{\partial\langle H\rangle}{\partial\lambda} = \frac{\hbar^2\lambda}{2m} - \frac{3A}{\lambda^5}$$

$$\lambda = \left(\frac{6mA}{\hbar^2}\right)^{1/6}$$

Substituting this value of λ, we obtain

$$\langle H\rangle = \frac{\hbar^2}{4m}\left(\frac{6mA}{\hbar^2}\right)^{1/3} + \frac{3A}{4}\left(\frac{\hbar^2}{6mA}\right)^{2/3}$$

$$= \frac{3^{1/3}}{2}\left(\frac{\hbar^2}{2m}\right)^{2/3}A^{1/3} + \frac{3^{1/3}}{2}\left(\frac{\hbar^2}{2m}\right)^{2/3}A^{1/3}$$

$$= \frac{3^{4/3}}{4}\left(\frac{\hbar^2}{2m}\right)^{2/3}A^{1/3} = 1.082\left(\frac{\hbar}{2m}\right)^{2/3}A^{1/3}$$

It may be noted that numerical integration gives a coefficient of 1.08, illustrating the usefulness of the variation method. It may also be noted that perturbation technique is not possible as there is no way to split H into an unperturbed part and a perturbed part.

9.10 The Hamiltonian of a system is given by

$$H = \frac{-\hbar^2}{2m}\frac{d^2}{dx^2} - a\delta(x)$$

where a is a constant and $\delta(x)$ is Dirac's delta function. Estimate the ground state energy of the system using a Gaussian trial function.

Solution. The normalized Gaussian trial function is given by $\phi(x) = (2b/\pi)^{1/4}\exp(-bx^2)$. Then,

$$\langle H\rangle = -\frac{\hbar^2}{2m}\left\langle\phi\left|\frac{d^2}{dx^2}\right|\phi\right\rangle - a\langle\phi|\delta(x)|\phi\rangle$$

$$\left\langle\phi\left|\frac{d^2}{dx^2}\right|\phi\right\rangle = \left(\frac{2b}{\pi}\right)^{1/2}2b\int_{-\infty}^{\infty}\exp(-2bx^2)\,dx + \left(\frac{2b}{\pi}\right)^{1/2}4b^2\int_{-\infty}^{\infty}x^2\exp(-2bx^2)\,dx$$

$$= -\left(\frac{2b}{\pi}\right)^{1/2}2b\left(\frac{\pi}{2b}\right)^{1/2} + \left(\frac{2b}{\pi}\right)^{1/2}4b^2\frac{1}{4b}\left(\frac{\pi}{2b}\right)^{1/2} = -b$$

$$\langle \phi | \delta(x) | \phi \rangle = \left(\frac{2b}{\pi}\right)^{1/2} \int_{-\infty}^{\infty} \delta(x) \exp(-2bx^2)\, dx$$

$$= \left(\frac{2b}{\pi}\right)^{1/2} \exp(-2bx^2)|_{x=0} = \left(\frac{2b}{\pi}\right)^{1/2}$$

$$\langle H \rangle = \frac{\hbar^2 b}{2m} - a\left(\frac{2b}{\pi}\right)^{1/2}$$

Minimizing $\langle H \rangle$ with respect to b, we get

$$b = \frac{2m^2a^2}{\pi\hbar^4} \quad \text{or} \quad \langle H \rangle_{\min} = -\frac{ma^2}{\pi\hbar^2}$$

9.11 Evaluate the ground state energy of hydrogen atom using a Gaussian trial function. Given

$$\int_0^{\infty} x^{2n} \exp(-\lambda x^2)\, dx = \frac{\pi^{1/2}(2n)!}{2^{2n+1} n!\, \lambda^{n+1/2}}$$

$$\int_0^{\infty} x^{2n+1} \exp(-\lambda x^2)\, dx = \frac{n!}{2\lambda^{n+1}}$$

Solution.

$$\text{Hamiltonian } H = -\frac{\hbar^2}{2\mu}\nabla^2 - \frac{e^2}{r}$$

The Gaussian trial function $\phi(r) = \exp(-br^2)$, where b is the variable parameter. Since ϕ depends only on r, only the radial derivative exists in ∇^2. However, the angular integration of $d\tau$ gives a factor of 4π. Hence,

$$\langle H \rangle = \frac{\left(-\frac{\hbar^2}{2m}\right)\left\langle \phi \left| \frac{d^2}{dr^2} \right| \phi \right\rangle - \left(\frac{\hbar^2}{2m}\right)\left\langle \phi \left| \frac{2}{r}\frac{d}{dr} \right| \phi \right\rangle - \left\langle \phi \left| \frac{e^2}{r} \right| \phi \right\rangle}{\langle \phi | \phi \rangle}$$

$$\langle \phi | \phi \rangle = 4\pi \int_0^{\infty} r^2 \exp(-2br^2)\, dr = \left(\frac{\pi}{2b}\right)^{3/2}$$

$$\left\langle \phi \left| \frac{d^2}{dr^2} \right| \phi \right\rangle = 4\pi(-2b) \int_0^{\infty} r^2 e^{-2br^2}\, dr + 4\pi \times 4b^2 \int_0^{\infty} r^4 e^{-2br^2}\, dr$$

$$= -\frac{\pi^{3/2}}{(2b)^{1/2}} + \frac{6\pi^{3/2}}{(2b)^{5/2}} = \left(\frac{\pi}{2b}\right)^{3/2} b$$

$$\left\langle \phi \left| \frac{2}{r}\frac{d}{dr} \right| \phi \right\rangle = -16\pi b \int_0^{\infty} r^2 e^{-2br^2}\, dr = \left(\frac{\pi}{2b}\right)^{3/2} (-4b)$$

$$\left\langle \phi \left| \frac{e^2}{r} \right| \phi \right\rangle = 4\pi e^2 \int_0^\infty r e^{-2br^2} dr = \frac{\pi e^2}{b}$$

$$\langle H \rangle = \frac{3\hbar^2 b}{2\mu} + 2e^2 b^{1/2} \left(\frac{2}{\pi} \right)^{1/2}$$

Minimizing $\langle H \rangle$ with respect to b given by $b = \dfrac{8\mu^2 e^4}{9\pi\hbar^4}$, we get

$$\langle H \rangle_{\min} = \frac{8}{3\pi} \left(\frac{-\mu e^4}{2\hbar^2} \right) = -11.59 \text{ eV}$$

9.12 A particle of mass m is moving in a one-dimensional box defined by the potential $V = 0$, $0 \le x \le a$ and $V = \infty$ otherwise. Estimate the ground state energy using the trial function $\psi(x) = Ax(a - x)$, $0 \le x < a$.

Solution. The normalization condition gives

$$\langle \psi | \psi \rangle = A^2 \int_0^a x^2 (a - x)^2 = 1$$

$$A^2 \left[\int_0^a a^2 x^2 dx - 2a \int_0^a x^3 dx + \int_0^a x^4 dx \right] = 1$$

$$\frac{A^2 a^5}{30} = 1 \quad \text{or} \quad A = \sqrt{\frac{30}{a^5}}$$

The normalized trial function is

$$\psi(x) = \sqrt{\frac{30}{a^5}} x(a - x), \qquad 0 \le x \le a$$

The Hamiltonian of the system is given by

$$H = -\frac{\hbar^2}{2m} \frac{d^2}{dx^2}$$

$$\langle H \rangle = -\frac{\hbar^2}{2m} \frac{30}{a^5} \int_0^a (ax - x^2) \frac{d^2}{dx^2} (ax - x^2)\, dx$$

$$= \frac{30\hbar^2}{ma^5} \int_0^a (ax - x^2)\, dx = \frac{5\hbar^2}{ma^2} = \frac{10\hbar^2}{2ma^2}$$

which is the ground state energy with the trial function. It may be noted that the exact ground state energy is $\pi^2\hbar^2/(2ma^2)$, which is very close to the one obtained here.

9.13 Evaluate, by the variation method, the energy of the first excited state of a linear harmonic oscillator using the trial function

$$\phi = Nx \exp(-\lambda x^2)$$

where is the λ variable parameter.

Solution. The Hamiltonian

$$H = -\frac{\hbar^2}{2m}\frac{d^2}{dx^2} + \frac{1}{2}kx^2$$

The trial function

$$\phi = Nx \exp(-\lambda x^2)$$

where λ is the variable parameter. The normalization condition gives

$$1 = N^2 \int_{-\infty}^{\infty} x^2 e^{-2\lambda x^2}\, dx = N^2 \times 2 \times \frac{\sqrt{\pi}}{4}\frac{1}{(2\lambda)^{3/2}}$$

$$N_2 = \frac{2^{5/2}\lambda^{3/2}}{\pi^{1/2}}$$

$$\langle H\rangle = -\frac{\hbar^2}{2m}\left\langle \phi \left| \frac{d^2}{dx^2} \right| \phi \right\rangle + \frac{1}{2}k\langle \phi | x^2 | \phi\rangle$$

$$\left\langle \phi \left| \frac{d^2}{dx^2} \right| \phi \right\rangle = N^2 \int_{-\infty}^{\infty} (-6\lambda x^2 + 4\lambda^2 x^4)\, e^{-2\lambda x^2}\, dx$$

$$= N^2\left(-\frac{3\pi^{1/2}}{2^{3/2}\lambda^{1/2}} + \frac{3\pi^{1/2}}{2^{5/2}\lambda^{1/2}}\right) = -\frac{3\pi^{1/2}}{2^{5/2}\lambda^{1/2}} N^2$$

Substituting the value of N^2, we get

$$\left\langle \phi \left| \frac{d^2}{dx^2} \right| \phi \right\rangle = -\frac{3\pi^{1/2}}{2^{5/2}\lambda^{1/2}}\frac{2^{5/2}\lambda^{3/2}}{\pi^{1/2}} = -3\lambda$$

$$\langle \phi | x^2 | \phi\rangle = N^2 \int_{-\infty}^{\infty} x^4 e^{-2\lambda x^2}\, dx = \frac{3\pi^{1/2}}{4(2\lambda)^{5/2}} N^2 = \frac{3}{4\lambda}$$

Substituting these values, we obtain

$$\langle H\rangle = \left(-\frac{\hbar^2}{2m}\right)(-3\lambda) + \frac{1}{2}k \times \frac{3}{4\lambda} = \frac{3\hbar^2\lambda}{2m} + \frac{3k}{8\lambda}$$

Minimizing $\langle H\rangle$ with respect to λ, we obtain

$$\frac{3\hbar^2}{2m} - \frac{3k}{8\lambda^2} = 0 \quad \text{or} \quad \lambda = \frac{\sqrt{km}}{2\hbar}$$

Substituting this value of λ in $\langle H\rangle$, we get

$$\langle H\rangle_{\min} = \frac{3}{2}\hbar\sqrt{\frac{k}{m}} = \frac{3}{2}\hbar\omega$$

9.14 Estimate the ground state energy of helium atom by taking the product of two normalized hydrogenic ground state wave functions as the trial wave function, the nuclear charge $Z'e$ being the variable parameter. Assume that the expectation value of the interelectronic repulsion term is (5/4) ZW_H, W_H = 13.6 eV.

Solution. The Hamiltonian of the helium atom having a nuclear charge Ze (Fig. 9.1) is given by

$$H = \left(-\frac{\hbar^2}{2m}\nabla_1^2 - \frac{kZe^2}{r_1}\right) + \left(-\frac{\hbar^2}{2m}\nabla_2^2 - \frac{kZe^2}{r_2}\right) + \frac{ke^2}{r_{12}} \tag{i}$$

where

$$k = \frac{1}{4\pi\varepsilon_0}$$

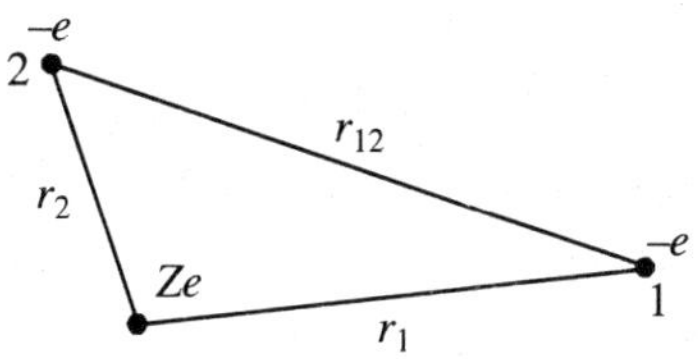

Fig. 9.1 The helium atom.

In terms of the variable parameter $Z'e$, it is convenient to write the Hamiltonian as

$$H = \left(-\frac{\hbar^2}{2m}\nabla_1^2 - \frac{kZ'e^2}{r_1}\right) + \left(-\frac{\hbar^2}{2m}\nabla_2^2 - \frac{kZ'e^2}{r_2}\right) + (Z' - Z)\,ke^2\left(\frac{1}{r_1} + \frac{1}{r_2}\right) + \frac{ke^2}{r_{12}} \tag{ii}$$

The product of the two normalized hydrogenic ground state wave functions is

$$\psi = \psi_1(r_1)\,\psi_2(r_2) = \frac{Z'^3}{\pi a_0^3}\exp\left[-\frac{Z'}{a_0}(r_1 + r_2)\right] \tag{iii}$$

where $\psi_1(r_1)$ and $\psi_2(r_2)$ are the normalized hydrogenic wave functions with Z replaced by Z'. The expectation value of H with the trial wave function, as seen from Eq. (iii), is

$$\langle H\rangle = \left\langle \psi_1\left| -\frac{\hbar^2}{2m}\nabla_1^2 - \frac{kZ'e^2}{r_1}\right|\psi_1\right\rangle + \left\langle \psi_2\left| -\frac{\hbar^2}{2m}\nabla_2^2 - \frac{kZ'e^2}{r_2}\right|\psi_2\right\rangle$$

$$+ (Z' - Z)\left\langle \psi_1\left|\frac{ke^2}{r_1}\right|\psi_1\right\rangle + (Z' - Z)\left\langle \psi_2\left|\frac{ke^2}{r_2}\right|\psi_2\right\rangle + \left\langle \psi_1\psi_2\left|\frac{ke^2}{r_{12}}\right|\psi_1\psi_2\right\rangle$$

The value of the first and second terms are equal and each is $-Z'^2 W_H$, where $W_H = k^2me^4/2\hbar^2$.

$$\left\langle \psi_1 \left| \frac{ke^2}{r_1} \right| \psi_1 \right\rangle = \frac{Z'^3 ke^2}{\pi a_0^3} \int_0^{2\pi} d\phi_1 \int_0^{\pi} \sin\theta_1 \, d\phi_1 \int_0^{\infty} r_1 \exp\left(-\frac{2Z' r_1}{a_0}\right) dr_1$$

$$= \frac{Z'^3 ke^2}{\pi a_0^3} 4\pi \frac{1}{(2Z'/a_0)^2}$$

$$= \frac{Z' ke^2}{a_0} = 2Z' W_H \qquad \text{(iv)}$$

where the value of a_0 is substituted. Given

$$\left\langle \psi_1\psi_2 \left| \frac{ke^2}{r_{12}} \right| \psi_1\psi_2 \right\rangle = \frac{5}{4} Z' W_H \qquad \text{(v)}$$

Summing up, we have

$$\langle H \rangle = -2Z'^2 W_H + 4(Z' - Z)\, Z' W_H + \frac{5}{4} Z' W_H \qquad \text{(vi)}$$

Minimizing $\langle H \rangle$ with respect to Z', we get

$$-4Z' W_H + 8Z' W_H - 4Z W_H + \frac{5}{4} W_H = 0$$

$$Z' = Z - \frac{5}{16} \qquad \text{(vii)}$$

With this value of Z', Eq. (vi) gives

$$E = \langle H \rangle = -2\left(Z - \frac{5}{16}\right)^2 W_H$$

Substitution of $W_H = 13.6$ eV leads to a ground state energy of –77.46 eV.

9.15 The attractive short range force between the nuclear particles in a deuteron is described by the Yukawa potential

$$V(r) = -V_0 \frac{e^{-r/\beta}}{r/\beta}$$

where V_0 and β are constants. Estimate the ground state energy of the system using the trial function

$$\phi = \left(\frac{\alpha^3}{\pi\beta^3}\right)^{1/2} e^{-\alpha r/\beta}$$

where α is a variable parameter.

Solution. The Hamiltonian for the ground state is

$$H = -\frac{\hbar^2}{2\mu}\nabla^2 + V(r) \qquad \text{(i)}$$

As the trial function depends only on r, we need to consider only the radial derivative in ∇^2:

$$\nabla^2 = \frac{1}{r^2}\frac{d}{dr}\left(r^2 \frac{d}{dr}\right) = \frac{d^2}{dr^2} + \frac{2}{r}\frac{d}{dr} \tag{ii}$$

Consequently,

$$\langle H \rangle = \langle \phi | H | \phi \rangle = -\frac{\hbar^2}{2\mu}\left\langle \phi \left| \frac{d^2}{dx^2} \right| \phi \right\rangle - \frac{\hbar^2}{2\mu}\left\langle \phi \left| \frac{2}{r} \right| \phi \right\rangle + \langle \phi | V | \phi \rangle \tag{iii}$$

While evaluating integrals in Eq. (iii), the factor $d\tau$ gives the angular contribution 4π. Using the integrals in the Appendix, we get

$$\begin{aligned}\left\langle \phi \left| \frac{d^2}{dx^2} \right| \phi \right\rangle &= \frac{\alpha^3}{\pi\beta^3}\frac{\alpha^2}{\beta^2} 4\pi \int_0^\infty r^2 \exp\left(-\frac{2\alpha r}{\beta}\right) dr \\ &= \frac{\alpha^3}{\pi\beta^3}\frac{\alpha^2}{\beta^2} 4\pi \frac{2}{(2\alpha/\beta)^3} = \frac{\alpha^2}{\beta^2}\end{aligned} \tag{iv}$$

$$\begin{aligned}\left\langle \phi \left| \frac{2}{r} \right| \phi \right\rangle &= \frac{\alpha^3}{\pi\beta^3}\left(-\frac{8\pi\alpha}{\beta}\right)\int_0^\infty r \exp\left(-\frac{2\alpha r}{\beta}\right) dr \\ &= -\frac{8\alpha^4}{\beta^4}\frac{\beta^2}{4\alpha^2} = -\frac{2\alpha^2}{\beta^2}\end{aligned} \tag{v}$$

$$\begin{aligned}\langle \phi | V(r) | \phi \rangle &= \frac{\alpha^3}{\pi\beta^3}(-4\pi\beta V_0)\int_0^\infty r \exp\left(-\frac{2\alpha+1}{\beta} r\right) dr \\ &= \frac{\alpha^3}{\pi\beta^3}(-4\pi V_0 \beta)\frac{\beta^2}{(2\alpha+1)^2} = -\frac{4V_0\alpha^3}{(2\alpha+1)^2}\end{aligned} \tag{vi}$$

Adding all the contributions, we here

$$\begin{aligned}\langle H \rangle &= -\frac{\hbar^2}{2\mu}\frac{\alpha^2}{\beta^2} + \frac{2\hbar^2\alpha^2}{2\mu\beta^2} - \frac{4V_0\alpha^3}{(2a+1)^2} \\ &= \frac{\hbar^2}{2\mu}\frac{\alpha^2}{\beta^2} - \frac{4V_0\alpha^3}{(2a+1)^2}\end{aligned} \tag{vii}$$

Minimizing with respect to α, we obtain

$$0 = \frac{\hbar^2\alpha}{\mu\beta^2} - \frac{4V_0\alpha^2(2\alpha+3)}{(2a+1)^3}$$

$$\frac{\hbar^2}{2\mu\beta^2} = \frac{2V_0\alpha(2\alpha+3)}{(2\alpha+1)^3} \tag{viii}$$

Repalcing $\hbar^2/2\mu\beta^2$ in Eq. (vii) using Eq. (viii), we get

$$\langle H\rangle = \frac{2V_0\alpha^3(2\alpha+3)}{(2\alpha+1)^3} - \frac{4V_0\alpha^3}{(2\alpha+1)^2}$$

$$= \frac{2V_0\alpha^3}{(2\alpha+1)^3}[(2\alpha+3) - 2(2\alpha+1)]$$

$$E = \frac{2V_0\alpha^3(2\alpha-1)}{(2\alpha+1)}$$

where α is given by Eq. (viii).

9.16 Consider a particle having momentum p moving inside the one-dimensional potential well shown in Fig. 9.2. If $E < V(x)$, show by the WKB method, that

$$2\int_{x_1}^{x_2} p\,dx = \left(n + \frac{1}{2}\right)h, \quad n = 0, 1, 2, \ldots$$

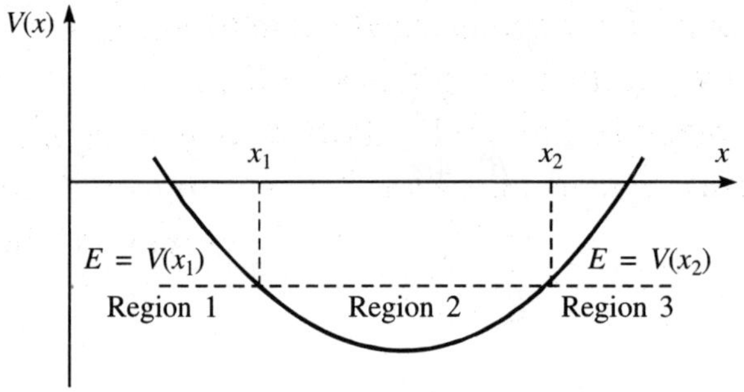

Fig. 9.2 A potential well with linear turning points at x_1 and x_2.

Solution. Classically, the particle will oscillate back and forth between the turning points x_1 and x_2. Quantum mechanically, the particles can penetrate into regions 1 and 2. The wave functions in regions 1 and 2 are exponentially decreasing. When we move from region 1 to region 2, the barrier is to the left of the turning point and, when we move from region 2 to region 3, the barrier is to the right of the turning point. The wave function in region 1 is

$$\psi_1 = \frac{1}{\sqrt{r}}\exp\left(-\int_{x_2}^{x_1}\gamma\,dx\right), \quad \gamma^2 = \frac{2m[V(x)-E]}{\hbar^2} \tag{i}$$

Applying Eq. (9.8), we get

$$\psi_2 = \frac{2}{\sqrt{k}}\cos\left(\int_{x_2}^{x_1} k\,dx - \frac{\pi}{4}\right), \quad k^2 = \frac{2m[E-V(x)]}{\hbar^2} \tag{ii}$$

The wave function that connects region 2 with the decreasing potential of region 3 being of the type

$$\cos\left(\int_{x_1}^{x_2} k\,dx - \frac{\pi}{4}\right)$$

Hence, Eq. (ii) should be modified as

$$\psi_2 = \frac{2}{\sqrt{k}}\cos\left(\int_{x_1}^{x_2} k\,dx + \int_{x_2}^{x} k\,dx - \frac{\pi}{4}\right) \tag{iii}$$

Since cos $(-\theta) = \cos\theta$ and $\sin(-\theta) = -\sin\theta$, Eq. (iii) can be rewritten as

$$\psi_2 = \frac{2}{\sqrt{k}}\cos\left(\int_{x_1}^{x_2} k\,dx\right)\cos\left(\int_{x}^{x_2} k\,dx + \frac{\pi}{4}\right) + \frac{2}{\sqrt{k}}\sin\left(\int_{x_1}^{x_2} k\,dx\right)\sin\left(\int_{x}^{x_2} k\,dx + \frac{\pi}{4}\right)$$

$$= \frac{2}{\sqrt{k}}\cos\left(\int_{x_1}^{x_2} k\,dx\right)\sin\left(\int_{x}^{x_2} k\,dx - \frac{\pi}{4}\right) + \frac{2}{\sqrt{k}}\sin\left(\int_{x_1}^{x_2} k\,dx\right)\cos\left(\int_{x}^{x_2} k\,dx - \frac{\pi}{4}\right) \tag{iv}$$

Comparison of Eqs. (iv) and (9.7) shows that the second term of Eq. (iv) is the one that connects with the decreasing exponential of region 3, while the first term connects with the increasing exponential. Since an increasing exponential in region 3 is not acceptable, the first term has to be zero. This is possible if

$$\cos\int_{x_1}^{x_2} k\,dx = 0 \quad \text{or} \quad \int_{x_1}^{x_2} k\,dx = \left(n + \frac{1}{2}\right)\pi, \qquad n = 0, 1, 2, \ldots \tag{v}$$

Substituting the value of k, we get

$$\left(\frac{2m}{\hbar^2}\right)^{1/2}\int_{x_1}^{x_2} [E - V(x)]^{1/2}\,dx = \left(n + \frac{1}{2}\right)\pi, \qquad n = 0, 1, 2, \ldots \tag{vi}$$

which gives the allowed energy value. Classically, since the linear momentum $p = [2m\,(E - V)]^{1/2}$, Eq. (vi) can be rewritten as

$$2\int_{x_1}^{x_2} p\,dx = \left(n + \frac{1}{2}\right)h, \qquad n = 0, 1, 2, \ldots \tag{vii}$$

The LHS is the value of the integral over a complete cycle.

9.17 Obtain the energy values of harmonic oscillator by the WKB method.

Solution. The classical turning points of the oscillator are those points at which the potential $V(x) = E$, i.e., $1/2m\omega^2x^2 = E$ or $x_1 = -(2E/m\omega^2)^{1/2}$ and $x_2 = (2E/m\omega^2)^{1/2}$. For a particle constrained to move between classical turning points x_1 and x_2 in a potential well, the energies can be obtained from the condition (vii) of Problem 9.16. We then have

$$E = \frac{p^2}{2m} + \frac{1}{2}m\omega^2x^2 \quad \text{or} \quad p = \left[2m\left(E - \frac{1}{2}m\omega^2x^2\right)\right]^{1/2}$$

Substituting this value of p in Eq. (vii) of Problem 9.16, we get

$$\int_{x_1}^{x_2} \left[2m\left(E - \frac{1}{2}m\omega^2 x^2\right)\right]^{1/2} dx = \left(n + \frac{1}{2}\right)\pi\hbar, \qquad n = 0, 1, 2, \ldots$$

Writing $\sin\theta = (m\omega^2/2E)^{1/2}x$, the above integral reduces to

$$\int_{-\pi/2}^{\pi/2} (2mE)^{1/2} \cos^2\theta \left(\frac{2E}{m\omega^2}\right)^{1/2} d\theta = \left(n + \frac{1}{2}\right)\pi\hbar$$

$$\left(\frac{2E}{\omega}\right) \int_{-\pi/2}^{\pi/2} \cos^2\theta \, d\theta = \left(n + \frac{1}{2}\right)\pi\hbar$$

$$\frac{2E}{\omega} \times \frac{\pi}{2} = \left(n + \frac{1}{2}\right)\pi\hbar \quad \text{or} \quad E = \left(n + \frac{1}{2}\right)\hbar\omega$$

9.18 Solve the following one-dimensional infinite potential well:

$$V(x) = 0 \quad \text{for } -a < x < a; \; V(x) = \infty \quad \text{for } |x| > a$$

using the WKB method and compare it with the exact solution.

Solution. $V(x) = 0$ for $-a < x < a$ and $V(x) = \infty$ for $|x| > a$. The turning points are $x_1 = -a$ and $x_2 = a$. The allowed energies can be obtained using the relation

$$\int_{-a}^{a} k\, dx = \left(n + \frac{1}{2}\right)\pi, \qquad k^2 = \frac{2mE}{\hbar^2}, \qquad n = 0, 1, 2, \ldots$$

$$\left(\frac{2mE}{\hbar^2}\right)^{1/2} \int_{-a}^{a} dx = \left(n + \frac{1}{2}\right)\pi$$

$$E_n = \frac{[n + (1/2)^2 \pi^2\hbar^2]}{8ma^2}, \qquad n = 0, 1, 2, \ldots$$

The exact solution gives

$$E_n = \frac{n^2\pi^2\hbar^2}{8ma^2}, \qquad n = 1, 2, 3, \ldots$$

The WKB solution has $n + (1/2)$ in place of n. Another major difference is in the allowed values of n.

9.19 Estimate the energy levels of a particle moving in the potential

$$V(x) = \begin{cases} \infty, & x < 0 \\ Ax, & x > 0 \end{cases}$$

A being a constant.

Solution. The classical turning points are at $x_1 = 0$ and at $x_2 = E/A$. Now,

$$\int_{x_1}^{x_2} k\, dx = \left(n + \frac{1}{2}\right)\pi, \qquad k^2 = \frac{2m}{\hbar^2}(E - V)$$

In the given case,

$$k = \left(\frac{2m}{\hbar^2}\right)^{1/2} (E - Ax)^{1/2}$$

$$\left(\frac{2m}{\hbar^2}\right)^{1/2} \int_0^{E/A} (E - Ax)^{1/2}\, dx = \left(n + \frac{1}{2}\right)\pi$$

$$-\left(\frac{2m}{\hbar^2}\right)^{1/2} \left[\frac{(E - Ax)^{3/2}}{A3/2}\right]^{E/A} = \left(n + \frac{1}{2}\right)\pi$$

$$E_n = \left(\frac{\hbar^2}{2m}\right)^{1/3} \left[\frac{3\pi A\,(2n + 1)}{4}\right]^{2/3}, \qquad n = 0, 1, 2, \ldots$$

9.20 Find the energy levels of a particle moving in the potential $V(x) = V_0\,|x|$, V_0 being a positive constant.

Solution. The turning points are given by

$$E = V_0\,|x| \quad \text{or} \quad |x| = E/V_0 \quad \text{or} \quad x = \pm E/V_0$$

$$\int_{x_1}^{x_2} k\, dx = \left(n + \frac{1}{2}\right)\pi, \qquad k = \left(\frac{2m}{\hbar^2}\right)^{1/2} (E - V_0\,|x|)^{1/2}$$

$$\left(\frac{2m}{\hbar^2}\right)^{1/2} \int_{-E/V_0}^{E/V_0} (E - V_0\,|x|)^{1/2}\, dx = \left(n + \frac{1}{2}\right)\pi$$

As the integrand is even,

$$\left(\frac{2m}{\hbar^2}\right)^{1/2} 2 \int_0^{E/V_0} (E - V_0\,|x|)^{1/2}\, dx = \left(n + \frac{1}{2}\right)\pi$$

$$2\left(\frac{2m}{\hbar^2}\right)^{1/2} \left[\frac{E - V_0\,|x|}{-3V_0/2}\right]_0^{E/V_0} = \left(n + \frac{1}{2}\right)\pi$$

$$E_n = \left[\frac{3}{4}\left(n + \frac{1}{2}\right)\pi V_0\right]^{2/3} \left(\frac{\hbar^2}{2m}\right)^{1/3}, \qquad n = 0, 1, 2, 3, \ldots$$

9.21 Consider a particle of mass m moving in a spherically symmetric potential $V = kr$, k being a positive constant. Estimate the ground state energy using a trial function of the type $\phi = \exp(-\alpha r)$, where α is the variable parameter.

Solution. The Hamiltonian operator

$$H = -\frac{\hbar^2}{2m}\nabla^2 + kr$$

As the trial wave function is not normalized,

$$\langle H\rangle = \frac{\langle\phi|H|\phi\rangle}{\langle\phi|\phi\rangle}, \qquad \phi = e^{-\alpha r}$$

$$\langle\phi|\phi\rangle = \int_0^\infty e^{-2\alpha r} r^2\, dr = \frac{2!}{(2\alpha)^3} = \frac{1}{4\alpha^3}$$

(see Appendix). Now,

$$\langle\phi|H|\phi\rangle = -\frac{\hbar^2}{2m}\int_0^\infty e^{-\alpha r}\frac{1}{r^2}\left[\frac{d}{dr}\left(r^2\frac{d}{dr}\right)e^{-\alpha r}\right]r^2 dr + k\int_0^\infty r^3 e^{-2\alpha r}\, dr$$

$$= -\frac{\hbar^2\alpha^2}{2m}\int_0^\infty r^2 e^{-2\alpha r} dr + \frac{\hbar^2\alpha}{m}\int_0^\infty r e^{-2\alpha r} dr + k\int_0^\infty r^3 e^{-2\alpha r}\, dr$$

Using the standard integral in the Appendix, we get

$$\langle\phi|H|\phi\rangle = -\frac{\hbar^2\alpha^2}{2m}\frac{2!}{(2\alpha)^3} + \frac{\hbar^2\alpha}{m}\frac{1}{(2\alpha)^2} + k\frac{3!}{(2\alpha)^4}$$

$$= \frac{\hbar^2}{8m\alpha} + \frac{3k}{8\alpha^4}$$

$$\langle H\rangle = \frac{\langle\phi|H|\phi\rangle}{\langle\phi|\phi\rangle} = \frac{\hbar^2\alpha^2}{2m} + \frac{3k}{2\alpha}$$

For $\langle H\rangle$ to be minimum, it is necessary that

$$\frac{\partial\langle H\rangle}{\partial\alpha} = 0$$

$$\frac{\hbar^2\alpha}{m} - \frac{3k}{2\alpha^2} = 0 \quad \text{or} \quad \alpha = \left(\frac{3km}{2\hbar^2}\right)^{1/3}$$

With this value of α, the ground state energy

$$E = \frac{\hbar^2}{2m}\left(\frac{3mk}{2\hbar^2}\right)^{2/3} + \frac{3k}{2}\left(\frac{2\hbar^2}{3km}\right)^{2/3} = \frac{3}{2}\left(\frac{9k^2\hbar^2}{4m}\right)^{1/3}$$

9.22 Using the WKB method, calculate the transmission coefficient for the potential barrier

$$V(x) = \begin{cases} V_0\left(1 - \dfrac{|x|}{\lambda}\right), & |x| < \lambda \\ 0, & |x| > \lambda \end{cases}$$

Solution. The transmission coefficient

$$T = \exp\left(-2\int_{x_1}^{x_2}\lambda\, dx\right), \qquad \lambda^2 = \frac{2m}{\hbar^2}[V(x) - E]$$

where x_1 and x_2 are the turning points. At the turning points,

$$E = V(x) = V_0\left(1 - \frac{|x|}{\lambda}\right) \quad \text{or} \quad \frac{E}{V_0} = 1 - \frac{|x|}{\lambda}$$

$$|x| = \lambda\left(\frac{V_0 - E}{V_0}\right) \quad \text{or} \quad x = \pm\lambda\left(\frac{V_0 - E}{V_0}\right)$$

$$x_1 = -\lambda\left(\frac{V_0 - E}{V_0}\right), \qquad x_2 = \lambda\left(\frac{V_0 - E}{V_0}\right)$$

$$-2\int_{x_1}^{x_2} \gamma\, dx = -2\sqrt{\frac{2m}{\hbar^2}} \int_{x_1}^{x_2} \left(V_0 - \frac{V_0 x}{\lambda} - E\right)^{1/2} dx$$

$$= -\frac{2\sqrt{2m}}{\hbar}\left(\frac{2}{3}\right)\left(-\frac{\lambda}{V_0}\right)\left[\left(V_0 - E - \frac{V_0 x}{\lambda}\right)^{3/2}\right]_{x_1}^{x_2}$$

$$= -\frac{16\sqrt{m}}{3\hbar}\frac{\lambda}{V_0}(V_0 - E)^{3/2}$$

$$T = \exp\left[-\frac{16\sqrt{m}}{3\hbar}\frac{\lambda}{V_0}(V_0 - E)^{3/2}\right]$$

9.23 Use the WKB method to calculate the transmission coefficient for the potential barrier

$$\text{V(x)} = \begin{cases} V_0 - ax, & x > 0 \\ 0, & x < 0 \end{cases}$$

Solution. The transmission coefficient

$$T = \exp\left(-2\int_{x_1}^{x_2} \gamma\, dx\right), \qquad \gamma^2 = \frac{2m}{\hbar^2}[V(x) - E]$$

From the value of $V(x)$, it is clear that the turning point $x_1 = 0$. To get the other turning point, it is necessary that

$$E = V(x) = V_0 - ax_2$$

$$x_2 = \frac{V_0 - E}{a}$$

$$\gamma = \frac{\sqrt{2m}}{\hbar}(V_0 - ax - E)^{1/2}$$

$$-2\int_{x_1}^{x_2} \gamma\, dx = -2\frac{\sqrt{2m}}{\hbar}\int_{0}^{x_2} (V_0 - E - ax)^{1/2}\, dx$$

$$= -2\frac{\sqrt{2m}}{\hbar}\frac{2}{3}\left(-\frac{1}{a}\right)\left[(V_0 - E - ax)^{3/2}\right]_0^{x_2}$$

$$= \frac{4\sqrt{2m}}{3\hbar a}\left[(V_0 - E - ax)^{3/2} - (V_0 - E)^{3/2}\right]$$

$$= -\frac{4\sqrt{2m}}{3\hbar a}[V_0 - E]^{3/2}$$

$$T = \exp\left[-\frac{4\sqrt{2m}}{3\hbar a}(V_0 - E)^{3/2}\right]$$

CHAPTER 10

Time-Dependent Perturbation

In certain systems, the Hamiltonian may depend on time, resulting in the absence of stationary states. The Hamiltonian can then be written as

$$H\,(\boldsymbol{r},\,t) = H^0(\boldsymbol{r}) + H'(\boldsymbol{r},\,t), \qquad H' \ll H^0 \tag{10.1}$$

where H^0 is time independent and H' is time dependent. The time-dependent Schrödinger equation to be solved is

$$i\hbar \frac{\partial \Psi\,(\boldsymbol{r},t)}{\partial t} = (H^0 + H')\,\Psi(\boldsymbol{r},\,t) \tag{10.2}$$

Let Ψ_n^0, $n = 1, 2, 3, \ldots$ be the stationary state eigenfunctions of H^0 forming a complete orthonormal set. Ψ_n^0's are of the form

$$\Psi_n^0 = \psi_n^0(\boldsymbol{r}) \exp\left(-\frac{iE_n t}{\hbar}\right) \qquad n = 1,\, 2,\, 3,\, \ldots \tag{10.3}$$

and obey the equation

$$i\hbar \frac{\partial}{\partial t}\,\Psi_n^0 = H^0 \Psi_n^0, \qquad n = 1,\, 2,\, 3,\, \ldots \tag{10.4}$$

10.1 First Order Perturbation

In the presence of H', the states of the system may be expressed as a linear combination of Ψ_n^0's as

$$\Psi(\boldsymbol{r},\,t) = \sum_n c_n(t)\,\Psi_n^0 = \sum_n c_n(t)\,\Psi_n^0(\boldsymbol{r}) \exp\left(-\frac{iE_n t}{\hbar}\right) \tag{10.5}$$

where $c_n(t)$'s are expansion coefficients. The system is initially in state n and the perturbation H' is switched on for a time t and its effect on the stationary states is studied. The first order contribution to the coefficient is

$$c_k^{(1)}(t) = \frac{1}{i\hbar}\int_0^t H'_{kn}(\mathbf{r}, t')\exp(i\omega_{kn}t')\,dt' \tag{10.6}$$

where

$$H'_{kn} = \langle\psi_k^0|H'|\psi_n^0\rangle = \langle k|H'|n\rangle$$

$$\omega_{kn} = \frac{E_k - E_n}{\hbar} \tag{10.7}$$

The perturbation H' has induced transition to other states and, after time t, the probability that a transition to state k has occurred is given by $|c_k^{(1)}(t)|^2$.

10.2 Harmonic Perturbation

A harmonic perturbation with an angular frequency ω has the form

$$H'(\mathbf{r},t) = 2H'(r)\cos\omega t = H'(\mathbf{r})\,(e^{i\omega t} + e^{-i\omega t}) \tag{10.8}$$

With this perturbation, we get

$$c_k^{(1)}(t) = -\frac{H'_{kn}}{\hbar}\left[\frac{\exp[i(\omega_{kn}+\omega)t]-1}{\omega_{kn}+\omega} + \frac{\exp[i(\omega_{kn}-\omega)t]-1}{\omega_{kn}-\omega}\right] \tag{10.9}$$

The first term on the RHS of Eq. (10.9) has a maximum value when $\omega_{kn} + \omega \cong 0$ or $E_k \cong E_n - \hbar\omega$ which corresponds to **induced** or **stimulated emission**. The second term is maximum when $E_k \cong E_n + \hbar\omega$ which corresponds to **absorption**. The probability for absorption is obtained as

$$P_{n\to k}(t) = |c_{kn}(t)|^2 = \frac{4|H'_{kn}|^2}{\hbar^2}\,\frac{\sin^2(\omega_{kn}-\omega)t/2}{(\omega_{kn}-\omega)^2} \tag{10.10}$$

10.3 Transition to Continuum States

Next we consider transitions from a discrete state n to a continuum of states around E_k, where the density of states is $\rho(E_k)$. The probability for transition into range dE_k is

$$P(t) = \frac{2\pi}{\hbar}t\,|H'_{kn}|^2\,\rho(E_k) \tag{10.11}$$

The **transition probability** ω is the number of transitions per unit time and is given by

$$\omega = \frac{2\pi}{\hbar}|H'_{kn}|^2\,\rho(E_k) \tag{10.12}$$

which is called **Fermi's Golden Rule**.

10.4 Absorption and Emission of Radiation

In dipole approximation, $kr \cong 1$, k being the wave vector $2\pi/\lambda$ of the incident plane electromagnetic wave. Under this approximation, the probability per unit time for absorption is given by

$$\omega = \frac{2\pi}{3\hbar^2} |\mu_{kn}|^2 \rho(\omega_{kn}) \tag{10.13}$$

where μ_{kn} is the **transition dipole moment** defined by

$$\mu_{kn} = \langle k | er_A | n \rangle \tag{10.14}$$

er being the dipole moment of the atom.

10.5 Einstein's A and B Coefficients

The transition probability per unit time for spontaneous emission, called Einstein's A coefficient, is defined by

$$A = \frac{4\omega_{kn}^3}{3\hbar c^3} |\mu_{kn}|^2 \tag{10.15}$$

The transition probability per unit time for stimulated emission or absorption, called Einstein's B coefficient, is defined by

$$B = \frac{2\pi}{3\hbar^2} |\mu_{kn}|^2 \tag{10.16}$$

From Eqs. (10.15) and (10.16),

$$\frac{A}{B} = \frac{2\hbar\omega_{kn}^3}{\pi c^3} = \frac{8\pi h\nu_{kn}^3}{c^3} \tag{10.17}$$

It can easily be proved that

$$\frac{\text{Spontaneous emission rate}}{\text{Stimulated emission rate}} = \exp\left(\frac{\hbar\omega}{kT}\right) - 1 \tag{10.18}$$

10.6 Selection Rules

Transitions between all states are not allowed. The selection rules specify the transitions that may occur on the basis of dipole approximation. Transitions for which μ_{kn} is nonzero are the allowed transitions and those for which it is zero are the forbidden transitions. The selection rules for hydrogenic atoms are

$$\Delta n = \text{any value}, \qquad \Delta l = \pm 1, \qquad \Delta m = 0, \pm 1 \tag{10.19}$$

The selection rule for electric dipole transitions of a linear harmonic oscillator is

$$\Delta n = \pm 1 \tag{10.20}$$

PROBLEMS

10.1 A system in an unperturbed state n is suddenly subjected to a constant perturbation $H'(r)$ which exists during time $0 \to t$. Find the probability for transition from state n to state k and show that it varies simple harmonically with angular frequency $(E_k - E_n)/2\hbar$ and amplitude $4|H'_{kn}|^2/(E_k - E_n)^2$.

Solution. Equation (10. 6) gives the value of $c_k^{(1)}(t)$. When the perturbation is constant in time, $H'_{kn}(r)$ can be taken outside the integral. Hence,

$$c_k^{(1)}(t) = \frac{H'_{kn}(r)}{i\hbar} \int_0^t \exp(i\omega_{kn}t')\, dt' = -\frac{H'_{kn}}{\hbar\omega_{kn}} [\exp(i\omega_{kn}t) - 1]$$

$$= -\frac{H'_{kn}}{\hbar\omega_{kn}} \exp(i\omega_{kn}t/2) [\exp(i\omega_{kn}t/2) - \exp(i\omega_{kn}t/2)]$$

$$= -\frac{2iH'_{kn}}{\hbar\omega_{kn}} \exp(i\omega_{kn}t/2) \sin(i\omega_{kn}t/2)$$

$$|c_k^{(1)}(t)|^2 = \frac{4}{\hbar^2} \frac{|H'_{kn}|^2}{\omega_{kn}^2} \sin^2(\omega_{kn}t/2)$$

which is the probability for transition from state n to state k. From the above expression it is obvious that the probability varies simple harmonically with angular frequency $\omega_{kn}/2 = (E_k - E_n)/2\hbar$. The amplitude of vibration is

$$\frac{4|H'_{kn}|^2}{\hbar^2\omega_{kn}^2} = \frac{4|H'_{kn}|^2}{(E_k - E_n)^2}$$

10.2 Calculate the Einstein B coefficient for the $n = 2$, $l = 1$, $m = 0$, $\to n = 1$, $l = 0$, $m = 0$ transition in the hydrogen atom.

Solution. Einstein's B coefficient is given by

$$B_{m\to n} = \frac{2\pi}{3\hbar^2} |\mu_m|^2 = \frac{2\pi e^2}{3\hbar^2} |\langle m|\mathbf{r}|n\rangle|^2$$

To get the value of $\langle 210|\mathbf{r}|100\rangle$, we require the values of $\langle 210|x|100\rangle$, $\langle 210|y|100\rangle$, $\langle 210|z|100\rangle$. In the spherical polar coordinates, $x = r\sin\theta\cos\phi$, $y = r\sin\theta\sin\phi$, $z = r\cos\theta$.

$$\psi_{210} = \left(\frac{1}{32\pi a_0^3}\right)^{1/2} \frac{r}{a_0} \exp\left(-\frac{r}{2a_0}\right) \cos\theta$$

$$\psi_{100} = \left(\frac{1}{\pi a_0^3}\right)^{1/2} \exp\left(-\frac{r}{a_0}\right)$$

$$\langle 210|x|100\rangle = \text{constant} \times r\text{-part} \times \theta\text{-part} \times \int_0^{2\pi} \cos\phi\, d\phi = 0$$

$$\langle 210|y|100\rangle = \text{constant} \times r\text{-part} \times \theta\text{-part} \times \int_0^{2\pi} \sin\phi\, d\phi = 0$$

$$\langle 210|z|100\rangle = \langle 210|r\cos\theta|100\rangle$$

$$= \frac{1}{4\sqrt{2}\pi a_0^4}\int_0^\infty r^4 e^{-3r/2a_0}\,dr\int_0^\pi \cos^2\theta\sin\theta\,d\theta\int_0^{2\pi} d\phi$$

$$= \frac{1}{4\sqrt{2}\pi a_0^4}\,\frac{4\,!}{(3/2a_0)^5}\,\frac{2}{3}\,2\pi = 4\sqrt{2}\left(\frac{2}{3}\right)^5 a_0$$

$$|\langle 210|\boldsymbol{r}|100\rangle|^2 = 32\left(\frac{2}{3}\right)^{10} a_0^2 = 0.1558\times 10^{-20}\ \text{m}^2$$

$$B = \frac{2\pi}{3}\,\frac{(1.6\times 10^{-19}\text{C})^2\,(0.1558\times 10^{-20}\text{m}^2)}{(1.054\times 10^{-34}\text{Js})^2} = 7.5\times 10^9\ \text{N}^{-1}\text{m}^2\text{s}^{-2}$$

10.3 Calculate the square of the electric dipole transition moment $|\langle 310|\mu|200\rangle|^2$ for hydrogen atom.

Solution.

$$\psi_{200} = \left(\frac{1}{32\pi a_0^3}\right)^{1/2}\left(2-\frac{r}{a_0}\right)\exp\left(-\frac{r}{2a_0}\right)$$

$$\psi_{310} = \frac{4}{27\,(2\pi)^{1/2}\,a_0^{5/2}}\,r\left(1-\frac{r}{6a_0}\right)\exp\left(-\frac{r}{3a_0}\right)\cos\theta$$

$$\langle 310|z|200\rangle = \langle 310|r\cos\theta|200\rangle$$

$$= \frac{1}{54\pi a_0^4}\int_0^\infty\left(2r^4-\frac{8r^5}{6a_0}+\frac{r^6}{6a_0^2}\right)\exp\frac{-5r}{6a_0}\,dr\int_{-\infty}^{\infty}\cos^2\theta\sin\theta\,d\theta\int_0^{2\pi} d\phi$$

Using standard integrals (see Appendix), we get

$$\langle 310|z|200\rangle = \frac{1}{54\pi a_0^4}\times\frac{144}{5}\left(\frac{6a_0}{5}\right)^5\times\frac{2}{3}\times 2\pi$$

$$= 1.7695a_o$$

$$\langle 310|\mu_z|200\rangle = -1.7695a_oe$$

$$|\langle 310|\mu_z|200\rangle|^2 = 3.13a_o^2e^2$$

Since the ϕ-part of the integral is given by $\langle 310|x|200\rangle = \langle 310|y|200\rangle = 0$ (refer Problem 10.2), we have

$$|\langle 310|\mu|200\rangle|^2 = 3.13a_o^2e^2$$

10.4 What are electric dipole transitions ? Show that the allowed electric dipole transitions are those involving a change in parity.

Solution. When the wavelength λ of the electromagnetic radiation is large, the matrix element H'_{kn} of the perturbation H' between the states k and n reduces to the dipole moment matrix $\langle k|er|n\rangle$ times the other factors. This approximation is called **dipole approximation**. Physically, when the wavelength of the radiation is large, it 'sees' the atom as a dipole and, when λ is small, the radiation 'sees' the individual charges of the dipole only.

The parity of an atomic orbital with quantum number l is $(-1)^l$. Hence, s ($l = 0$) and d ($l = 2$) orbitals have even parity, whereas p ($l = 1$) and f($l = 3$) orbitals have odd parity. A transition is allowed if the dipole matrix element $\mu_{kn} = \langle \psi_k | er | \psi_n \rangle$ is nonvanishing. For that to happen, the integrand of the dipole moment matrix must have even parity. The parity of the integrand is governed by

$$(-1)^{l_k}(-1)(1)^{l_n} = (-1)^{l_k + l_n + 1}$$

If $l_k + l_n + 1$ is odd, the integrand of μ_{kn} will be odd and μ_{kn} vanishes. Hence, for μ_{kn} to be nonvanishing, $l_k + l_n + 1$ = even or $l_k + l_n$ = odd. That is, for μ_{kn} to be finite, the two orbitals must have opposite parity. This is often referred to as **Laporte selection rule.**

10.5 For hydrogenic atoms, the states are specified by the quantum numbers n, l, m. For a transition to be allowed, show that

$$\Delta n = \text{any value}, \qquad \Delta l = \pm 1, \qquad \Delta l = 0, \pm 1$$

Solution. The form of the radial wave functions are such that the radial part of the integral $\langle n'l'm' | er | nlm \rangle$ is nonvanishing, whatever be the values of n', l', n and l. Hence,

$$\Delta n = \text{any value is allowed.}$$

By the Laporte selection rule (see Problem 10.4) , for a transition to be allowed, it is neccessary that

$$l_k + l_n = \text{odd}$$

Therefore,

$$l_k - l_n = \Delta l = \pm 1$$

To obtain the selection rule for the quantum number m, the matrix element may be written as

$$\langle n'l'm' | \mathbf{r} | nlm \rangle = \hat{i}\langle n'l'm' | x | nlm \rangle + \hat{j}\langle n'l'm' | y | nlm \rangle + \hat{k}\langle n'l'm' | z | nlm \rangle$$

If the radiation is plane polarized with the electric field in the z-direction, the z-component is the only relevant quantity, which is $\langle n'l'm' | r \cos\theta | nlm \rangle$. The ϕ-part of this integral is

$$\int_0^{2\pi} \exp\,[i(m - m')\phi]\, d\phi$$

which is finite only when

$$m - m' = 0 \quad \text{or} \quad \Delta m = 0$$

If the radiation is polarized in the xy-plane, it is convenient to find the matrix elements of $x \pm iy$ since it is always possible to get the values for x and y by the relations

$$x = \frac{1}{2}[(x + iy) + (x - iy)], \qquad y = \frac{1}{2i}[(x + iy) - (x - iy)]$$

In the polar coordinates,

$$x \pm iy = r \sin\theta \cos\phi \pm ir \sin\theta \sin\phi = r \sin\theta\, e^{\pm i\phi}$$

The matrix elements of $x \pm iy$ are

$$\langle n'l'm' | r \sin\theta e^{\pm i\phi} | nlm \rangle = f(r, \theta) \int_0^{2\pi} \exp\,[i(m - m' \pm 1)\phi]\, d\phi$$

This integral is nonvanishing only when

$$m - m' \pm 1 = 0 \quad \text{or} \quad m' - m = \pm 1 \quad \text{or} \quad \Delta m = \pm 1$$

For arbitrary polarization, the general selection rule is

$$\Delta m = 0, \pm 1$$

Thus, the selection rules for hydrogenic atoms are

$$\Delta n = \text{any value}, \qquad \Delta l = \pm 1, \qquad \Delta m = 0, \pm 1$$

10.6 Find the condition under which stimulated emission equals spontaneous emission. If the temperature of the source is 500 K, at what wavelength will both the emissions be equal? Comment on the result.

Solution. Stimulated emission equals spontaneous emission when (Eq. 10.18). Hence,

$$e^{h\nu/kT} - 1 = 1 \quad \text{or} \quad e^{h\nu/kT} = 2$$

Taking logarithm on both sides, we get

$$\frac{h\nu}{kT} = \ln 2 = 0.693 \quad \text{or} \quad \frac{\nu}{T} = \frac{0.693\,K}{h}$$

$$\frac{\nu}{T} = \frac{0.693 \times 1.38 \times 10^{-23}\ \text{J/K}}{6.626 \times 10^{-34}\ \text{Js}}$$

$$= 1.44 \times 10^{10}\ \text{K}^{-1}\text{s}^{-1}$$

When $T = 500$ K,

$$\nu = (1.44 \times 10^{10}\ \text{K}^{-1}\text{s}^{-1})\ 500\ \text{K}$$

$$= 7.2 \times 10^{12}\ \text{s}^{-1}$$

$$\lambda = \frac{c}{\nu} = \frac{3 \times 10^{8}\ \text{ms}^{-1}}{7.2 \times 10^{12}\ \text{s}^{-1}}$$

$$= 4.17 \times 10^{-5}\ \text{m}$$

Wavelength of the order of 10^{-5} m corresponds to the near infrared region of the electromagnetic spectrum.

10.7 Spontaneous emission far exceeds stimulated emission in the visible region, whereas reverse is the situation in the microwave region. Substantiate.

Solution. Visible region: Wavelength ~ 5000 Å. So,

$$\frac{\text{Spontaneous emission rate}}{\text{Stimulated emission rate}} = e^{h\nu/kT} - 1$$

$$\frac{h\nu}{kT} = \frac{hc}{\lambda kT} \cong \frac{(6.626 \times 10^{-34}\ \text{J s})(3 \times 10^{8}\text{m s}^{-1})}{(5000 \times 10^{-10}\ \text{m})(1.38 \times 10^{-23}\text{J/K})\ 300\ k}$$

$$= 96.03$$

$$\text{Spontaneous emission rate} = (e^{96.03} - 1) \times \text{stimulated emission rate}$$

$$= 4.073 \times\ \text{stimulated emission rate}$$

Microwave region: Wavelength ≈ 1cm. Therefore,

$$\frac{h\nu}{kT} = \frac{(6.626 \times 10^{-34}\ \text{J s})(3 \times 10^{8}\ \text{m s}^{-1})}{0.01\text{m}\ (1.38 \times 10^{-23}\ \text{J/K})\ 300\ k}$$

$$= 0.004$$

$$e^{0.004} - 1 = 1.004 - 1 = 0.004$$

Spontaneous emission rate = 0.004 × stimulated emission rate

Hence the required result.

10.8 Obtain the selection rule for electric dipole transitions of a linear harmonic oscillator.

Solution. Consider a charged particle having a charge e executing simple harmonic motion along the x-axis about a point where an opposite charge is situated. At a given instant, the dipole moment is ex, where x is the displacement from the mean position. The harmonic oscillator wave function is

$$\psi_n(y) = N_n H_n(y)\ \exp\left(-\frac{y^2}{2}\right), \qquad y = \left(\frac{m\omega}{\hbar}\right)^{1/2} x$$

The dipole matrix element is given by

$$\langle k|y|n\rangle = \text{constant} \int H_k(y)\ yH_n(y)\ \exp(-y^2)\, dy$$

For Hermite polynomials,

$$y\ H_n(y) = nH_{n-1}(y) + \frac{1}{2}H_{n+1}(y)$$

Substituting this value of $y\ H_n(y)$, we get

$$\langle k|y|n\rangle = \text{constant} \int H_k(y)\left[nH_{n-1}(y) + \frac{1}{2}\ H_{n+1}(y)\right] \exp(-y^2)\, dy$$

In view of the orthogonality relation, we have

$$\int H_k(y)\ H_n(y)\ \exp(-y^2)\, dy = \text{constant}\ \delta_{kn}$$

$\langle k|y|n\rangle$ is finite only when $k = n - 1$ or $k = n + 1$, i.e., the harmonic oscillator selection rule is

$$k - n = \pm 1 \quad \text{or} \quad \Delta n = \pm 1$$

10.9 Which of the following transitions are electric diploe allowed?
(i) 1s → 2s; (ii) 1s → 2p; (iii) 2p → 3d; (iv) 3s → 5d.

Solution.

(i) 1s → 2s: The allowed electric dipole transitions are those involving a change in parity. The quantum number $l = 0$ for both 1s and 2s. Hence both the states have the same parity and the transition is not allowed.

(ii) 1s → 2p: The quantum number l for 1s is zero and for 2p it is 1. Hence the transition is allowed.

(iii) 2p → 3d: The l value for 2p is 1 and for 3d it is 2. The transition is the refue allowed.

(iv) 3s → 5d: The l value for 3s is zero and for 5d it is 2. As both states have same parity, the transition is not allowed.

10.10 A hydrogen atom in the 2p state is placed in a cavity. Find the temperature of the cavity at which the transition probabilities for stimulated and spontaneous emissions are equal.

Solution. The probability for stimulated emission = $B\rho(\nu)$. The probability for spontaneous emission = A. When the two are equal,

$$A = B\rho(\nu)$$

$$\rho(\nu) = \frac{A}{B} = \frac{8\pi h\nu_{21}^3}{c^3}$$

The radiation density $\rho(\nu)$ is given by Eq.(1.3). Hence,

$$\frac{8\pi h\nu_{21}^3}{c^3}\,\frac{1}{\exp(h\nu_{21}/kT)_{-1}} = \frac{8\pi h\nu_{21}^3}{c^3}$$

$$\frac{1}{\exp(h\nu_{21}/kT)_{-1}} = 1 \quad \text{or} \quad \exp\left(\frac{h\nu_{21}}{kT}\right) = 2$$

$$T = \frac{h\nu_{21}}{k\ln 2}$$

$$h\nu_{21} = (10.2\text{ eV})\,(1.6\times 10^{-19}\text{ J/eV}) = 16.32\times 10^{-19}\text{ J}$$

$$T = \frac{16.32\times 10^{-19}}{(1.38\times 10^{-23}\text{ J/K})\,0.693} = 17.1\times 10^4\text{ K}$$

10.11 A particle of mass m having charge e, confined to a three-dimensional cubical box of side $2a$, is acted on by an electric field

$$E = E_0 e^{-\alpha t}, \qquad t > 0$$

where α is a constant, in the x-direction. Calculate the prbability that the charged particle in the ground state at $t = 0$ is excited to the first excited state by the time $t = \infty$.

Solution. The energy eigenfunctions and eigenvalues of a partcile in a cubical box of side $2a$ are given by

$$E_{jkl} = \frac{\pi^2\hbar^2}{8ma^2}(j^2 + k^2 + l^2), \qquad j, k, l = 1, 2, 3, \ldots$$

$$\Psi_{jkl} = \frac{1}{\sqrt{a^3}}\sin\frac{j\pi x}{2a}\sin\frac{k\pi y}{2a}\sin\frac{l\pi z}{2a} = |jkl\rangle$$

The ground state is $|111\rangle$ and the first excited states are $|211\rangle$, $|121\rangle$, $|112\rangle$. Since the electric field is along the x-axis, the dipole moment $\mu = ex$ and the perturbation are given by

$$H' = -\mu\cdot E = -eE_0 x e^{-at}$$

The transition probability for a transition from state n to state m is obtained as

$$P = |C_m^{(1)}|^2 = \frac{1}{\hbar^2}\left|\int_0^\infty H'_{mn}\exp(i\omega_{mn}t)\,dt\right|^2$$

where $\omega_{mn} = (E_m - E_n)/\hbar$, $c_m^{(1)}$ and H'_{mn} is the transition moment.

$$H'_{mn} = \langle 111|H'|211\rangle = \langle 111|-eE_0 x e^{-\alpha t}|211\rangle$$

$$= -eE_0 e^{-\alpha t}\langle 111|x|211\rangle$$

$$= \frac{-eE_0 e^{-\alpha t}}{a^3}\int_0^{2a} x\sin\frac{\pi x}{2a}\sin\frac{\pi x}{a}\,dx \int_0^{2a}\sin^2\frac{\pi y}{2a}\,dy\int_0^{2a}\sin^2\frac{\pi z}{2b}\,dz$$

$$= \frac{-eE_0 e^{-\alpha t}}{a^3}\left(-\frac{32a^2}{9\pi^2}\right)\times a\times a = \frac{32aeE_0 e^{-\alpha t}}{9\pi^2}$$

$$\langle 111|H'|121\rangle = \frac{-eE_0 e^{-\alpha t}}{a^3}\int_0^{2a} x\sin^2\frac{\pi x}{2b}\,dx\int_0^{2a}\sin\frac{\pi y}{2b}\sin\frac{\pi y}{b}\,dy\int_0^{2a}\sin^2\frac{\pi z}{2b}\,dz = 0$$

Similarly,

$$\langle 111|H'|112\rangle = 0$$

$$\omega_{21} = \frac{E_2 - E_1}{\hbar} = \frac{\pi^2\hbar^2}{8ma^2}(2^2 + 1^2 + 1^2 - 1^2 - 1^2) = \frac{3\pi^2\hbar^2}{8ma^2}$$

Consequently,

$$P = \left(\frac{32aeE_0}{9\pi^2\hbar}\right)^2\left|\int_0^{\alpha}\exp(-\alpha t + i\omega_{21}t)\,dt\right|^2$$

$$= \left(\frac{32aeE_0}{9\pi^2\hbar}\right)^2\frac{1}{\alpha^2 + \omega_{21}^2}$$

10.12 Calculate the electric dipole transition moment $\langle 2p_z|\mu_z|2s\rangle$ for the $2s \to 2p_z$ transition in a hydrogen atom.

Solution.

$$\psi_{2s} = \left(\frac{1}{32\pi a_0^3}\right)^{1/2}\left(2 - \frac{r}{a_0}\right)e^{-r/2a_0}$$

$$\psi_{2p_z} = \left(\frac{1}{32\pi a_0^3}\right)^{1/2}\frac{r}{a_0}e^{-r/2a_0}\cos\theta$$

$$\langle 2p_z|\mu_z|2s\rangle = \langle 2p_z|-er\cos\theta|2s\rangle$$

$$= \frac{-e}{32\pi a_0^4}\left[\int_0^{\infty}2r^4e^{-r/a_0}\,dr - \frac{1}{a_0}\int_0^{\infty}r^5e^{-r/a_0}\,dr\right]\int_0^{\infty}\cos^2\theta\sin\theta\,d\theta\int_0^{2\pi}d\phi$$

$$= \frac{-e}{32\pi a_0^4}\left[\frac{2\times 4!}{(1/a_0)^5} - \frac{1}{a_0}\frac{5!}{(1/a_0)^6}\right]\frac{2}{3}\times 2\pi = 3ea_0$$

10.13 Calculate Einsten's A coefficient for the $n = 2, l = 1, m = 0 \to n = 1, l = 0, m = 0$ transition in the hydrogen atom.

Solution.

$$\text{Einstein's } A \text{ coefficient} = \frac{4\omega_{mn}^3}{3\text{hc}^3} \,|\,\mu_{mn}\,|^2 = \frac{4\omega_{mn}^3 e^2}{3\text{hc}^3} |\langle m|\mathbf{r}|n\rangle|^2$$

$$\psi_{100} = \left(\frac{1}{\pi a_0^3}\right)^{1/2} e^{-r/a_0}, \quad \psi_{210} = \left(\frac{1}{32\pi a_0^3}\right)^{1/2} \frac{r}{a_0} e^{-r/2a_0} \cos\theta$$

To evaluate $\langle 210\,|\boldsymbol{r}|\,100\rangle$, we require the values of $\langle 210\,|x|\,100\rangle$, $\langle 210\,|y|\,100\rangle$, and $\langle 210\,|z|\,100\rangle$. In the spherical polar coordinates, $x = r \sin\theta \cos\phi$, $y = r \sin\theta \sin\phi$ and $z = r \cos\theta$. The x- and y-components of the matrix element vanish since

$$\int_0^{2\pi} \cos\phi \, d\phi = 0 \quad \text{and} \quad \int_0^{2\pi} \sin\phi \, d\phi = 0$$

$$\langle 210\,|z|\,100\rangle = \langle 210\,|r\cos\theta|\,100\rangle = \frac{1}{4\sqrt{2}\pi a_0^4} \int_0^\infty r^4 e^{-3r/2a_0}\, dr \int_0^\infty \cos^2\theta \sin\theta\, d\theta \int_0^{2\pi} d\phi$$

$$= \frac{1}{4\sqrt{2}\pi a_0^4} \frac{4!}{(3/2a_0)^5} \frac{4\pi}{3} = 4\sqrt{2}\left(\frac{2}{3}\right)^5 a_0$$

$$|\langle 210\,|\boldsymbol{r}|\,100\rangle|^2 = 32 \times \left(\frac{2}{3}\right)^{10} a_0^2 = 0.1558 \times 10^{-20}\ \text{m}^2$$

For $n = 2 \to n = 1$ transition,

$$\nu = \frac{E_2 - E_1}{h} = \frac{10.2\ \text{eV}}{h} = 2.463 \times 10^{15}\ \text{Hz}$$

$$\omega = 2\pi\nu = 15.482 \times 10^{15}\ \text{Hz}$$

$$e^2 = \frac{e'^2}{4\pi\varepsilon_0} = \frac{1.6 \times 10^{-19} \times 1.6 \times 10^{-19}}{4\pi \times 8.854 \times 10^{-12}} = 2.3 \times 10^{-28}\ \text{Nm}^2$$

$$A = \frac{4 \times (15.482 \times 10^{15}\ \text{s}^{-1})^3}{3 \times 1.055 \times 10^{-34}\ \text{Js} \times (3 \times 10^8\,\text{ms}^{-1})^3} \times 2.3 \times 10^{-28}\ \text{Nm}^2 \times 0.1558 \times 10^{-20}\,\text{m}^2$$

$$= 6.2 \times 10^8\ \text{s}^{-1}$$

10.14 Prove the following:

(i) If the source temperature is 1000 K, in the optical region (λ = 5000 Å), the emission is predominantly due to spontaneous transitions.

(ii) If the source temperature is 300 K, in the microwave region (λ = 1 cm), the emission is predominantly due to stimulated emission. The Boltzmann constant is 1.38×10^{-23} JK^{-1}.

Solution.

$$\frac{\text{Spontaneous emission}}{\text{Stimulated emission}} = \exp\left(\frac{\hbar\nu}{kT}\right) - 1$$

(i) In the optical region,

$$\nu = \frac{c}{\lambda} = \frac{3 \times 10^8}{5000 \times 10^{-10}} = 6 \times 10^{14}\ \text{Hz}$$

$$\frac{\hbar\nu}{kT} = \frac{6.626 \times 10^{-34}}{1.38 \times 10^{-23}} \times \frac{6 \times 10^{14}}{1000} = 28.8$$

$$\exp(28.8) - 1 = 3.22 \times 10^{12}$$

Thus, spontaneous emission is predominant.

(ii) In the microwave region,

$$\nu = \frac{c}{\lambda} = \frac{3 \times 10^8}{10^{-2}} = 3 \times 10^{10}\ \text{Hz}$$

$$\frac{\hbar\nu}{kT} = \frac{6.626 \times 10^{-34} \times 3 \times 10^{10}}{1.38 \times 10^{-23} \times 300} = 4.8 \times 10^{-3}$$

$$\exp(4.8 \times 10^{-3}) - 1 = 0.0048$$

Therefore, stimulated emission is predominant.

10.15 Obtain Einstein's A coefficient for a one-dimensional harmonic oscillator of angular frequency ω in its nth state.

Solution.

$$A_{n\to k} = \frac{4\omega_{kn}^3}{3\hbar c^3} \, |\mu_{kn}|^2 = \frac{4e^2\omega_{kn}^3}{3\hbar c^3} \, |\langle k|x|n\rangle|^2$$

For linear harmonic oscillator, $\langle k|x|n\rangle$ is finite only when $k = n - 1$ or $k = n + 1$. For emission from state n, k must be $n - 1$. Hence,

$$\langle k|x|n\rangle = \langle n-1|x|n\rangle = \left\langle n-1 \left| \left(\frac{\hbar}{2m\omega}\right)^{1/2} (a + a^\dagger) \right| n \right\rangle$$

$$= \left(\frac{\hbar}{2m\omega}\right)^{1/2} [(n-1)|a|n\rangle + \langle(n-1)|a^\dagger|n\rangle$$

$$= \left(\frac{\hbar}{2m\omega}\right)^{1/2} \left[\sqrt{n} + 0\right] = \left(\frac{n\hbar}{2m\omega}\right)^{1/2}, \qquad k = n - 1$$

Substituting this value of $\langle k|x|n\rangle$,

$$A_{n\to k} = \frac{4e^2\omega^3}{3\hbar c^3} \frac{n\hbar}{2m\omega} = \frac{2e^2\omega^2 n}{3mc^3}$$

10.16 Calculate the rates of stimulated and spontaneous emission for the transition $3p - 2s$ (H_α line) of hydrogen atom, essuming the atoms are at a temperature of 1000 K.

Solution.

$$\text{Stimulated emission rate} = B_{m\to n}\ \rho(\upsilon) = \frac{2\pi}{3\hbar^2}|\mu_{mn}|^2\ \rho(\upsilon)$$

From Problem 10.3, $|\langle 200|\mu|310\rangle|^2 = 3.13a_0^2e^2$

Since $e^2 = 2.3 \times 10^{-28}$ N m^2

$$|\langle 200|\mu|310\rangle|^2 = 3.13\,(0.53\times 10^{-10}\text{m})^2 \times 2.3\times 10^{-28}\text{N m}^2 = 2.0222\times 10^{-48}\text{ N m}^4$$

$$\nu = \frac{E_3 - E_2}{h} = \frac{1.89\times 1.6\times 10^{-19}}{6.626\times 10^{-34}} = 4.564\times 10^{14}\text{ Hz}$$

$$\frac{1}{e^{h\nu/kt}-1} = \frac{1}{e^{21.914}-1} = \frac{1}{3.289\times 10^9}$$

$$\rho = \frac{8\pi h\nu^3}{c^3}\frac{1}{e^{h\nu/kt}-1} = \frac{8\pi\times 6.626\times 10^{-34}}{(3\times 10^8)^3}\times\frac{(4.564\times 10^{14})^3}{3.289\times 10^9}$$

$$= 178.3\times 10^{-25}\text{ J m}^{-3}\text{ s}$$

$$\text{Stimulated emission rate} = \frac{2\pi\times 2.0222\times 10^{-48}\text{N m}^4\times 178.3\times 10^{-25}\text{ J m}^{-3}\text{s}}{3\times(1.055\times 10^{-34}\text{ Js})^2}$$

$$= 6.79\times 10^{-3}\text{ s}^{-1}$$

$$\text{Spontaneous emission rate } A = \frac{4\omega_{mn}^3}{3\hbar c^3}|\mu_{mn}|^2 = \frac{32\pi^3\nu^3}{3\hbar c^3}|\mu_{mn}|^2$$

$$A = \frac{32\pi^3\times(4.564\times 10^{14})^3}{3\times 1.055\times 10^{-34}\times(3\times 10^8)^3}\times 2.0222\times 10^{-48} = 2.235\times 10^7\text{ s}^{-1}$$

10.17 A harmonic oscillator in the ground state is subjected to a perturbation

$$H' = -x\exp\left(-\frac{t^2}{t_0^2}\right)\quad\text{from } t = 0 \text{ to } t = \infty.$$

Calculate the probability for transition from the ground state, given that

$$\int_0^\infty \exp(-\alpha t^2 + i\omega t)\,dt = -i\sqrt{\frac{\pi}{\alpha}}\exp\left(\frac{-\omega^2}{4\alpha}\right)$$

Solution. The probability that a transition to state k has occurred is $|c_k^{(1)}(t)|^2$

$$c_k^{(1)}(t) = \frac{1}{i\hbar}\int_0^t H'_{kn}\exp(i\omega_{kn}t')\,dt',\qquad H' = -x\exp\left(-\frac{t^2}{t_0^2}\right)$$

Since the only transition possible is $0 \to 1$,

$$c_k^{(1)}(\infty) = -\frac{1}{i\hbar}\int_0^{\infty} \langle 0|x|1\rangle\, e^{i\omega t} \exp\left(\frac{-t'^2}{t_0^2}\right) dt$$

$$\langle 0|x|1\rangle = \sqrt{\frac{\hbar}{2m\omega}}, \qquad w_{kn} = w$$

$$c_k^{(1)}(\infty) = -\frac{1}{i\hbar}\sqrt{\frac{\hbar}{2m\omega}}\int_0^{\infty} e^{i\omega t} \exp\left(-\frac{t'^2}{t_0^2}\right) dt$$

$$= \frac{1}{\sqrt{2m\hbar\omega}}\sqrt{\pi t_0^2}\, \exp\left(-\frac{\omega^2 t_0^2}{4}\right)$$

The probability for the $0 \to 1$ transition is

$$|c_k^{(1)}|^2 = \frac{\pi t_0^2}{2m\hbar\omega} \exp\left(-\frac{\omega^2 t_0^2}{2}\right)$$

10.18 The time varying Hamiltonain $H'(t)$ induces transitions between states $|j\rangle$ and $|k\rangle$. Using time-dependent perturbation theory, show that the probability for a transition from state $|j\rangle$ to state $|k\rangle$ is the same as the probability for a transition from state $|k\rangle$ to state $|j\rangle$.

Solution. The probability for a transition from state $|j\rangle$ to state $|k\rangle$ at time t is

$$P_{j\to k}(t) = |C_{j\to k}(t)|^2$$

The relation for $C_{j\to k}$ is

$$C_{j\to k}(t) = \frac{1}{i\hbar}\int_0^t \langle k|H'|j\rangle \exp(i\omega_{kj}t)\, dt$$

See Eq. (10.6). The coefficient for transition from state $|k\rangle$ to state $|j\rangle$ is given by

$$C_{j\to k}(t) = \frac{1}{i\hbar}\int_0^t \langle j|H'|k\rangle \exp(i\omega_{jk}t)\, dt$$

Since H' is Hermitian, $\langle k|H'|j\rangle = \langle j|H'|k\rangle$. Also, it follows that $\hbar\omega_{kj} = E_k - E_j = -\hbar\omega_{jk}$. As the integrand of the second integral is the complex conjugate of that of the first one, we have

$$|C_{j\to k}(t)|^2 = |C_{k\to j}(t)|^2$$

i.e.,

$$P_{j\to k}(t) = P_{k\to j}(t)$$

10.19 A quantum mechanical system is initially in the ground state $|0\rangle$. At $t = 0$, a perturbation of the form $H'(t)$, $H_0 e^{-\alpha t}$, where α is a constant, is applied. Show that the probability that the system is in state $|1\rangle$ after long time is

$$P_{10} = \frac{|\langle 0|H_0|1\rangle|^2}{\hbar^2(\alpha^2 + \omega_{10})^2}, \qquad \omega_{10} = \frac{E_1 - E_0}{\hbar}$$

Solution. In the first-order perturbation, the transition probability amplitude is given by Eq. (10.6). So,

$$C_k^{(1)}(t) = \frac{1}{i\hbar}\int_0^t H'_{kn} \exp(i\omega_{kn}t')\, dt'$$

where

$$H'_{kn} = \langle k|H'|n\rangle, \qquad \omega_{kn} = \frac{E_k - E_n}{\hbar}$$

Substituting the value of H' and allowing $t \to \infty$, we get

$$C_k^{(1)}(t) = \frac{1}{i\hbar}\int_0^\infty \exp(i\omega_{10}t)\, e^{-\alpha t}\, \langle 1|H_0|0\rangle\, dt$$

$$= \frac{\langle 1|H_0|0\rangle}{i\hbar}\left[\frac{\exp[-(\alpha - i\omega_{10})t]}{-(\alpha - i\omega_{10})}\right]_0^\infty$$

$$= \frac{\langle 1|H_0|0\rangle}{i\hbar}\,\frac{1}{\alpha - i\omega_{10}}$$

The probability for a transition from state $|0\rangle$ to state $|1\rangle$ after a long time is

$$P_{10} = |C_k^{(1)}|^2 \frac{|\langle 0|H_0|1\rangle|^2}{\hbar^2(a^2 + \omega_{10}^2)^2}$$

10.20 A hydrogen atom in the ground state is subjected to an electric field

$$E = E_0 e^{-t/\tau}, \qquad t > 0,\ \tau \text{ being constant}$$

along the z-axis. Calculate the probability for transition to the (200) and (210) states when it is very large.

Solution. The interaction Hamiltonian

$$H' = -\boldsymbol{\mu}\cdot\mathbf{E} = -\mu E\cos\theta = erE_0 e^{-t/\tau}\cos\theta$$

$$\psi_{100} = \left(\frac{1}{\pi a_0^3}\right)^{1/2} e^{-r/a_0}$$

$$\psi_{200} = \frac{1}{\pi^{1/2}}\left(\frac{1}{2a_0}\right)^{3/2}\left(1 - \frac{r}{2a_0}\right)e^{-r/a_0}$$

$$\psi_{210} = \frac{1}{\pi^{1/2}}\left(\frac{1}{2a_0}\right)^{5/2} r\, e^{-r/2a_0}\cos\theta$$

The probability for transition from $n \to k$ state is

$$P_{n\to k} = \frac{1}{\hbar^2}\left|\int_0^t H'_{kn}(r, t)\exp(i\omega_{kn}t)\, dt\right|^2 \qquad \omega_{kn} = \frac{E_k - E_n}{\hbar}$$

(100) to (200) transition:

$$H'_{21}(t) = \langle 200|H'|100\rangle = \int \psi_{200}(erE_0 e^{-t/\tau} \cos\theta)\, \psi_{100}\, d\tau$$

The θ-part of the integral is

$$\int_0^\pi \cos\theta \sin\theta\, d\theta = 0$$

Hence, H'_{21} is zero. Therefore, the probability $P_{n\to k} = 0$.

(100) to (210) transition:

$$H'_{21}(t) = \langle 200|H'|100\rangle = \frac{eE_0 e^{-t/\tau}}{\pi 2^{5/2} a_0^4} \int_0^\infty r^4 e^{-3r/2a_0} dr \int_0^\pi \cos^2\theta \sin\theta\, d\theta \int_0^{2\pi} d\phi$$

Writing $y = \cos\theta$, $dy = -\sin\theta\, d\theta$,
we have

$$\int_0^\pi \cos^2\theta \sin\theta\, d\theta = -\int_1^{-1} y^2 dy = \frac{2}{3}$$

$$H'_{21}(t) = \frac{eE_0 e^{-t/\tau}}{\pi 2^{5/2} a_0^4} \cdot \frac{4\,!}{(3/2a_0)^5} \times \frac{2}{3} \times 2\pi$$

$$= \frac{256 eE_0 a_0 e^{-t/\tau}}{243 \times \sqrt{2}} = Ae^{-t/\tau}$$

where

$$A = \frac{256 eE_0 a_0}{243 \times \sqrt{2}}$$

$$\int_0^t H'_{21}\, e^{i\omega_{21}t}\, dt = A\int_0^t e^{-t/\tau} e^{i\omega_{21}t}\, dt = A\int_0^t e^{-t/\tau}\, (\cos\omega_{21}t + i\sin\omega_{21}t)\, dt$$

As t is very large, we can assume the limits of integral as 0 to ∞. Then,

$$\int_0^\infty H'_{21}\, e^{i\omega_{21}t}\, dt = A\left(\frac{1/\tau}{(1/\tau^2) + \omega_{21}^2} + i\frac{\omega_{21}}{(1/\tau^2) + \omega_{21}^2}\right)$$

$$P_{1\to 2} = \frac{A^2}{\hbar^2}\left(\frac{1/\tau}{(1/\tau^2) + \omega_{21}^2} + i\frac{\omega_{21}}{(1/\tau^2) + \omega_{21}^2}\right)\left(\frac{1/\tau}{(1/\tau^2) + \omega_{21}^2} - i\frac{\omega_{21}}{(1/\tau^2) + \omega_{21}^2}\right)$$

$$= \frac{A^2}{\hbar^2}\left(\frac{(1/\tau)^2}{[(1/\tau^2) + \omega_{21}^2]^2} + i\frac{\omega_{21}^2}{[(1/\tau^2) + \omega_{21}^2]^2}\right)$$

$$= \frac{A^2}{\hbar^2}\left(\frac{1}{(1/\tau^2) + \omega_{21}^2}\right)$$

CHAPTER 11

Identical Particles

Systems of identical particles are of considerable importance for the understanding of structures of atoms, molecules and nuclei.

11.1 Indistinguishable Particles

Particles that can be substituted for each other with no change in the physical situation are said to be **indistinguishable** or **identical**. For example, n electrons are strictly indistinguishable. Since the interchange of coordinates of any two electrons does not change the Hamiltonian, we have

$$H(1, 2, \ldots, i, j, \ldots, n) = H(1, 2, \ldots, i, j, \ldots, n) \tag{11.1}$$

A **particle exchange operator** P_{ij} is defined such that when it operates on a state, the coordinates of particles i and j are interchanged. The eigenvalue of the particle exchange operator is either +1 or –1, i.e.,

$$P_{ij}\psi(1, 2, \ldots, i, j, \ldots, n) = \pm 1\,\psi(1, 2, \ldots, j, i, \ldots, n) \tag{11.2}$$

Consequently, the indistinguishability requires that the wave function must be either symmetric or antisymmetric with respect to the interchange of any pair of particles. The symmetry character of a wave function does not change with time.

The solution of the Schrödinger equation of an n-identical particle system gives ψ which is a function of the coordinates of the n particles. This leads to n ! solutions from one solution since n ! permutations of the n arguments are possible. All these n ! solutions correspond to the same energy. The degeneracy arising due to this interchange is called **exchange degeneracy**.

11.2 The Pauli Principle

From simple considerations, Pauli has shown that the symmetry of a system is related to the spin of the identical particles:

1. Systems of identical particles with half odd integer spins (spin 1/2, 3/2, 5/2, ...) are described by antisymmetric wave functions. Such particles obey Fermi-Dirac statistics and are called **fermions**.
2. Systems of identical particles with integer spins (spin 0, 1, 2, ...) are described by symmetric wave functions. Such particles obey Bose-Einstein statistics and are called **bosons**.

One form of Pauli's exclusion principle is that two identical fermions cannot occupy the same state. For electrons, this is stated as "No two electrons can have the same set of quantum numbers". For a system having n particles, if $u_a(1)$, $u_b(2)$, ..., $u_n(n)$ are the $n1$ particle eigenfunctions, the normalized antisymmetric combination is given by the Slater determinant

$$\psi_{\rm as}(1, 2, 3, \ldots, n) = \frac{1}{\sqrt{n\,!}} \begin{vmatrix} u_a(1) & u_a(2) & \ldots & u_a(n) \\ u_b(1) & u_b(2) & \ldots & u_b(n) \\ \vdots & \vdots & \vdots & \vdots \\ u_n(1) & u_n(2) & \ldots & u_n(n) \end{vmatrix} \tag{11.3}$$

The factor $1/\sqrt{n\,!}$ is the normalization constant.

11.3 Inclusion of Spin

The spin can be included in the formalism by taking the single particle eigenfunctions of both position wave function $\phi(\boldsymbol{r})$ and spin function $\chi(m_s)$, i.e.,

$$\psi(\boldsymbol{r}, m_s) = \phi(\boldsymbol{r})\chi(m_s) \tag{11.4}$$

The spin functions of spin –1/2 system are discussed in problem

Boson states:
$$\psi_{\rm s} = \begin{cases} \psi_{\rm s}(\text{spatial}) & \chi_{\rm s}(\text{spin}) \\ \psi_{\rm as}(\text{spatial}) & \chi_{\rm as}(\text{spin}) \end{cases} \tag{11.5}$$

Fermion states:
$$\psi_{\rm as} = \begin{cases} \psi_{\rm s}(\text{spatial}) & \chi_{\rm as}(\text{spin}) \\ \psi_{\rm as}(\text{spatial}) & \chi_{\rm s}(\text{spin}) \end{cases} \tag{11.6}$$

Here, s refers to symmetric and as refers to antisymmetric.

For a system with two identical electrons, the possible spin product functions alongwith the eigenvalues are given in Table 11.1.

Table 11.1 Two Electron Spin Product Functions

Spin product functions	*Symmetry character*	*Eigenvalue of* $S_z = S_{1z} + S_{2z}$	*Eigenvalue of* $S^2 = (S_1 + S_2)^2$
$\alpha\alpha$	Symmetric	$\hbar$	$2\hbar^2$
$\frac{1}{\sqrt{2}}(\alpha\beta + \beta\alpha)$	Symmetric	0	$2\hbar^2$
$\beta\beta$	Symmetric	$-\hbar$	$2\hbar^2$
$\frac{1}{\sqrt{2}}(\alpha\beta - \beta\alpha)$	Antisymmetric	0	0

PROBLEMS

11.1 Consider a system having three identical particles. Its wave function $\psi(1,2,3)$ is 3 ! fold degenerate due to exchange degeneracy. (i) Form symmetric and antisymmetric combinations of the degenerate functions. (ii) If the Hamiltonian $H(1,2,3) = H(1) + H(2) + H(3)$ and $\psi(1,2,3) = u_a(1)\, u_b(2)\, u_c(3)$, where $u_a(1)\, u_b(2)$ and $u_c(3)$ are the eigenfunctions of H_1, H_2, H_3 respectively, what are the symmetric and antisymmetric combinations?

Solution.

(i) In the three-particle system the wave function $\psi(1,2,3)$ = 6-fold degenerate. The six functions are $\psi(123)$, $\psi(132)$, $\psi(321)$, $\psi(213)$, $\psi(231)$, and $\psi(312)$.

The symmetric combination is the sum of all functions:

$$\psi_s = \psi(123) + \psi(132) + \psi(321) + \psi(213) + \psi(231) + \psi(312)$$

The antisymmetric combination is the sum of all functions with even number of interchanges–the sum of all functions with odd number of interchanges.

$$\psi_{as} = \psi(123) + \psi(231) + \psi(312) - \psi(213) + \psi(132) + \psi(321)$$

(ii) $\psi(1,2,3) = u_a(1)\, u_b(2)\, u_c(3)$

The six product functions are

$$u_a(1)\, u_b(2)\, u_c(3), \qquad u_a(1)\, u_b(3)\, u_c(2), \qquad u_a(2)\, u_b(1)\, u_c(3)$$

$$u_a(2)\, u_b(3)\, u_c(1), \qquad u_a(3)\, u_b(2)\, u_c(1), \qquad u_a(3)\, u_b(1)\, u_c(2)$$

The symmetric combination of these is simply the sum. The antisymmetric combination

$$\psi_{as} = u_a(1)\, u_b(2)\, u_c(3) + u_a(2)\, u_b(3)\, u_c(1) + u_a(3)\, u_b(1)\, u_c(2)$$
$$- u_a(1)\, u_b(3)\, u_c(2) - u_a(2)\, u_b(1)\, u_c(3) - u_a(3)\, u_b(2)\, u_c(1)$$

$$= \frac{1}{\sqrt{3\,!}} \begin{vmatrix} u_a(1) & u_a(2) & u_a(3) \\ u_b(1) & u_b(2) & u_b(3) \\ u_c(1) & u_c(2) & u_c(3) \end{vmatrix}; \quad \frac{1}{\sqrt{3\,!}} \text{ is the normalization constant}$$

11.2 Consider a one-dimensional infinite square well of width 1 cm with free electrons in it. If its Fermi energy is 2 eV, what is the number of electrons inside the well?

Solution. In an infinite square well, energy

$$E_n = \frac{\pi^2 \hbar^2 n^2}{2ma^2}, \qquad n = 1, 2, 3, \ldots$$

Each level accommodates two electrons, one spin up and the other spin down. If the highest filled level is n, then the Fermi energy $E_F = E_n$.

$$n^2 = \frac{E_F\, 2ma^2}{\pi^2 \hbar^2}$$

$$= \frac{(2 \times 1.6 \times 10^{-19}\ \text{J}) \times 2 \times (9.1 \times 10^{-31}\ \text{kg})(0.01\ \text{m})^2}{\pi^2 (1.05 \times 10^{-34}\ \text{J s})^2}$$

$$= 5.3475 \times 10^4$$

$$n = 2.312 \times 10^7$$

The number of electrons inside the well = $2n = 4.62 \times 10^7$.

11.3 N noninteracting bosons are in an infinite potential well defined by $V(x) = 0$ for $0 < x < a$; $V(x) = \infty$ for $x < 0$ and for $x > a$. Find the ground state energy of the system. What would be the ground state energy if the particles are fermions.

Solution. The energy eigenvalue of a particle in the infinite square well (Problem 4.1) is given by

$$E_n = \frac{\pi^2 \hbar^2 n^2}{2ma^2}, \qquad n = 1, 2, 3, \ldots$$

As the particles are bosons, all the N particles will be in the $n = 1$ state. Hence the total energy

$$E = \frac{N\pi^2\hbar^2}{2ma^2}$$

If the particles are fermions, a state can have only two of them, one spin up and the other spin down. Therefore, the lowest $N/2$ states will be filled. The total ground state energy will be

$$E = 2\frac{\pi^2\hbar^2}{2ma^2}[1^2 + 2^2 + 3^3 + \cdots + (N/2)^2]$$

$$= \frac{\pi^2\hbar^2}{ma^2}\frac{1}{6}\left[\frac{N}{2}\left(\frac{N}{2}+1\right)\left(2\frac{N}{2}+1\right)\right]$$

$$= \frac{\pi^2\hbar^2}{24ma^2} N(N+1)(N+2)$$

11.4 Consider two noninteracting electrons described by the Hamiltonian

$$H = \frac{p_1^2}{2m} + \frac{p_2^2}{2m} + V(x_1) + V(x_2)$$

where $V(x) = 0$ for $0 < x < a$; $V(x) = \infty$ for $x < 0$ and for $x > a$. If both the electrons are in the same spin state, what is the lowest energy and eigenfunction of the two-electron system?

Solution. As the electrons are noninteracting, the wave function of the system $\psi(1, 2)$ can be written as

$$\psi(1, 2) = \psi(1)\,\psi(2)$$

With this wave function, the Schrödinger equation for the system breaks into two one-particle equations:

$$-\frac{\hbar^2}{2m}\frac{d^2}{dx_1^2}\psi(1) + V(x_1)\,\psi(1) = E^{(1)}\,\psi(1)$$

$$-\frac{\hbar^2}{2m}\frac{d^2}{dx_2^2}\psi(2) + V(x_2)\,\psi(1) = E^{(2)}\,\psi(1)$$

where $E^{(1)} + E^{(2)} = E$, which is the total energy of the system. The energy eigenvalues and eigenfunctions for a single particle in such a potential (see Problem 4.1) are

$$E_{n_1}^{(1)} = \frac{\pi^2 \hbar^2 n_1^2}{2ma^2}, \qquad \psi_{n_1}(1) = \sqrt{\frac{2}{a}} \sin \frac{n_1 \pi x_1}{a}, \qquad n_1 = 1, 2, 3, \ldots$$

$$E_{n_2}^{(1)} = \frac{\pi^2 \hbar^2 n_2^2}{2ma^2}, \qquad \psi_{n_2}(2) = \sqrt{\frac{2}{a}} \sin \frac{n_2 \pi x_2}{a}, \qquad n_2 = 1, 2, 3, \ldots.$$

As both the electrons are in the same spin state, the possible combinations of spin functions are $\alpha(1)$ $\alpha(2)$ or $\beta(1)\ \beta(2)$, both being symmetric. Hence the space function must be antisymmetric. As the electrons are either spin up ($\alpha\alpha$) or spin down ($\beta\beta$), $n_1 = n_2 = 1$ is not possible. The next possibility is $n_1 = 1$, $n_2 = 2$.

$$\text{Energy of the state } (n_1 = 1,\ n_2 = 2) = \frac{\pi^2 \hbar^2}{2ma^2} + \frac{4\pi^2 \hbar^2}{2ma^2} = \frac{5\pi^2 \hbar^2}{2ma^2}$$

$$\text{Energy eigenfunction } \psi(1, 2) = \frac{2}{a} \sin \frac{\pi x_1}{a} \sin \frac{2\pi x_2}{a}$$

When the two electrons are interchanged, the eigenfunction

$$\psi(2, 1) = \frac{2}{a} \sin \frac{\pi x_2}{a} \sin \frac{2\pi x_1}{a}$$

Since both the states have the same energy, the space wave function of the system must be a linear combination of the two functions. The antisymmetric combination is

$$\psi(1, 2) - \psi(2, 1)$$

To get the complete energy eigenfunction, this space part has to be multiplied by $\alpha\alpha$ or $\beta\beta$. Since the energy depends only on the space part,

$$\text{Energy eigenvalue } E = \frac{5\pi^2 \hbar^2}{2ma^2}$$

11.5 Show that for a system of two identical particles of spin I, the ratio of the number of states which are symmetric under spin interchange to the number of states which are antisymmetric under spin interchange is $(I + 1)/I$.

Solution. We shall denote the m_I values of the two spins by m_I and m_I'. The spin states of the combined system are given by $|m_I(1)\rangle\ |m_I'(2)\rangle$. The products $|m_I(1)\rangle\ |m_I(2)\rangle$ corresponding to $m_I = m_I'$ will be symmetric and we will have $(2I + 1)$ such product functions. The number of product functions corresponding to $m_I \neq m_I'$ will be $2I\ (2I + 1)$. With these we have to form combinations of the type

$$|m_I(1)\rangle\ |m_I'(2)\rangle \pm |m_I'(1)\rangle\ |m_I(2)\rangle$$

where the plus sign gives symmetric and the minus sign gives antisymmetric functions. As we take two product functions to form such a combination, we will have $(1/2)\ 2I\ (2I + 1)$ symmetric and $(1/2)\ 2I\ (2I + 1)$ antisymmetric combinations. The total number of symmetric combinations = $(2I + 1) + (1/2)\ 2I\ (2I + 1) = (I + 1)\ (2I + 1)$. Hence,

$$\frac{\text{No. of symmetric combinations}}{\text{No. of antisymmetric combinations}} = \frac{(I + 1)(2I + 1)}{I(2I + 1)} = \frac{I + 1}{I}$$

11.6 Show that if a wave function $\psi(1, 2, 3, \ldots, n)$ is an energy eigenfunction of a symmetric Hamiltonian that corresponds to a nondegenerate eigenvalue, it is either symmetric or antisymmetric.

Solution. The eigenvalue equation of the Hamiltonian is

$$H(1, 2, \ldots, i, j, \ldots, n)\, \psi(1, 2, \ldots, i, j, \ldots, n) = E\psi(1, 2, \ldots, i, j, \ldots, n)$$

Interchange of the indistinguishable particles i and j does not change the energy. Hence,

$$H(1, 2, \ldots, j, i, \ldots, n)\, \psi(1, 2, \ldots, j, i, \ldots, n) = E\psi(1, 2, \ldots, j, i, \ldots, n)$$

Since H is symmetric,

$$H(1, 2, \ldots, i, j, \ldots, n)\, \psi(1, 2, \ldots, j, i, \ldots, n) = E\psi(1, 2, \ldots, j, i, \ldots, n)$$

$$H(1, 2, \ldots, i, j, \ldots, n)\, P_{ij}\psi(1, 2, \ldots, i, j, \ldots, n) = EP_{ij}\psi(1, 2, \ldots, i, j, \ldots, n)$$

$$= P_{ij}H(1, 2, \ldots, i, j, \ldots, n)\, \psi(1, 2, \ldots, i, j, \ldots, n)$$

$$(HP_{ij} - P_{ij}H)\, \psi = 0 \quad \text{or} \quad [H, P_{ij}] = 0$$

Since P_{ij} commutes with the Hamiltonian, $\psi(1, 2, \ldots, i, j, \ldots, n)$ is an eigenfunction of P_{ij} also.

$$P_{ij}\psi(1, 2, \ldots, i, j, \ldots, n) = p\psi(1, 2, \ldots, i, j, \ldots, n)$$

$$\psi(1, 2, \ldots, j, i, \ldots, n) = p\psi(1, 2, \ldots, i, j, \ldots, n)$$

Operating both sides by P_{ij}, we get

$$\psi(1, 2, \ldots, i, j, \ldots, n) = p^2\psi(1, 2, \ldots, i, j, \ldots, n)$$

Hence, $p^2 = 1$ or $p = \pm 1$, i.e.,

$$P_{ij}\psi(1, 2, \ldots, i, j, \ldots, n) = \pm\psi(1, 2, \ldots, i, j, \ldots, n)$$

which means that the wavefunction must be either symmetric or antisymmetric with respect to interchange of two identical particles.

11.7 Sixteen noninteracting electrons are confined in a potential $V(x) = \infty$ for $x < 0$ and $x > 0$; $V(x) = 0$, for $0 < x < a$.

(i) What is the energy of the least energetic electron in the ground state?

(ii) What is the energy of the most energetic electron in the ground state?

(iii) What is the Fermi energy E_f of the system?

Solution.

(i) The least energetic electron in the ground state is given by $E_1 = \dfrac{\pi^2\hbar^2}{2ma^2}$.

(ii) In the given potential, the energy eigenvalue

$$E_n = \frac{\pi^2\hbar^2 n^2}{2ma^2}, \qquad n = 1, 2, 3, \ldots$$

As two electrons can go into each of the states $n = 1, 2, 3, \ldots$, the highest filled level will have $n = 8$ and its energy will be

$$E_8 = \frac{\pi^2\hbar^2 8^2}{2ma^2} = \frac{32\pi^2\hbar^2}{ma^2}$$

(iii) The energy of the highest filled state is the Fermi energy E_F. Hence,

$$E_F = \frac{32\pi^2\hbar^2}{ma^2}$$

11.8 What is the ground state energy and wave function for two identical particles in the potential defined in Problem 11.7 if the two particles are (i) bosons, and (ii) fermions?

Solution. The solution of the Schrödinger equation of a particle in the given potential gives

$$E_n = \frac{\pi^2\hbar^2 n^2}{2ma^2}, \qquad \psi_n(x) = \sqrt{\frac{2}{a}} \sin\frac{n\pi x}{a}, \qquad n = 1, 2, 3, \ldots$$

(i) *Bosons:* Both the particles can be in the same state. Hence,

$$E_1(1) = \frac{\pi^2\hbar^2}{2ma^2}, \qquad \psi_1(x_1) = \sqrt{\frac{2}{a}} \sin\frac{\pi x_1}{a}$$

$$E_1(2) = \frac{\pi^2\hbar^2}{2ma^2}, \qquad \psi_2(x_2) = \sqrt{\frac{2}{a}} \sin\frac{\pi x_2}{a}$$

The energy and wave function of the combined system are

$$E = E_1(1) + E_1(2) = \frac{\pi^2\hbar^2}{ma^2}, \qquad \psi = \left(\frac{2}{a}\right) \sin\frac{\pi x_1}{a} \sin\frac{\pi x_2}{a}$$

Interchange does not change ψ. Hence it is symmetric. Therefore, the spin function of the two-particle system must be symmetric. The wave function of the system including spin is

$$\psi(x, m_s) = \left(\frac{2}{a}\right) \sin\frac{\pi x_1}{a} \sin\frac{\pi x_2}{a} \begin{cases} \alpha\alpha \\ \beta\beta \\ (\alpha\beta + \beta\alpha)/\sqrt{2} \end{cases}$$

(ii) *Fermions:* In the ground state, one particle has to be spin up and the other spin down. Hence the energy and wave functions are

$$E = \frac{\pi^2\hbar^2}{ma^2}, \qquad \psi(x, m_s) = \left(\frac{2}{a}\right) \sin\frac{\pi x_1}{a} \sin\frac{\pi x_2}{a} \frac{1}{\sqrt{2}} (\alpha\beta - \beta\alpha)$$

11.9 Consider two identical particles described by the Hamiltonian

$$H = \frac{p_1^2(x_1)}{2m} + \frac{p_2^2(x_2)}{2m} + \frac{1}{2}m\omega^2 x_1^2 + \frac{1}{2}m\omega^2 x_2^2$$

Obtain the energy spectrum of this system. Discuss its degeneracy.

Solution. The Schrödinger equation of the system splits into two equations:

$$\left(-\frac{\hbar^2}{2m}\frac{d^2}{dx_1^2} + \frac{1}{2}m\omega^2 x_1^2\right)\psi(x_1) = E_1\psi(x_1)$$

$$\left(-\frac{\hbar^2}{2m}\frac{d^2}{dx_2^2}+\frac{1}{2}m\omega^2 x_2^2\right)\psi(x_2)=E_2\psi(x_2)$$

The solution of these equations is

$$E_{n_1}=\left(n_1+\frac{1}{2}\right)\hbar\omega;\quad \psi_{n_1}(x_1)=NH_n(y_1)\,e^{-y_1^2/2},\quad y_1=\left(\frac{m\omega}{\hbar}\right)^{1/2}x_1$$

$$E_{n_2}=\left(n_2+\frac{1}{2}\right)\hbar\omega;\quad \psi_{n_2}(x_2)=NH_n(y_2)\,e^{-y_2^2/2},\quad y_2=\left(\frac{m\omega}{\hbar}\right)^{1/2}x_2$$

where $\quad n_1 = 0, 1, 2, \ldots;\quad n_2 = 0, 1, 2, 3, \ldots$

$$\text{Total energy } E_n = E_{n_1} + E_{n_2} = (n_1 + n_2)\,\hbar\omega + \hbar\omega = (n+1)\,\hbar\omega$$

$$\text{Wave function of the system } \psi_n(x_1, x_2) = \psi_{n_1}(x_1)\,\psi_{n2}(x_2)$$

Each level is $(n + 1)$-fold degenerate.

11.10 Prove that the three column vectors

$$\begin{pmatrix}1\\0\\0\end{pmatrix},\quad \begin{pmatrix}0\\1\\0\end{pmatrix},\quad \begin{pmatrix}0\\0\\1\end{pmatrix}$$

are the spin eigenfunctions of S_z of a spin $s = 1$ system. Also prove that they are mutually orthogonal.

Solution. The S_z matrix of a spin $s = 1$ system is given by

$$S_z=\begin{pmatrix}1\hbar & 0 & 0\\ 0 & 0 & 0\\ 0 & 0 & -1\hbar\end{pmatrix}$$

$$\begin{pmatrix}1\hbar & 0 & 0\\ 0 & 0 & 0\\ 0 & 0 & -1\hbar\end{pmatrix}\begin{pmatrix}1\\0\\0\end{pmatrix}=\begin{pmatrix}1\hbar\\0\\0\end{pmatrix}=1\hbar\begin{pmatrix}1\\0\\0\end{pmatrix}$$

$$\begin{pmatrix}1\hbar & 0 & 0\\ 0 & 0 & 0\\ 0 & 0 & -1\hbar\end{pmatrix}\begin{pmatrix}0\\1\\0\end{pmatrix}=\begin{pmatrix}0\\0\\0\end{pmatrix}=0\hbar\begin{pmatrix}0\\1\\0\end{pmatrix}$$

$$\begin{pmatrix}1\hbar & 0 & 0\\ 0 & 0 & 0\\ 0 & 0 & -1\hbar\end{pmatrix}\begin{pmatrix}0\\0\\1\end{pmatrix}=\begin{pmatrix}0\\0\\-1\hbar\end{pmatrix}=-1\hbar\begin{pmatrix}0\\0\\1\end{pmatrix}$$

As expected, the eigenvalues of S_z are $1\hbar$, 0 and $-1\hbar$. Thus,

$$(1\ \ 0\ \ 0)\begin{pmatrix}0\\1\\0\end{pmatrix} = 0, \qquad (0\ \ 1\ \ 0)\begin{pmatrix}0\\0\\1\end{pmatrix} = 0, \qquad (0\ \ 0\ \ 1)\begin{pmatrix}1\\0\\0\end{pmatrix} = 0$$

Hence the result.

11.11 Give the zeroth order wave functions for helium atom (i) in the ground state ($1s^2$), and (ii) in the excited state 1s 2s. Also, express them in the form of Slater determinants.

Solution.

(i) The ground state of helium is $1s^2$. As both the electrons are in the ψ_{100} state, the space part of the wave function is $\psi_{100}(r_1)\psi_{100}(r_2)$. The spin part that multiplies this must be antisymmetric so that the total wave function is antisymmetric. Hence, the zeroth order wave function for helium atom in the $1s^2$ state is

$$1s(1)\ 1s\ (2)\ \frac{1}{\sqrt{2}}\ [\alpha(1)\ \beta(2) - \beta(1)\ \alpha(2)]$$

In terms of the Slater determinant, this takes the form

$$\frac{1}{\sqrt{2}}\begin{vmatrix} 1s\,(1)\ \alpha(1) & 1s\,(2)\ \alpha(2) \\ 1s\,(1)\ \beta(1) & 1s\,(2)\ \beta(2) \end{vmatrix}$$

(ii) For the 1s 2s state, taking exchange degeneracy into account, the possible product functions are

$$1s(1)2s(2) \quad \text{and} \quad 1s(2)\ 2s(1)$$

The symmetric combination ψ_s and the antisymmetric combination ψ_{as} are given by

$$\psi_s = \frac{1}{\sqrt{2}}\ [1s(1)\ 2s(2) + 1s(2)\ 2s(1)]$$

$$\psi_{as} = \frac{1}{\sqrt{2}}\ [1s(1)\ 2s(2) - 1s(2)\ 2s(1)]$$

Combining these with the spin wave function for a two-electron system, with the condition that the total wave function must be antisymmetric, we get

$$\psi_1 = \frac{1}{\sqrt{2}}\ [1s(1)\ 2s(2) + 1s(2)\ 2s(1)]\ [\alpha(1)\ \beta(2) - \beta(1)\ \alpha(2)]\frac{1}{\sqrt{2}}$$

$$\psi_2 = \frac{1}{\sqrt{2}}\ [1s(1)\ 2s(2) - 1s(2)\ 2s(1)]\ \alpha(1)\ \alpha(2)$$

$$\psi_3 = \frac{1}{\sqrt{2}}\ [1s(1)\ 2s(2) - 1s(2)\ 2s(1)]\ [\alpha(1)\ \beta(2) + \beta(1)\ \alpha(2)]\frac{1}{\sqrt{2}}$$

$$\psi_4 = \frac{1}{\sqrt{2}}\ [1s(1)\ 2s(2) - 1s(2)\ 2s(1)]\ \beta(1)\ \beta(2)$$

For 1s 2s configuration, we have the following spin orbital combinations: 1sα, 1sβ, 2sα and 2$s\beta$, leading to the four Slater determinants (the normalization factor $1/\sqrt{2}$ not included.):

$$D_1 = \begin{vmatrix} 1s(1)\,\alpha(1) & 1s(2)\,\alpha(2) \\ 2s(1)\,\alpha(1) & 2s(2)\,\alpha(2) \end{vmatrix}, \qquad D_2 = \begin{vmatrix} 1s(1)\,\alpha(1) & 1s(2)\,\alpha(2) \\ 2s(1)\,\beta(1) & 2s(2)\,\beta(2) \end{vmatrix}$$

$$D_3 = \begin{vmatrix} 1s(1)\,\beta(1) & 1s(2)\,\beta(2) \\ 2s(1)\,\alpha(1) & 2s(2)\,\alpha(2) \end{vmatrix}, \qquad D_4 = \begin{vmatrix} 1s(1)\,\beta(1) & 1s(2)\,\beta(2) \\ 2s(1)\,\beta(1) & 2s(2)\,\beta(2) \end{vmatrix}$$

A comparison of the above wave functions with these determinants shows that ψ_1, ψ_2, ψ_3, ψ_4 are equal to the determinants $(D_2 - D_3)/2$, $D_1/\sqrt{2}$, $(D_2 + D_3)/2$ and $D_4/\sqrt{2}$, respectively.

11.12 Prove that it is impossible to construct a completely antisymmetric spin function for three electrons.

Solution. Let a, b, c stand for three functions and 1, 2, 3 for three identical particles. In the function $a(1)\; b\,(2)\; c\,(3)$, particle 1 is in a, particle 2 is in b, and particle 3 is in c. Let us proceed without specifying that these functions correspond to space or spin functions. The third-order Slater determinant for the case is

$$\frac{1}{\sqrt{6}}\begin{vmatrix} a(1) & a(2) & a(3) \\ b(1) & b(2) & b(3) \\ c(1) & c(2) & c(3) \end{vmatrix}$$

This is completely antisymmetrized as interchange of two spins amounts to interchanging two columns of the determinant, which multiplies it by –1. Let us now specify the functions a, b, c as that due to electron spins. Let $a = \alpha$, $b = \beta$ and $c = \beta$ in the above determinant. The determinant reduces to

$$\frac{1}{\sqrt{6}}\begin{vmatrix} \alpha(1) & g(2) & \alpha(3) \\ \beta(1) & \beta(2) & \beta(3) \\ \beta(1) & \beta(2) & \beta(3) \end{vmatrix}$$

As the second and third rows of the determinant are identical, its value is zero. In whatever way we select a, b, c, the two rows of the determinant will be equal. Therefore, we cannot construct a completely antisymmetric three-electron spin function.

11.13 Two particles of mass m are in a three-dimensional box of sides a, b, c ($a < b < c$). The potential representing the interaction between the particles is $V = A\delta(\mathbf{r}_1 - \mathbf{r}_2)$, where δ is the Dirac delta function. Using the first-order perturbation theory, calculate the lowest possible energy of the system if it is equal to (i) spin zero identical particles, (ii) spin half identical particles with spins parallel. Given

$$\int_0^a \sin^4 \frac{\pi x}{a}\, dx = \frac{3}{8}a.$$

Solution. The energy eigenvalues and eigenfunctions of a particle in a rectangular box of side a, b, c are given by (Problem 5.1)

$$E = \frac{\pi^2\hbar^2}{2m}\left(\frac{n_x^2}{a^2} + \frac{n_y^2}{b^2} + \frac{n_z^2}{c^2}\right), \qquad n_x,\, n_y,\, n_z = 1, 2, 3, \ldots$$

$$\psi(x, y, z) = \sqrt{\frac{8}{abc}} \sin \frac{n_x \pi x}{a} \sin \frac{n_y \pi y}{b} \sin \frac{n_z \pi z}{c}$$

(i) For a system of spin zero particles, the total wave function must be symmetric for interchange of any pair of particles. Hence, for the two-particle system, the unperturbed wave function can be taken as the product of two single-particle wave function which is symmetric, i.e.,

$$\psi_s(\mathbf{r}_1, \mathbf{r}_2) = \psi(\mathbf{r}_1)\, \psi(\mathbf{r}_2)$$

$$= \frac{8}{abc} \sin \frac{\pi x_1}{a} \sin \frac{\pi y_1}{b} \sin \frac{\pi z_1}{c} \sin \frac{\pi x_2}{a} \sin \frac{\pi y_2}{b} \sin \frac{\pi z_2}{c}$$

The unperturbed energy

$$E_0 = \frac{\pi^2 \hbar^2}{m} \left(\frac{1}{a^2} + \frac{1}{b^2} + \frac{1}{c^2} \right)$$

The Hamiltonian representing the interaction between the two particles is

$$H' = A\delta(\mathbf{r}_1 - \mathbf{r}_2)$$

where A is a constant, can be taken as the perturbation. The first order correction to the energy

$$E_s^{(1)} = \int \psi_s^*(\mathbf{r}_1, \mathbf{r}_2)\, A\delta(\mathbf{r}_1 - \mathbf{r}_2)\, \psi_s(\mathbf{r}_1, \mathbf{r}_2)\, d\tau_1\, d\tau_2$$

$$= A \int |\psi_s(\mathbf{r}_1, \mathbf{r}_1)|^2\, d\tau_1$$

$$= A \left(\frac{8}{abc} \right)^2 \int_0^a \int_0^b \int_0^c \left(\sin \frac{\pi x_1}{a} \sin \frac{\pi y_1}{b} \sin \frac{\pi z_1}{c} \right)^4 dx_1\, dy_1\, dz_1$$

$$= A \left(\frac{8}{abc} \right)^2 \int_0^a \sin^4 \frac{\pi x_1}{a}\, dx_1 \int_0^b \sin^4 \frac{\pi y_1}{b}\, dy_1 \int_0^c \sin^4 \frac{\pi z_1}{c}\, dz_1$$

$$= A \left(\frac{8}{abc} \right)^2 \frac{3a}{8} \frac{3b}{8} \frac{3c}{8} \frac{27A}{8abc}$$

Consequently, the energy corrected to first order is

$$E_s = \frac{\pi^2 \hbar^2}{m} \left(\frac{1}{a^2} + \frac{1}{b^2} + \frac{1}{c^2} \right) + \frac{27A}{8abc}$$

(ii) For a system of spin half particles, the total wave function must be antisymmetric for interchange of any pair of particles. As the spins are parallel, the spin wave function is symmetric and, therefore, the space part must be antisymmetric. One of the particles will be in the ground state ψ_{111}, and the other will be in the first excited state ψ_{211} since $1/a^2 < 1/b^2 < 1/c^2$. The antisymmetric combination is then given by

$$\psi_a(\mathbf{r}_1, \mathbf{r}_2) = \frac{1}{\sqrt{2}} [\psi_{111}(\mathbf{r}_1)\, \psi_{211}(\mathbf{r}_2) - \psi_{111}(\mathbf{r}_2)\, \psi_{211}(\mathbf{r}_1)]$$

The unperturbed energy

$$E_a = \frac{\pi^2\hbar^2}{2m}\left(\frac{1}{a^2}+\frac{1}{b^2}+\frac{1}{c^2}+\frac{4}{a^2}+\frac{1}{b^2}+\frac{1}{c^2}\right)$$

$$= \frac{\pi^2\hbar^2}{2m}\left(\frac{5}{a^2}+\frac{2}{b^2}+\frac{2}{c^2}\right) = \frac{\pi^2\hbar^2}{m}\left(\frac{5}{2a^2}+\frac{1}{b^2}+\frac{1}{c^2}\right)$$

The first-order correction to the energy is

$$E_a^{(1)} = \int \psi_a^*(\mathbf{r}_1, \mathbf{r}_2)\, A\delta(\mathbf{r}_1 - \mathbf{r}_2)\, \psi_a(\mathbf{r}_1, \mathbf{r}_2)\, d\tau_1\, d\tau_2$$

which reduces to zero when ψ_a^* and ψ_a are substituted. Hence,

$$E_a = \frac{\pi^2\hbar^2}{m}\left(\frac{5}{2a^2}+\frac{1}{b^2}+\frac{1}{c^2}\right)$$

11.14 A one-dimensional infinite potential well of width a contains two spinless particles, each of mass m. The potential representing the interaction between the particles $V = a\delta(x_1 - x_2)$. Calculate the ground state energy of the system corrected to first order in A.

Solution. The energy eigenvalues and eigenfunctions of a particle in an infinite square well of width a are given by

$$E_n = \frac{\pi^2\hbar^2 n^2}{2ma^2}, \qquad n = 1, 2, 3, \ldots$$

$$\psi_n(x) = \sqrt{\frac{2}{a}}\,\sin\frac{n\pi x}{a}$$

For the two-particle system, the unperturbed wave function

$$\psi_{nk}(x_1, x_2) = \psi_n(x_1)\,\psi_k(x_2) = \frac{2}{a}\sin\frac{n\pi x_1}{a}\sin\frac{k\pi x_2}{a}$$

$$E_{nk} = \frac{\pi^2\hbar^2}{2ma^2}(n^2 + k^2), \qquad n, k = 1, 2, 3, \ldots$$

For the ground state, $n = k = 1$. The unperturbed ground state energy is, then,

$$E_{11} = \frac{\pi^2\hbar^2}{ma^2}$$

Next we consider the perturbation $H' = A\delta(x_1 - x_2)$. The first-order correction to the ground state energy

$$E^{(1)} = A\left(\frac{2}{a}\right)^2 \int_0^a\!\!\int_0^a \sin^2\frac{\pi x_1}{a}\sin^2\frac{\pi x_2}{a}\,\delta(x_1 - x_2)\,dx_1\,dx_2$$

$$= A\frac{4}{a^2}\int_0^a \sin^4\frac{\pi x_1}{a}\,dx_1 = \frac{3A}{2a}$$

Hence, the first-order corrected ground state energy

$$E_{11} = \frac{\pi^2 \hbar^2}{ma^2} + \frac{3A}{2a}$$

11.15 Two identical bosons, each of mass m, move in the one-dimensional harmonic potential $V = (1/2)\, m\omega^2 x^2$. They also interact with each other via the potential

$$V_{\text{int}} = a \exp\left[-\beta(x_1 - x_2)^2\right]$$

where α and β are positive parameters. Compute the ground state energy of the system to first order in the parameter α.

Solution. Since the particles are bosons, both of them can remain in the ground state. The V_{int} term can be treated as a perturbation. The ground state wavefunction of a harmonic oscillator is

$$\left(\frac{m\omega}{\hbar\pi}\right)^{1/4} \exp\left(-\frac{m\omega x^2}{2\hbar}\right)$$

Hence the unperturbed wavefunction of the ground state for this two-particle system is

$$\psi_0\,(x_1, x_2) = \left(\frac{m\omega}{\hbar\pi}\right)^{1/4} \exp\left(-\frac{m\omega x_1^2}{2\hbar}\right)\left(\frac{m\omega}{\hbar\pi}\right)^{1/4} \exp\left(-\frac{m\omega x_2^2}{2\hbar}\right)$$

$$= \left(\frac{m\omega}{\hbar\pi}\right)^{1/4} \exp\left[-\frac{m\omega}{2\hbar}\,(x_1^2 + x_2^2)\right]$$

The first-order correction to the energy

$$E^{(1)} = \frac{m\omega\alpha}{\hbar\pi} \int\int_{-\infty}^{\infty} \exp\left[-\frac{m\omega}{\hbar}\,(x_1^2 + x_2^2) - \beta(x_1^2 + x_2^2)^2\right] dx_1\, dx_2$$

$$= \frac{m\omega\alpha}{\hbar\pi} \frac{1}{\sqrt{(m\omega/\hbar) + 2\beta}}$$

The ground state energy of the system is, therefore,

$$E = \hbar\omega + \frac{m\omega\alpha}{\hbar\pi} \frac{1}{\sqrt{(m\omega/\hbar) + 2\beta}}$$

11.16 Consider the rotation of the hydrogen molecule H_2. How does the identity of the two nuclei affect the rotational spectrum? Discuss the type of transition that occurs between the rotational levels.

Solution. The rotational energy levels of hydrogen molecule are given by

$$E = \frac{\hbar^2 l(l+1)}{2I}, \quad l = 0, 1, 2, \ldots$$

The total wave function of the molecule ψ is the product of electronic (ψ_e), vibrational (ψ_v), rotational (ψ_r) and nuclear (ψ_n) wave functions.

$$\psi = \psi_e \psi_v \psi_r \psi_n$$

The spin of proton is half . Hence the total wave function ψ must be antisymmetric to nuclear exchange. Since ψ_e and ψ_v are symmetric to nuclear exchange, the product $\psi_r\psi_n$ must be antisymmetric. For $l = 0, 2, 4, \ldots$, the rotational wave function ψ_r is symmetric with respect to nuclear exchange and for $l = 1, 3, 5, \ldots$, it is antisymmetric. Hence, the antisymmetric ψ_n combines with ψ_r of even l states and the symmetric ψ_n combines with ψ_r of odd l states. As there is no interconversion between symmetric and antisymmetric nuclear spin states, transitions can take place between odd l and even l values. Since three symmetric nuclear spin functions and one anitsymmetric functions are possible (similar to electron product functions), the transitions between odd l values are considered to be strong. In other words, there will be an alternation in intensity of the rotational spectrum of H_2 molecule.

Note: The hydrogen molecules corresponding to antisymmetric nuclear spin states are called **para-hydrogen**, and those corresponding to symmetric spin states are called **ortho-hydrogen**.

11.17 Obtain the zeroth-order wave function for the state $1s^2$ 2s of lithium atom.

Solution. The 1s orbital accomodates two electrons with opposite spins and 2s orbital the third electron. The third-order Slater determinant is given by

$$\frac{1}{\sqrt{6}}\begin{vmatrix} a(1) & a(2) & a(3) \\ b(1) & b(2) & b(3) \\ c(1) & c(2) & c(3) \end{vmatrix}$$

where a, b, c stands for the three functions and 1, 2, 3 for the three identical particles. Identifying a, b, c with the spin-orbitals: $a(1) = 1s(1)\,\alpha(1)$, $b(1) = 1s(1)\,\beta(1)$, $c(1) = 2s(1)\,\alpha(1)$, the above determinant becomes

$$\frac{1}{\sqrt{6}}\begin{vmatrix} 1s(1)\,a(1) & 1s(2)\,\alpha(2) & 1s(3)\,\alpha(3) \\ 1s(1)\,\beta(1) & 1s(2)\,\beta(2) & 1s(3)\,b(3) \\ 2s(1)\,\alpha(1) & 2s(2)\,\alpha(2) & 2s(3)\,\alpha(3) \end{vmatrix}$$

An equally good ground state is when we take $c(1) = 2s(1)$ b(1).

11.18 Consider a system of two identical particles occupying any of three energy levels A, B and C having energies E, $2E$ and $3E$, respectively. The level A is doubly degenerate (A_1 and A_2) and the system is in thermal equilibrium. Find the possible configurations and the corresponding energy in the following cases:

(i) the particles obey Fermi statistics;
(ii) the particles obey Bose statistics; and
(iii) the particles are distinguishable and obey Boltzmann statistics.

Solution. Denote the two states with energy E by A_1 and A_2 and the states with $2E$ and $3E$ by B and C, respectively.

If particle 1 is in A_1 and particle 2 is in A_2, the configuration is marked as (A_1, A_2). Thus, the symbol (B, C) indicates that one particle is in B and the other is in C.

(i) If the particles obey Fermi statistics, the system has the following configuration and energy:

Configuration:	(A_1, A_2),	(A_1, B),	(A_2, B),	(A_1, C),	(A_2, C),	(B, C)
Energy:	2E	3E	3E	4E	4E	5E

(ii) If the particles obey Bose statistics, the additional configurations: (A_1, A_1), (A_2, A_2), (B, B) and (C, C) are also possible. Hence the configuration and energy are

(A_1, A_2),	(A_1, B),	(A_2, B),	(A_1, C),	(A_2, C),	(B, C),	(A_1, A_1),	(A_2, A_2),	(B, B),	(C, C)
2E,	3E,	3E,	4E,	4E,	5E,	2E,	2E,	4E,	6E

(iii) Since the particles are distinguishable, the following configurations are also possible:

Configuration:	(A_2, A_1),	(B, A_1),	(B, A_2),	(C, A_1),	(C, A_2),	(C, B)
Energy:	2E,	3E,	3E,	4E,	4E,	5E

11.19 Consider the rotational spectrum of the homonuclear diatomic molecule $^{14}N_2$. Show that the ratio of intensities of adjacent rotational lines is approximately 2 : 1.

Solution. The rotational energy levels of N_2 molecule are given by

$$E_l = \frac{\hbar^2 l(l+1)}{2I}, \qquad l = 0, 1, 2, \ldots$$

The spin of ^{14}N is 1; hence it is a boson. The possible values of the total nuclear spin I of N_2 molecule are 0, 1, 2, making it a boson. The total wave function must be symmetric to nuclear exchange. The rotational functions corresponding to $l = 0, 2, 4, \ldots$ combine with the symmetric spin functions ($I = 0, 2$), and the functions for $l = 1, 3, 5, \ldots$ combine with antisymmetric spin function $I = 1$. The total degeneracy of symmetric spin functions $= (2 \times 0 + 1) + (2 \times 2 + 1) = 6$, and of antisymmetric spin functions $= (2 \times 1 + 1) = 3$. Since transitions are allowed only between symmetric or antisymmetric rotational states, $\Delta l = 2$. The first line will be $l = 0 \to l = 2$ and the second one $l = 1 \to l = 2$. The nuclear spin I usually remains unchanged in optical transitions.

The energy difference between adjacent rotational levels is very small, the effect due to this in intensity can be neglected. Hence, the intensity of the lines will be in the ratio 6:3 or 2:1.

11.20 Ignoring the interaction between the electrons and considering exchange degeneracy and spin effects, write the wave functions for the ground and the excited states $(1s)^1\,(2p)^1$ of helium atom.

Solution. The Hamiltonian

$$H = \left(-\frac{\hbar^2}{2m}\nabla_1^2 - \frac{Ze^2}{4\pi\varepsilon_0 r_1}\right) + \left(-\frac{\hbar^2}{2m}\nabla_2^2 - \frac{Ze^2}{4\pi\varepsilon_0 r_2}\right)$$

where ∇_1 and ∇_2 refer to the coordinates of electron 1 and 2, respectively. Distances r_1 and r_2 are those of electron 1 and electron 2. The electrostatic repulsion between the two electrons is neglected.

Ground state. The ground state of helium is $1s^2$. As both the electrons are in the $|100\rangle$ state, the space part of the wave function is

$$\psi_{\text{space}} = |100\rangle_1\,|100\rangle_2$$

The subscripts 1, 2 refer to the two electrons. Exchange degeneracy does not exist as both the electrons are in the same state. Since the system is of fermions, the total wave function must be antisymmetric. The space part of the wave function is symmetric. Hence the spin part must be antisymmetric. Multiplying ψ_{space} by the antisymmetric spin combination, the wave function of the ground state is obtained as

$$\psi = |100\rangle_1\,|100\rangle_2\,\frac{1}{\sqrt{2}}\,[\alpha(1)\,\beta(2) - \beta(1)\,\alpha(2)]$$

$(1s)^1$ $(2p)^1$ **state:** Since $l = 1$, $m = 1, 0, -1$. Therefore, the states obtained are

$$|100\rangle_1\ |211\rangle_2, \qquad |100\rangle_1\ |210\rangle_2, \qquad |100\rangle_1\ |21, -1\rangle_2$$

Taking exchange degeneracy into account, the symmetric and antisymmetric combinations of the space part are

$$\psi_{s1} = \frac{1}{\sqrt{2}} = [|100\rangle_1\,|211\rangle_2 + |100\rangle_2\,|211\rangle_1]$$

$$\psi_{as1} = \frac{1}{\sqrt{2}} = [|100\rangle_1\,|211\rangle_2 - |100\rangle_2\,|211\rangle_1]$$

$$\psi_{s2} = \frac{1}{\sqrt{2}} = [|100\rangle_1\,|210\rangle_2 + |100\rangle_2\,|210\rangle_1]$$

$$\psi_{as2} = \frac{1}{\sqrt{2}} = [|100\rangle_1\,|210\rangle_2 - |100\rangle_2\,|210\rangle_1]$$

$$\psi_{s3} = \frac{1}{\sqrt{2}} = [|100\rangle_1\,|21, -1\rangle_2 + |100\rangle_2\,|21, -1\rangle_1]$$

$$\psi_{as3} = \frac{1}{\sqrt{2}} = [|100\rangle_1\,|21, -1\rangle_2 - |100\rangle_2\,|21, -1\rangle_1]$$

Combining these with the spin functions, we get

$$\psi(s_1) = \psi_{s1}\chi_{as} \qquad \psi(t_1) = \psi_{as1}\chi_s$$

$$\psi(s_2) = \psi_{s2}\chi_{as} \qquad \psi(t_2) = \psi_{as2}\chi_s$$

$$\psi(s_3) = \psi_{s3}\chi_{as} \qquad \psi(t_3) = \psi_{as3}\chi_s$$

where S_1, S_2, S_3 refer to singlet states and t_1, t_2, t_3 refer to triplet states.

11.21 The excited electronic state $(1s)^1$ $(2s)^1$ of helium atom exists as either a singlet or a triplet state. Which state has the higher energy? Explain why. Find out the energy separation between the singlet and triplet states in terms of the one-electron orbitals $\psi_{1s}(\boldsymbol{r})$ and $\psi_{2s}(\boldsymbol{r})$.

Solution. The electrostatic repulsion between the electrons $e^2/(4\pi\varepsilon_0 r_{12})$ can be treated as perturbation on the rest of the Hamiltonian. Here, r_{12} is the distance between the electrons. Taking exchange degeneracy into account, the two unperturbed states are

$$\psi_{1s}(\boldsymbol{r}_1)\ \psi_{2s}(\boldsymbol{r}_2) \quad \text{and} \quad \psi_{1s}(\boldsymbol{r}_2)\ \psi_{2s}(\boldsymbol{r}_1) \tag{i}$$

As the spin part of the wave function does not contribute to the energy, the perturbation for these two degerate states can easily be evaluated [refer Eqs. (8.5) and (8.6)]. The energy eigenvalues of the perturbation matrix can be evaluated from the determinant

$$\begin{vmatrix} J - E^{(1)} & K \\ K & J - E^{(1)} \end{vmatrix} = 0 \tag{ii}$$

where

$$J = \iint \psi^*_{1s}(\boldsymbol{r}_1)\,\psi^*_{2s}(\boldsymbol{r}_2)\frac{e^2}{4\pi\varepsilon_0 r_{12}}\,\psi_{1s}(\boldsymbol{r}_1)\,\psi_{2s}(\boldsymbol{r}_2)\,d\tau_1\,d\tau_2 \tag{iii}$$

$$K = \iint \psi_{1s}^*(r_1)\, \psi_{2s}^*(r_2) \frac{e^2}{4\pi\varepsilon_0 r_{12}} \psi_{1s}(r_2)\, \psi_{2s}(r_1)\, d\tau_1\, d\tau_2 \tag{iv}$$

Both J and K are positive. The solution of the determinant gives

$$(J - E^{(1)})^2 - K^2 = 0$$

$$(J - E^{(1)} + K)\,(J - E^{(1)} - K)$$

$$E^{(1)} = J + K \quad \text{or} \quad E^{(1)} = J - K \tag{v}$$

These energies correspond to the normalized eigenfunctions

$$\psi_s = \frac{1}{\sqrt{2}} [\psi_{1s}(r_1)\psi_{2s}(r_2) + \psi_{1s}(r_2)\psi_{2s}(r_1)] \tag{vi}$$

$$\psi_{as} = \frac{1}{\sqrt{2}} [\psi_{1s}(r_1)\psi_{2s}(r_2) - \psi_{1s}(r_2)\psi_{2s}(r_1)] \tag{vii}$$

The total wave function must be antisymmetric. Hence ψ_s combines with the antisymmetric spin part and ψ_{as} combines with the symmetric spin part, i.e.,

$$\psi(s) = \frac{\psi_s\,(\alpha\beta - \beta\alpha)}{\sqrt{2}} \tag{viii}$$

$$\psi(t) = \psi_{as} \begin{cases} \alpha\alpha \\ \dfrac{\alpha\beta + \beta\alpha}{\sqrt{2}} \\ \beta\beta \end{cases} \tag{ix}$$

The Eq. (viii) is the wave function for the singlet state as $S = 0$ for it. The Eq. (ix) refers to the triplet state as $S = 1$ for the state. The energy of ψ_s is $J + K$ and that of $\psi(t)$ is $J - K$. Hence the singlet lies above the triplet. The energy difference

$$\Delta E = (J + K) - (J - K) = 2K$$

where the value of K is given by Eq. (iv).

11.22 The first two wave functions of an electron in an infinite potential well are $U_1(x)$ and $U_2(x)$ Write the wave function for the lowest energy state of three electrons in this potential well.

Solution. By Pauli's exclusion principle, two electrons can go into the $n = 1$ state and the third electron must go in the $n = 2$ state. The spin of the third can be in an up or down state with the same energy. We shall assume it to be in the spin up state. The antisymmetric combination of the two electrons in the $n = 1$ state multiplied by the function of the third electron gives

$$[U_{1\uparrow}(x_1)\, U_{1\downarrow}(x_2) - U_{1\downarrow}(x_1)\, U_{1\uparrow}(x_2)]\, U_{2\uparrow}(x_3) \tag{i}$$

This product would not be antisymmetric under the interchange of any pair of electrons. To make the product function antisymmetric, we take the product in Eq. (i) and subtract from it the same expression with x_2 and x_3 interchanged, as well as a second expression with x_1 and x_3 interchanged. We then get

$$[U_{1\uparrow}(x_1)\, U_{1\downarrow}(x_2) - U_{1\downarrow}(x_1)\, U_{1\uparrow}(x_2)]\, U_{2\uparrow}(x_3) - [U_{1\uparrow}(x_1)\, U_{1\downarrow}(x_3) - U_{1\downarrow}(x_1)\, U_{1\uparrow}(x_3)]\, U_{2\uparrow}(x_2)$$
$$- [U_{1\uparrow}(x_3)\, U_{1\downarrow}(x_2) - U_{1\downarrow}(x_3)\, U_{1\uparrow}(x_2)]\, U_{2\uparrow}(x_1)$$

Multiplying, we obtain

$$U_{1\uparrow}(x_1)\, U_{1\downarrow}(x_2)\, U_{2\uparrow}(x_3) - U_{1\downarrow}(x_1)\, U_{1\uparrow}(x_2)\, U_{2\uparrow}(x_3) - U_{1\uparrow}(x_1)\, U_{1\downarrow}(x_3)\, U_{2\uparrow}(x_2)$$
$$+ U_{1\downarrow}(x_1)\, U_{1\uparrow}(x_3)\, U_{2\uparrow}(x_2) - U_{1\uparrow}(x_3)\, U_{1\downarrow}(x_2)\, U_{2\uparrow}(x_1) + U_{1\downarrow}(x_3)\, U_{1\uparrow}(x_2)\, U_{2\uparrow}(x_1)$$

This expression changes sign under the interchange of any two electrons.

11.23 Consider two identical fermions, both in the spin up state in a one-dimensional infinitely deep well of width $2a$. Write the wave function for the lowest energy state. For what values of position, does the wave function vanish?

Solution. The wave function and energies of a particle in an infinite potential well of side $2a$ is

$$\psi_n = \frac{1}{\sqrt{a}} \sin \frac{n\pi x}{2a}, \qquad -a \le x \le a$$

$$E_n = \frac{\pi^2 \hbar^2 n^2}{8ma^2}, \qquad n = 1, 2, 3$$

In the given case, both the fermions are in the spin up states. Hence, one will be in $n = 1$ state and the other will be in the $n = 2$ state. Taking exchange degeneracy into account, the two product functions are

$$\psi_1(1)\ \psi_2(2) \quad \text{and} \quad \psi_1(2)\ \psi_2(1)$$

For fermions, the function must be antisymmetric. The antisymmetric combination of these two functions is

$$\psi_a = \frac{1}{\sqrt{2}} [\psi_1(1)\, \psi_2(2) - \psi_1(2)\, \psi_2(1)]$$

$$= \frac{1}{\sqrt{2a}} \left[\sin \frac{\pi x_1}{2a} \sin \frac{\pi x_2}{a} - \sin \frac{\pi x_2}{2a} \sin \frac{\pi x_1}{a} \right]$$

The function ψ_a will be zero at $x = 0$, $a/2$, a.

11.24 Consider a system of two spin half particles in a state with total spin quantum number $S = 0$. Find the eigenvalue of the spin Hamiltonian $H = A\mathbf{S}_1 \cdot \mathbf{S}_2$, where A is a positive constant in this state.

Solution. The total spin angular momentum $\mathbf{S}$ of this two spin-half system is

$$\mathbf{S} = \mathbf{S}_1 + \mathbf{S}_2$$

$$S^2 = S_1^2 + S_2^2 + 2\mathbf{S}_1 \cdot \mathbf{S}_2$$

$$\mathbf{S}_1 \cdot \mathbf{S}_2 = \frac{S^2 - S_1^2 - S_2^2}{2}$$

Hence,

$$H = \frac{A}{2} (S^2 - S_1^2 - S_2^2)$$

Let the simultaneous eigenkets of S^2, S_z, S_1^2 and S_2^2 be $|sm_s\rangle$. Then,

$$H|sm_s\rangle = \frac{A}{2}(S^2 - S_1^2 - S_2^2)|sm_s\rangle$$

$$= \frac{A\left(0 - \frac{3}{4} - \frac{3}{4}\right)\hbar^2}{2} = -\frac{3}{4}A\hbar^2$$

The eigenvalue of the spin Hamiltonian H' is $-(3/4)A\hbar^2$.

11.25 The valence electron in the first excited state of an atom has the electronic configuration $3s^1\ 3p^1$.

(i) Under L-S coupling what values of L and S are possible?

(ii) Write the spatial part of their wavefunctions using the single particle functions $\psi_s(\boldsymbol{r})$ and $\psi_p(\boldsymbol{r})$.

(iii) Out of the levels, which will have the lowest energy and why?

Solution.

(i) Electronic configuration $3s^1 3p^1$. Hence,

$$l_1 = 0,\ l_2 = 1, \qquad s_1 = (1/2), \qquad s_2 = (1/2)$$

$$L = 1, \qquad S = 0,\ 1$$

(ii) Taking exchange degeneracy into account, the two possible space functions are

$$\psi_s(\boldsymbol{r}_1)\ \psi_p(\boldsymbol{r}_2) \quad \text{and} \quad \psi_s(\boldsymbol{r}_2)\ \psi_p(\boldsymbol{r}_1)$$

The symmetric combination

$$\psi_s = N_s\,[\psi_s(\boldsymbol{r}_1)\ \psi_p(\boldsymbol{r}_2) + \psi_s(\boldsymbol{r}_2)\ \psi_p(\boldsymbol{r}_1)]$$

Antisymmetric combination

$$\psi_{as} = N_{as}\,[\psi_s(\boldsymbol{r}_1)\ \psi_p(\boldsymbol{r}_2) - \psi_s(\boldsymbol{r}_2)\ \psi_p(\boldsymbol{r}_1)]$$

where N_s and N_{as} are normalization constants.

(iii) Since the system is of fermions, the total wave function must be antisymmetric. Including the spin part of the wave function, the total wave function for the singlet ($S = 0$) and triplet ($S = 1$) states are

$$\psi_{sing} = N_s\,[\psi_s(r_1)\psi_p(r_2) + \psi_s(r_2)\psi_p(r_1)][\alpha(1)\beta(2) - \beta(1)\alpha(2)]\frac{1}{\sqrt{2}}$$

$$\psi_{trip} = N_{as}\,[\psi_s(r_1)\,\psi_p(r_2) - \psi_s(r_2)\,\psi_p(r_1)]\begin{cases}\alpha(1)\,\alpha(2)\\ [\alpha(1)\,\beta(2) - \beta(1)\,\alpha(2)]\dfrac{1}{\sqrt{2}}\\ \beta(1)\,\beta(2)\end{cases}$$

The spin function associated with the antisymmetric space function is symmetric with $S = 1$. When the space part is antisymmetric for the interchange of the electron $1 \leftrightarrow 2$, the probability for the two electrons gets closer, is very low and, therefore, the Coulomb repulsive energy is very small, giving a lower total energy. Thus, the triplet state ($S = 1$) is the lower of the two.

11.26 A one-dimensional potential well has the single-particle energy eigenfunctions $\psi_1(x)$ and $\psi_2(x)$ corresponding to energies E_1 and E_2 for the two lowest states. Two noninteracting particles are placed in the well. Obtain the two lowest total energies of the two-particle system with the wavefunction and degeneracy if the particles are (i) distinguishable spin-half particles, (ii) identical spin half particles, and (iii) identical spin zero particles.

Solution.

(i) Distinguishable spin-half particles. The particles have spin = half. Hence the total spin $S = 0, 1$ when $S = 0$, $M_s = 0$ and when $S = 1$, $M_s = 1, 0, -1$. Let us denote the spin wave functions by the corresponding $|SM_s\rangle$. As the particles are distinguishable, the two particles can be in ψ_1 even when $S = 1$. The different wave functions and energies are

$$\psi_1(x_1)\,\psi_1(x_2)\,|00\rangle, \qquad E_1 + E_1 = 2E_1$$

$$\psi_1(x_1)\,\psi_1(x_2)\,|1\,M_s\rangle, \qquad M_s = 1, 0, -1, \qquad E_1 + E_1 = 2E_1$$

The degeneracy is 1 + 3 = 4.

(ii) Two identical spin-half particles. Again, the total spin $S = 0$ or 1. When $S = 0$, the two praticles are in ψ_1 with their spins in the opposite directions. The total wave function must be antisymmetric. The space part of the wave function is symmetric. Hence the spin part must be antisymmetric. The wave function of the system is

$$\psi_1(x_1)\,\psi_1(x_2)\frac{1}{\sqrt{2}}[\alpha(1)\,\beta(2) - \beta(1)\,\alpha(2)]$$

with energy $E_1 + E_1 = 2E_1$.

When $S = 1$, one particle will be in level 1 and the other will be in level 2. Hence, the symmetric and antisymmetric combinations of space functions are

$$\psi_s = \frac{1}{\sqrt{2}}[\psi_1(x_1)\,\psi_2(x_2) + \psi_1(x_2)\,\psi_2(x_1)]$$

$$\psi_{as} = \frac{1}{\sqrt{2}}[\psi_1(x_1)\,\psi_2(x_2) - \psi_1(x_2)\,\psi_2(x_1)]$$

As the total wave function has to be antisymmetric, the wave functions including the spin are

$$\psi(s) = \psi_s\frac{1}{\sqrt{2}}[\alpha(1)\,\beta(2) - \beta(1)\,\alpha(2)]$$

$$\psi(t) = \psi_{as}\begin{cases}\alpha(1)\,\alpha(2)\\[4pt] \dfrac{1}{\sqrt{2}}[\alpha(1)\,\beta(2) - \beta(1)\,\alpha(2)]\\[4pt] \beta(1)\,\beta(2)\end{cases}$$

The first equation corresponds to a singlet state and the second equation to a triplet state. As the energy does not depend on spin function, the energy of both are equal to $E_1 + E_2$.

11.27 Consider two identical linear harmonic oscillators, each of mass m and frequency ω having interaction potential $\lambda x_1 x_2$, where x_1 and x_2 are oscillator variables. Find the energy levels.

Solution. The Hamiltonian of the system is

$$H = -\frac{\hbar^2}{2m}\frac{\partial^2}{\partial x_1^2} - \frac{\hbar^2}{2m}\frac{\partial^2}{\partial x_2^2} + \frac{1}{2}m\omega^2 x_1^2 + \frac{1}{2}m\omega^2 x_2^2 + \lambda x_1 x_2$$

Setting

$$x_1 = \frac{1}{\sqrt{2}}(X + x), \qquad x_2 = \frac{1}{\sqrt{2}}(X - x)$$

In terms of X and x,

$$H = -\frac{\hbar^2}{2m}\frac{\partial^2}{\partial X^2} - \frac{\hbar^2}{2m}\frac{\partial^2}{\partial x^2} + \frac{1}{2}(m\omega^2 + \lambda)X^2 + \frac{1}{2}(m\omega^2 - \lambda)\,x^2$$

Hence the system can be regarded as two independent harmonic oscillators of coordinates X and x. Therefore, the energy

$$E_{n_1,n_2} = \left(n_1 + \frac{1}{2}\right)\hbar\sqrt{\left(\omega^2 + \frac{\lambda}{m}\right)} + \left(n_2 + \frac{1}{2}\right)\hbar\sqrt{\left(\omega^2 - \frac{\lambda}{m}\right)}$$

where $n_1,\ n_2 = 0, 1, 2, \ldots$

11.28 What is the Slater determinant? Express it in the form of a summation using a permutation operator.

Solution. For the Slater determinant, refer Eq. (11.3). The determinant can also be written as

$$\psi_{\text{as}} = \frac{1}{n!}\sum_{1}^{n!} (-1)^p\, Pu_a(1)\, u_b(2) \ldots u_n(n)$$

where P represents the permutation operator and p is the number of interchanges (even or odd) involved in the particular permutation.

CHAPTER 12

Scattering

In scattering, a beam of particles is allowed to pass close to a scattering centre and their energies and angular distributions are measured. In the process, the scattering centre may remain in its original state (elastic scattering) or brought to a different state (inelastic scattering). We are mainly interested in the angular distribution of the scattered particles which in turn is related to the wave function.

12.1 Scattering Cross-section

Let N be the number of incident particles crossing unit area normal to the incident beam in unit time and n be the number of particles scattered into solid angle $d\Omega$ in the direction (θ, ϕ) in unit time, θ being the angle of scattering. The **differential scattering cross-section** is

$$\sigma(\theta, \phi) = \frac{n/d\Omega}{N} \tag{12.1}$$

The solid angle $d\Omega$ in the directon (θ, ϕ) is

$$\frac{r \sin\theta \, d\phi \, r d\theta}{r^2} = \sin\theta \, d\theta \, d\phi$$

$$\text{Total cross-section } \sigma = \int \sigma(\theta, \phi) \, d\Omega = \int_0^{\pi} \int_0^{2\pi} \sigma(\theta, \phi) \sin\theta \, d\theta \, d\phi \tag{12.2}$$

For spherically symmetric potential, $\sigma(\theta, \phi)$ becomes $\sigma(\theta)$.

12.2 Scattering Amplitude

If the potential V depends only on the relative distance between the incident particle and scattering centre, the Schrodinger equation to be solved is

$$-\frac{\hbar^2}{2\mu} \nabla^2 \psi + V(r) \psi = E\psi, \qquad \mu = \frac{mM}{m + M} \tag{12.3}$$

where m is the mass of the incident particle and M is the mass of the scattering centre. For incident particles along the z-axis, the wave function is represented by the plane wave

$$\psi_i \xrightarrow[r \to \infty]{} Ae^{ikz} \tag{12.4}$$

The spherically diverging scattered wave can be represented by

$$\psi_s \xrightarrow[r \to \infty]{} Af(\theta, \phi)\, \frac{e^{ikr}}{r} \tag{12.5}$$

where $f(\theta, \phi)$ is the **scattering amplitude**.

12.3 Probability Current Density

The probability current density corresponding to ψ_i and ψ_s can be calculated separately as

$$j_i = \frac{\hbar k |A|^2}{\mu} = \frac{p|A|^2}{\mu} = \mathbf{v}|A|^2 \tag{12.6}$$

$$j_s = \frac{k\hbar |A|^2}{\mu} |f(\theta)|^2 \frac{1}{r^2} = \frac{\mathbf{v}|A|^2 |f(\theta)|^2}{r^2} \tag{12.7}$$

$$\sigma(\theta) = \frac{j_s \text{ per unit solid angle}}{j_s \text{ of the incident wave}} = \frac{\mathbf{v}|A|^2 |f(\theta)|^2}{\mathbf{v}|A|^2} = |f(\theta)|^2 \tag{12.8}$$

Partial waves. The incident plane wave is equivalent to the superposition of an infinite number of *spherical waves*, and the individual spherical waves are called the *partial waves*. The waves with $l = 0, 1, 2, \ldots$ are respectively called the s-waves, p-waves, d-waves, and so on.

12.4 Partial Wave Analysis of Scattering

As the incident particles are along the z-axis, the scattering amplitude is given by

$$f(\theta) = \frac{1}{2ik} \sum_l (2l + 1)(\exp 2i\delta_l - 1)\, P_l(\cos\theta) \tag{12.9}$$

$$f(\theta) = \frac{1}{k} \sum_{l=0}^{\infty} (2l + 1) \exp i\delta_l\, P_l(\cos\theta) \sin\delta_l \tag{12.10}$$

The scattering cross-section $\sigma(\theta)$ is given by

$$\sigma(\theta) = |f(\theta)|^2 = \frac{1}{k^2} \left| \sum_{l=0}^{\infty} (2l + 1) \exp i\delta_l\, P_l(\cos\theta) \sin\delta_l \right|^2 \tag{12.11}$$

$$k^2 = \frac{2\mu E}{\hbar^2} \tag{12.12}$$

P_l (cos θ) are Legendre polynomials and δ_l are the phase shifts of the individual partial waves. The total cross-section

$$\sigma = 2\pi \int_0^{\pi} \sigma(\theta) \sin\theta \, d\theta = \frac{4\pi}{k^2} \sum_{l=0}^{\infty} (2l+1) \sin^2 \delta_l \tag{12.13}$$

Expression for phase shifts. For weak potentials,

$$\sin \delta_l \cong \delta_l = -\frac{2\mu k}{\hbar^2} \int_0^{\infty} V(r)\, j_l(kr)\, r^2 dr \tag{12.14}$$

where j_l (kr) are the spherical Bessel functions.

12.5 The Born Approximation

The wave function $\psi(r)$ is in the form of an integral equation in which ψ appears inside the integral. In the first Born approximation, $\psi(r')$ in the integral is replaced by the incoming plane wave, exp $(i\mathbf{k} \cdot \mathbf{r}')$. This leads to an improved value for the wave function $\psi(r)$ which is used in the integral in the second Born approximation. This iterative procedure is continued till both input and output ψ's are almost equal. The theory leads to

$$f(\theta) = -\frac{\mu}{2\pi\hbar^2} \int_0^{\infty} \exp(iq\mathbf{r}')\, V(\mathbf{r}')\, d\tau' \tag{12.15}$$

$$f(\theta) = -\frac{2\mu}{\hbar^2} \int_0^{\infty} \frac{\sin qr'}{qr'} V(\mathbf{r}')\, r'^2 d\tau' \tag{12.15a}$$

where

$$|q| = 2|k| \sin\frac{\theta}{2} \tag{12.16}$$

PROBLEMS

12.1 A beam of particles is incident normally on a thin metal foil of thickness t. If N_0 is the number of nuclei per unit volume of the foil, show that the fraction of incident particles scattered in the direction (θ, ϕ) is $\sigma(\theta, \phi) N_0 t \, d\Omega$, where $d\Omega$ is the small solid angle in the direction (θ, ϕ).

Solution. From Eq. (12.1), the differential scattering cross-section is

$$\sigma(\theta, \phi) = \frac{n/d\Omega}{N}$$

where n is the number scattered into solid angle $d\Omega$ in the direction (θ, ϕ) in unit time and N is the incident flux. Hence,

$$n = \sigma(\theta, \phi) N \, d\Omega$$

This is the number scattered by a single nucleus. The number of nuclei in a volume $At = N_0 At$. The number scattered by $N_0 At$ nuclei $= \sigma(\theta, \phi) N \, d\Omega \, N_0 At$. Thus, Number of particles striking an area $A = NA$.

$$\text{Fraction scattered in the direction } (\theta, \phi) = \frac{\sigma(\theta, \phi) N \, d\Omega \, N_0 At}{NA}$$

$$= \sigma(\theta, \phi) N_0 t \, d\Omega$$

12.2 Establish the expansion of a plane wave in terms of an infinite number of spherical waves.

Solution. Free particles moving parallel to the z-axis can be described by the plane wave

$$\psi_k = e^{ikz} = e^{ikr \cos\theta}$$

When the free particles are along the z-axis, the wave function must be independent of the angle ϕ. This reduces the associated Legendre polynomials in $Y_{lm}(\theta, \phi)$ to the Legendre polynomials $P_l(\cos\theta)$. Equating the two expressions for wave function, we get

$$\sum_{l=0}^{\infty} A_l j_l(kr) P_l(\cos\theta) = e^{ikr\cos\theta}$$

Multiplying both sides by $P_l(\cos\theta)$ and integrating over $\cos\theta$, we obtain

$$A_l j_l(kr) \frac{2}{2l+1} = \int_{-1}^{+1} e^{ikr\cos\theta} P_l(\cos\theta) \, d(\cos\theta)$$

$$= \left[\frac{P_l(\cos\theta) \, e^{ikr\cos\theta}}{ikr} \right]_{-1}^{+1} - \int_{-1}^{+1} \frac{e^{ikr\cos\theta}}{ikr} P_l'(\cos\theta) \, d(\cos\theta)$$

The second term on the RHS leads to terms in $1/r^2$ and, therefore, it vanishes as $r \to \infty$. Since $P_l(1) = 1$, $P_l(-1) = (-1)^l P_l(1) = e^{il\pi}$ as $r \to \infty$,

$$A_l \frac{2}{2l+1} j_l(kr) = \frac{1}{ikr} (e^{ikr} - e^{-ikr} e^{il\pi})$$

$$A_l \frac{2}{2l+1} \frac{1}{kr} \sin\left(kr - \frac{l\pi}{2}\right) = \frac{e^{il\pi/2}}{ikr} \left[\exp i\left(kr - \frac{l\pi}{2}\right) - \exp -i\left(kr - \frac{l\pi}{2}\right) \right]$$

$$A_l = (2l + 1)\, e^{il\pi/2} = (2l + 1)\, i^l$$

Consequently,

$$e^{ikz} = \sum_{l=0}^{\infty} (2l + 1)\, i^l j_l(kr) P_l (\cos \theta)$$

This is **Bauer's formula.**

12.3 In the theory of scattering by a fixed potential, the asymptotic form of the wave function is

$$\psi \xrightarrow[r\to\infty]{} A\left[e^{ikz} + f(\theta, \phi)\frac{e^{ikr}}{r}\right]$$

Obtain the formula for scattering cross-section in terms of the scattering amplitude $f(\theta, \phi)$.

Solution. The probability current density $\boldsymbol{j}(\boldsymbol{r}, t)$ is given by

$$\boldsymbol{j}(\boldsymbol{r}, t) = \frac{i\hbar}{2\mu}(\psi\nabla\psi^* - \psi^*\nabla\psi) \tag{i}$$

If $\boldsymbol{j}(\boldsymbol{r}, t)$ is calculated with the given wave function, we get interference terms between the incident and scattered waves. In the experimental arrangements, these do not appear. Hence we calculate the incident and scattered probability current densities $\boldsymbol{j}_i$ and $\boldsymbol{j}_s$ separately. The value of $\boldsymbol{j}_i$ due to exp (ikz) is

$$\boldsymbol{j}_i = \frac{i\hbar}{2\mu}[|A|^2(-ik) - |A|^2(-ik)] = \frac{\hbar k|A|^2}{\mu} \tag{ii}$$

The scattered probability current density

$$\boldsymbol{j}_s = \frac{i\hbar}{2\mu}|A|^2|f(\theta, \phi)|^2\left[-\frac{ik}{r^2} - \frac{1}{r^3} - \frac{ik}{r^2} + \frac{1}{r^3}\right]$$

$$= \frac{\hbar k}{\mu}|A|^2|f(\theta, \phi)|^2\frac{1}{r^2} \tag{iii}$$

In the above equation, $1/r^2$ is the solid angle subtended by unit area of the detector at the sacttering centre. The differential scattering cross-section

$$\sigma(\theta) = \frac{\text{Probability current density of the scattered wave per unit solid angle}}{\text{Probability current density of the incident wave}}$$

$$= \frac{(\hbar k/\mu)\,|A|^2\,|f[\theta(\phi)]|^2}{(\hbar k/\mu)\,|A|^2}$$

$$= |f(\theta, \phi)|^2$$

12.4 In the partial wave analysis of scattering, the scattering amptitude

$$f(\theta) = \frac{1}{k}\sum_{l=0}^{\infty}(2l + 1)\exp(i\delta_l)\, P_l(\cos\theta)\sin\delta_l, \qquad k^2 = \frac{2\mu E}{\hbar^2}$$

Obtain an expression for the total cross-section σ. Hence show that

$$\sigma = \frac{4\pi}{k} \text{ Im } f(0)$$

where Im $f(0)$ is the imaginary part of scattering amplitude $f(\theta)$ at $\theta = 0$.

Solution. The differential scattering cross section

$$\sigma(\theta) = |f(\theta)|^2 = \frac{1}{k^2} \left| \sum_{l=0}^{\infty} (2l+1) \exp(i\delta_l)\, P_l(\cos\theta) \sin\delta_l \right|^2 \tag{i}$$

$$\sigma = \int \sigma(\theta)\, d\Omega, \qquad d\Omega = \sin\theta\, d\theta\, d\phi$$

$$= \int_0^{\pi} \int_0^{2\pi} \sigma(\theta) \sin\theta\, d\theta\, d\phi = 2\pi \int_0^{\pi} \sigma(\theta) \sin\theta\, d\theta$$

$$= \frac{2\pi}{k^2} \int_0^{\pi} \left[\sum_{l=0}^{\infty} (2l+1) e^{i\delta_l} P_l(\cos\theta) \sin\delta_l \right]$$

$$\times \left[\sum_{l'=0}^{\infty} (2l'+1) e^{-i\delta_{l'}} P_{l'}(\cos\theta) \sin\delta_{l'} \right] \sin\theta\, d\theta \tag{ii}$$

For Legendre polynomials, we have the orthogonality relation

$$\int_{-1}^{+1} P_l(x) P_m(x)\, dx = \frac{2}{2l+1} \delta_{lm}$$

Changing the variable of integration from θ to x by defining $\cos\theta = x$ and using the orthogonal property of Legendre polynomials, Eq. (ii) reduces to

$$\sigma = \frac{4\pi}{k^2} \sum_{l=0}^{\infty} (2l+1) \sin^2\delta_l \tag{iii}$$

For $\theta = 0$, $P_l(1) = 1$ and the scattering amptitude

$$f(0) = \frac{1}{k} \sum_{l=0}^{\infty} (2l+1) \exp(i\delta_l) \sin\delta_l \tag{iv}$$

The imaginary part of $f(0)$ is

$$\text{Im } f(0) = \frac{1}{k} \sum_{l=0}^{\infty} (2l+1) \sin^2\delta_l \tag{v}$$

From Eqs. (iii) and (v),

$$\sigma = \frac{4\pi}{k} \text{ Im } f(0) \tag{vi}$$

Note: Equation (vi) is referred to as the **optical theorem**.

12.5 Write the radial part of the Schrodinger equation that describes scattering by the square well potential

$$V(r) = \begin{cases} -V_0, & 0 < r < a \\ 0, & r > a \end{cases}$$

and solve the same. Assuming that the scattering is mainly due to s-waves, derive an expression for the s-wave phase shift.

Solution. The radial part of the Schrodinger equation is

$$\frac{1}{r^2}\frac{d}{dr}\left(r^2\frac{dR}{dr}\right) + \frac{2\mu}{\hbar^2}(E - V_0)R - \frac{l(l+1)}{r^2}R = 0 \tag{i}$$

Writing

$$R = \frac{u}{r} \tag{ii}$$

we get

$$\frac{dR}{dr} = \frac{1}{r}\frac{du}{dr} - \frac{u}{r^2}, \qquad r^2\frac{dR}{dr} = r\frac{du}{dr} - u$$

$$\frac{d}{dr}\left(r^2\frac{dR}{dr}\right) = r\frac{d^2u}{dr^2}$$

For s-waves, $l = 0$. Equation (i) now takes the form

$$\frac{d^2u}{dr^2} + \frac{2\mu}{\hbar^2}(E + V_0)\,u = 0$$

$$\frac{d^2u}{dr^2} + k_1^2 u = 0, \quad k_1^2 = \frac{2\mu}{\hbar^2}(E + V_0), \quad r < a \tag{iii}$$

$$\frac{d^2u}{dr^2} + k^2 u = 0, \quad k^2 = \frac{2\mu E}{\hbar^2}, \quad r > a \tag{iv}$$

The solutions of Eq. (iii) and (iv) are

$$u = A\sin k_1 r + B\cos k_1 r, \qquad r < a \tag{v}$$

$$u = C\sin kr + D\cos kr, \qquad r > a \tag{vi}$$

In the region $r < a$, the solution $R = u/r = (1/r)\cos k_1 r$ can be left out as it is not finite at $r = 0$. The solution in the region $r > a$ can be written as

$$u = B\sin(kr + \delta_0) \qquad r > a \tag{vii}$$

$$u = A\sin k_1 r, \qquad r < a \tag{viii}$$

where we have replaced the constants C and D by constants B and δ_0. The constant δ_0 is the s-wave phase shift. As the wave function and its derivative are continuous at $r = a$.

$$A\sin k_1 a = B\sin(ka + \delta_0)$$

$$Ak_1\cos k_1 a = Bk\cos(ka + \delta_0)$$

Dividing one by the other, we get

$$\tan (ka + \delta_0) = \frac{k}{k_1} \tan k_1 a \quad \text{(ix)}$$

$$\delta_0 = \tan^{-1}\left(\frac{k}{k_1} \tan k_1 a\right) - ka \quad \text{(x)}$$

12.6 In a scattering problem, the scattering length a is defined by

$$a = \lim_{E \to 0} [-f(\theta)]$$

Show that (i) the zero energy cross-section $\sigma_0 = 4\pi a^2$, and (ii) for weak potentials $\delta_0 = -ka$.

Solution. When E is very low, only s-state is involved in the scattering. Consequently, from Eq. (12.10), the scattering amplitude

$$f_0(\theta) = \frac{1}{k} e^{i\delta_0} \sin \delta_0$$

(i) In the limit $E \to 0$,

$$a = -\frac{1}{k} e^{i\delta_0} \sin \delta_0$$

$$\sin \delta_0 = -kae^{-i\delta_0}$$

From Eq. (12.13) we have

$$\sigma_0 = \frac{4\pi}{k^2} \sin^2 \delta_0 = \frac{4\pi}{k^2} k^2 a^2 = 4\pi a^2$$

(ii) If the potential $V(r)$ is weak, δ_0 will be small. Then $\exp (i\delta_0) \cong 1$ and $\sin \delta_0 \cong \delta_0$. Hence,

$$f(\theta) = \frac{\delta_0}{k}$$

$$a = -\frac{\delta_0}{k} \quad \text{or} \quad \delta_0 = -ka$$

12.7 Consider the scattering of a particle having charge $Z'e$ by an atomic nucleus of charge Ze. If the potential representing the interaction is

$$V(r) = -\frac{ZZ'e^2}{r} e^{-\alpha r}$$

where α is a parameter. Calculate the scattering amplitude. Use this result to derive Rutherford's scattering formula for scattering by a pure Coulomb potential.

Solution. In the first Born approximation, the scattering amplitude $f(\theta)$ is given by Eq. (12.15). Substituting the given potential

$$f(\theta) = \frac{2ZZ'e^2\mu}{q\hbar^2} \int_0^\infty \sin qre^{-\alpha r}\, dr \quad \text{(i)}$$

The value of this integral is evaluated in Problem 12.7. Substituting the value of the integral, we get

$$f(\theta) = \frac{2\mu ZZ'e^2}{q\hbar^2} \frac{q}{q^2 + \alpha^2} = \frac{2\mu ZZ'e^2}{\hbar^2(q^2 + \alpha^2)} \tag{ii}$$

The momentum transfer

$$|\dot{q}| = 2\,|k|\sin\frac{\theta}{2} \tag{iii}$$

If the momentum transfer $q \gg \alpha$, then

$$q^2 + a^2 \cong q^2 = 4k^2 \sin^2\frac{\theta}{2} \tag{iv}$$

With this value of q^2, the differential scattering cross-section is

$$\sigma(\theta) = |f(\theta)|^2 = \frac{\mu^2 Z^2 Z'^2 e^4}{4\hbar^4 k^4 \sin^4(\theta/2)} \tag{v}$$

which is Rutherford's scattering formula for Coulomb scattering.

12.8 In a scattering experiment, the potential is spherically symmetric and the particles are scattered at such energy that only s and p waves need be considered.

(i) Show that the differential cross-section $\sigma(\theta)$ can be written in the form $\sigma(\theta) = a + b\cos\theta + c\cos^2\theta$.

(ii) What are the values of a, b, c in terms of phase shifts?

(iii) What is the value of total cross-section in terms of a, b, c?

Solution.

(i) The scattering amplitude

$$f(\theta) = \frac{1}{k}\sum_{l=0}^{\infty} (2l + 1)\, e^{i\delta_1} P_l(\cos\theta) \sin\delta_1$$

$$= \frac{1}{k}[e^{i\delta_0}\sin\delta_0 + 3e^{i\delta_1}\cos\theta\sin\delta_1]$$

since

$$P_0(\cos\theta) = 1, \qquad P_1(\cos\theta) = \cos\theta$$

$$s(\theta) = |f(\theta)|^2 = \frac{1}{k^2}[\sin^2\delta_0 + 6\sin\delta_0\sin\delta_1\cos(\delta_0 - \delta_1)\cos\theta + 9\sin^2\delta_1\cos^2\theta]$$

$$\sigma(\theta) = a + b\cos\theta + c\cos^2\theta$$

(ii) $a = \dfrac{\sin^2\delta_0}{k^2}$, $b = \dfrac{6}{k^2}\sin\delta_0\sin\delta_1\cos(\delta_0 - \delta_1)$, $c = \dfrac{9}{k^2}\sin^2\delta_1$

(iii) Total cross-section $\quad \sigma = \dfrac{4\pi}{k^2}(\sin^2\delta_0 + 3\sin^2\delta_1)$

$$= 4\pi a + \frac{4\pi c}{3} = 4\pi\left(a + \frac{c}{3}\right)$$

12.9 Consider scattering by a central potential by the methods of partial wave analysis and Born approximation. When δ_l is small, prove that the expressions for scattering amplitude in the two methods are equivalent. Given

$$\sum_l (2l+1)\, P_l(\cos\theta)\, j_l^2(kr) = \frac{\sin qr}{qr}$$

where $q = 2k \sin(\theta/2)$.

Solution. In the case of partial wave analysis, the scattering amplitude is given by Eq. (12.9), and hence

$$f(\theta) = \frac{1}{2ik} \sum_l (2l+1)\,(e^{2i\delta_l} - 1)\, P_l(\cos\theta)$$

Since δ_l is very small, $e^{2i\delta_l} - 1 \cong 2i\delta_l$, and, therefore,

$$f(\theta) \cong \frac{1}{k} \sum_l (2l+1)\, \delta_l P_l\,(\cos\theta)$$

Substituting the value of δ_l from Eq. (14.75), we get

$$f(\theta) = -\frac{2\mu}{\hbar^2} \sum_l (2l+1)\, P_l\,(\cos\theta) \int_0^\infty V(r)\, j_l^2(kr)\, r^2 dr$$

Using the given result in the question, we obtain

$$f(\theta) = -\frac{2\mu}{\hbar^2} \int_0^\infty \frac{\sin qr}{qr}\, V(r) r^2\, dr$$

which is the expression for the scattering amplitude under Born approximation (12.15).

12.10 Evaluate the scattering amplitude in the Born approximation for scattering by the Yukawa potential

$$V(r) = V_0 \exp\frac{-\alpha r}{r}$$

where V_0 and α are constants.

Also show that $\sigma(\theta)$ peaks in the forward direction ($\theta = 0$) except at zero energy and decreases monotonically as θ varies from 0 to π.

Solution. Substituting the given potential in the expression for $f(\theta)$, we get

$$f(\theta) = -\frac{2\mu}{q\hbar^2} \int_0^\infty V(r)\, r \sin qr\, dr, \qquad q = 2k \sin \theta/2$$

$$f(\theta) = -\frac{2\mu V_0}{q\hbar^2} \int_0^\infty e^{-\alpha r} \sin qr\, dr$$

Writing $I = \int_0^\infty e^{-\alpha r} \sin qr\, dr$ and integrating by parts, we obtain

$$I = -\left(e^{-\alpha r} \frac{\cos qr}{q}\right)_0^\infty - \frac{\alpha}{q} \int_0^\infty \cos qr\, e^{-\alpha r}\, dr$$

$$= \frac{1}{q} - \frac{\alpha}{q}\left(e^{-\alpha r} \frac{\sin qr}{q}\right)_0^\infty - \frac{\alpha^2}{q^2} \int_0^\infty \sin qr\, e^{-\alpha r}\, dr$$

$$I = \frac{1}{q} - \frac{\alpha^2}{q^2} I \quad \text{or} \quad I = \frac{q}{q^2 + \alpha^2}$$

$$f(\theta) = -\frac{2\mu V_0}{\hbar^2 (q^2 + \alpha^2)} = -\frac{2\mu V_0}{\hbar^2 (\alpha^2 + 4k^2 \sin^2 \theta/2)}$$

$$\sigma(\theta) = |f(\theta)|^2 = \frac{4\mu^2 V_0^2}{\hbar^4 (\alpha^2 + 4k^2 \sin^2 \theta/2)^2}$$

$\sigma(\theta)$ is maximum when $4k^2 \sin^2 \theta/2 = 0$, i.e., when $\theta = 0$ except at k or E is zero. $\sigma(\theta)$ decreases from this maximum value as $\theta \to \pi$.

12.11 Obtain an expression for the phase shift δ_0 for s-wave scattering by the potential

$$V(r) = \begin{cases} \infty & \text{for } 0 \le r \le a \\ 0 & \text{for } r > a \end{cases}$$

Assuming that the scattering is dominated by the $l = 0$ term, show that the total cross-section $\sigma_0 \cong 4\pi a^2$.

Solution. For the s-state, as $V = \infty$, the wave function $= 0$ for $r \le a$. For $r > a$, from Eq. (iv) of Problem (12.5),

$$\frac{d^2 u}{dr^2} + \frac{2mEu}{\hbar^2} = 0, \qquad R = \frac{u}{r}$$

$$u = B \sin (kr + \delta_0), \qquad k^2 = \frac{2mE}{\hbar^2}, \quad r > a$$

As $u = 0$ at $r = a$,

$$B \sin (ka + \delta_0) = 0, \quad \text{or} \quad \sin (ka + \delta_0) = 0$$

$$ka + \delta_0 = n\pi, \qquad (n \text{ being an integer})$$

$$\delta_0 = n\pi - ka$$

When scattering is dominated by $l = 0$, E/k is very small and, therefore, $\sin ka \cong ka$. The total cross-section

$$\sigma_0 = \frac{4\pi}{k^2} \sin^2 \delta_0 = \frac{4\pi}{k^2} \sin^2 (n\pi - ka)$$

$$= \frac{4\pi}{k^2} \sin^2 ka \cong 4\pi a^2$$

12.12 Using Born approximation, calculate the differential and total cross-sections for scattering of a particle of mass m by the δ-function potential $V(\mathbf{r}) = g\delta(\mathbf{r})$, g-constant.

Solution. From Eq. (12.15), the scattering amplitude

$$f(\theta) = -\frac{m}{2\pi\hbar^2}\int \exp(i\mathbf{q}\cdot\mathbf{r}')\, V(\mathbf{r}')\, d\tau'$$

where $\mathbf{q} = \mathbf{k} - \mathbf{k}'$ and $|\theta| = 2k \sin \theta/2$. Here, $\mathbf{k}$ and $\mathbf{k}'$ are, respectively, the wavevectors of the incident and scattered waves. Substituting the value of $V(r)$, we get

$$f(\theta) = -\frac{mg}{2\pi\hbar^2}\int \exp(i\mathbf{q}\cdot\mathbf{r}')\, \delta(\mathbf{r}')\, d\tau'$$

Using the definition of δ-function given in the Appendix, we get

$$f(\theta) = -\frac{mg}{2\pi\hbar^2}$$

The differential scattering cross-section is

$$\sigma(\theta) = |f(\theta)|^2 = \frac{m^2 g^2}{4\pi^2\hbar^4}$$

Since the distribution is isotropic, the total cross-section is given by

$$\sigma = 4\pi\sigma(\theta) = \frac{m^2 g^2}{\pi\hbar^4}$$

12.13 For the attractive square well potential,

$$V(r) = -V_0 \quad \text{for } 0 < r < r_0$$

$V(r) = 0$ for $r > r_0$. Find the energy dependence of the phase shift δ_0 by Born approximation. Hence show that at high energies,

$$\delta_0(k) \to \frac{mr_0V_0}{\hbar^2 k}, \qquad k^2 = \frac{2mE}{\hbar^2}$$

Solution. In the Born approximation for phase shifts, the phase shift δ_l is given by Eq.(12.14). Then the phase shift

$$\delta_0 = \frac{2mk}{\hbar^2} V_0 \int_0^{r_0} j_0^2(kr)\, r^2 dr, \qquad k^2 = \frac{2mE}{\hbar^2}$$

since $j_0(kr) = \sin(kr)/kr$. Now,

$$\delta_0 = \frac{2mkV_0}{\hbar^2 k^2}\int_0^{r_0} \sin^2(kr)\, dr = \frac{2mkV_0}{\hbar^2 k^2}\int_0^{r_0} \frac{1-\cos(2kr)}{2}\, dr$$

$$= \frac{2mkV_0}{\hbar^2 k^2}\left[\frac{r_0}{2} - \frac{\sin(2kr_0)}{4k}\right]$$

$$= \frac{mV_0}{\hbar^2 k^2}\left[kr_0 - \frac{1}{2}\sin(2kr_0)\right]$$

which is the energy dependence of the phase shift δ_0. At high energies, $k \to \infty$. When $k \to \infty$, the second term

$$\frac{mV_0}{2\hbar^2 k^2} \sin(2kr_0) \to 0$$

Hence at high energies,

$$\delta_0(k) \to \frac{mr_0 V_0}{\hbar^2 k}$$

12.14 In the Born approximation, calculate the scattering amplitude for scattering from the square well potential $V(r) = -V_0$ for $0 < r < r_0$ and $V(r) = 0$ for $r > r_0$.

Solution. In the Born approximation, from Eq. (12.15a), the scattering amplitude

$$f(\theta) = -\frac{2\mu}{\hbar^2} \int_0^\infty V(r)\, r \frac{\sin qr}{q}\, dr$$

where $q = 2k \sin(\theta/2)$, $k^2 = 2\mu E/\hbar^2$, θ is the scattering angle. Substituting $V(r)$ in the above equation, we get

$$f(\theta) = \frac{2\mu V_0}{\hbar^2 q} \int_0^{r_0} r \sin qr\, dr$$

$$= \frac{2\mu V_0}{\hbar^2 q} \left\{ \left[\frac{r \cos qr}{-q} \right]_0^{r_0} + \frac{1}{q} \int_0^{r_0} \cos qr\, dr \right\}$$

$$= \frac{2\mu V_0}{\hbar^2 q} \left(-\frac{r_0 \cos qr_0}{q} + \frac{\sin qr_0}{q^2} \right)$$

$$= \frac{2\mu V_0}{\hbar^2 q^3} (\sin qr_0 - qr_0 \cos qr_0)$$

12.15 In Problem 12.14, if the geometrical radius of the scatterer is much less than the wavelength associated with the incident particles, show that the scattering will be isotropic.

Solution. When the wavelength associated with the incident particle is large, wave vector k is small and, therefore, $kr_0 \ll 1$ or $qr_0 \ll 1$. Expanding $\sin qr_0$ and $qr_0 \cos qr_0$, we get

$$f(\theta) = \frac{2\mu V_0}{\hbar^2 q^3} \left[qr_0 - \frac{(qr_0)^3}{6} - r_0 q \left(1 - \frac{q^2 r_0^2}{2} \right) \right]$$

$$= \frac{2\mu V_0 r_0^3}{3\hbar^2}$$

which is independent of θ. Thus, the scattering will be isotropic.

12.16 Consider scattering by the attractive square well potential of Problem 12.14. Obtain an expression for the scattering length. Hence, show that, though the bombarding energy tends to zero, the s-wave scattering cross-section σ_0 tends to a finite value.

Solution. From Eq. (ix) of Problem 12.5,

$$\tan\,(kr_0 + \delta_0) = \frac{k}{k_1}\tan k_1 r_0$$

where

$$k^2 = \frac{2\mu E}{\hbar^2}, \qquad k_1^2 = \frac{2\mu}{\hbar^2}(E + V_0)$$

Expanding $\tan\,(kr_0 + \delta_0)$ and rearranging, we get

$$\tan\delta_0 = \frac{k\tan k_1 r_0 - k_1 \tan kr_0}{k_1 + k\tan kr_0 \tan k_1 r_0}$$

In the zero energy limit, $k \to 0$, $kr_0 \to kr_0$. Hence,

$$k_1 r_0 \to \left(\frac{2\mu V_0}{\hbar^2}\right)^{1/2} r_0 = k_0 r_0, \qquad k_0^2 = \frac{2\mu V_0}{\hbar^2}$$

$$k\tan kr_0 \tan k_1 r_0 \to k_2 r_0 \tan k_0 r_0$$

which may be neglected in comparison with k_0. Therefore,

$$\tan\,\delta_0 = \frac{k\tan k_0 r_0 - k_0 k r_0}{k_0} \quad \text{or} \quad \delta_0 \cong \frac{k}{k_0}\tan k_0 r_0 - kr_0$$

$$\text{The scattering length } a = -\frac{\delta_0}{k} = r_0 - \frac{\tan k_0 r_0}{k_0}$$

$$\sigma_0 = 4\pi a^2 = 4\pi r_0^2 \left(1 - \frac{\tan k_0 r_0}{k_0 r_0}\right)^2$$

That is, the s-wave scattering cross-section σ_0 tends to a finite value.

12.17 Use the Born approximation to calculate the differential cross section for scattering by the central potential $V(r) = \alpha/r^2$, where α is a constant. Given

$$\int_0^\infty \left(\sin\frac{x}{x}\right) dx = \frac{\pi}{2}$$

Solution. In the Born approximation,

$$f(\theta) = -\frac{2\mu}{\hbar^2}\int_0^\infty \frac{\sin qr}{qr} V(r)\, r^2\, dr, \qquad q = 2k\sin\frac{\theta}{2}$$

$$= -\frac{2\mu\alpha}{\hbar^2}\int_0^\infty \frac{\sin qr}{qr}\, dr = -\frac{2\mu\alpha}{q\hbar^2}\int_0^\infty \frac{\sin x}{x}\, dx, \qquad x = qr$$

$$= -\frac{2\mu\alpha}{q\hbar^2}\frac{\pi}{2} = \frac{-\pi\mu\alpha}{q\hbar^2}$$

$$\sigma(\theta) = |f(\theta)|^2 = \frac{\pi^2\mu^2\alpha^2}{q^2\hbar^4} = \frac{\pi^2\mu^2\alpha^2}{4k^2\hbar^4 \sin^2\theta/2}$$

12.18 Consider scattering by the Yukawa potential $V(r) = V_0 \exp(-\alpha r)/r$, where V_0 and α are constants. In the limit $E \to 0$, show that the differential scattering cross-section is independent of θ and ϕ.

Solution.

$$f(\theta) = -\frac{2\mu}{\hbar^2}\int_0^\infty \frac{\sin qr}{qr} V(r)\, r^2\, dr$$

$$= -\frac{2\mu V_0}{q\hbar^2}\int_0^\infty e^{-\alpha r}\sin qr\, dr = \frac{-2\mu V_0}{q\hbar^2}\,\frac{q}{q^2+\alpha^2} = -\frac{2\mu V_0}{\hbar^2(q^2+\alpha^2)}$$

As $E \to 0$, $k \to 0$ and $q = 2k \sin\theta/2 \to 0$. Hence,

$$\sigma(\theta) = |f(\theta)|^2 = \frac{4\mu^2V_0^2}{\hbar^4\alpha^4}$$

which is independent of θ and ϕ.

12.19 Consider the partial wave analysis of scattering by a potential $V(r)$ and derive an expression for the phase shift δ_l in terms of $V(r)$ and the energy E of the incident wave.

Solution. The radial part of the Schrodinger equation that describes the scattering is

$$\frac{1}{r^2}\frac{d}{dr}\left(r^2\frac{dR_l}{dr}\right) + \left[\frac{2\mu E}{\hbar^2} - \frac{2\mu V}{\hbar^2} - \frac{l(l+1)}{r^2}\right]R_l = 0 \tag{i}$$

Writing

$$R_l = \frac{u_l}{r} \tag{ii}$$

we get

$$\frac{d^2u_l}{dr^2} + \left[\frac{2\mu E}{\hbar^2} - \frac{2\mu V}{\hbar^2} - \frac{l(l+1)}{r^2}\right]u_l = 0 \tag{iii}$$

In the incident wave region $V = 0$ and, therefore,

$$\frac{d^2u_l}{dr^2} + \left[k^2 - \frac{l(l+1)}{r^2}\right]u_l(r) = 0, \quad k^2 = \frac{2mE}{\hbar^2} \tag{iv}$$

whose solution is

$$u_l(kr) = krj_l(kr) \tag{v}$$

Assymptotically,

$$u_l(kr) \xrightarrow[r\to\infty]{} \sin\left(kr - \frac{l\pi}{2}\right) \tag{vi}$$

Similarly, the approximate solution of

$$\frac{d^2\mathbf{v}_l}{dr^2} + \left[k^2 - \frac{2\mu V(r)}{\hbar^2} - \frac{l(l+1)}{r^2}\right]\mathbf{v}_l = 0 \tag{vii}$$

$$v_l(kr) \xrightarrow[r\to\infty]{} \sin\left(kr - \frac{l\pi}{2} + \delta_l\right) \tag{viii}$$

Multiplying Eq. (iv) by v_l, Eq. (vii) by u_l and subtracting, we get

$$v_l \frac{d^2u_l}{dr^2} - u_l \frac{d^2v_l}{dr^2} = -\frac{2\mu V}{\hbar^2} u_l v_l \tag{ix}$$

Integrating from 0 to r and remembering that $u_l(0) = v_l(0) = 0$, we obtain

$$v_l \frac{du_l}{dr} - u_l \frac{dv_l}{dr} = -\frac{2\mu}{\hbar^2} \int_0^r V(r')\, u_l(r')\, v_l(r')\, dr'$$

Allowing $r \to \infty$ and substituting the values of $u_l(r)$ and $v_l(r)$, we have

$$k \sin\left(kr - \frac{l\pi}{2} + \delta_l\right) \cos\left(kr - \frac{l\pi}{2}\right) - k \sin\left(kr - \frac{l\pi}{2}\right) \cos\left(kr - \frac{l\pi}{2} + \delta_l\right)$$

$$= -\frac{2\mu}{\hbar^2} \int_0^\infty V(r)\, u_l(kr)\, v_l(kr)\, dr$$

Since

$$\left(kr - \frac{l\pi}{2} + \delta_l\right) - \left(kr - \frac{l\pi}{2}\right) = \delta_l$$

the equation reduces to

$$k \sin \delta_l = -\frac{2\mu}{\hbar^2} \int_0^\infty V(r)\, u_l(kr)\, v_l(kr)\, dr$$

which is the equation for the phase shift δ_l.

12.20 Show that an attractive potential leads to positive phase shifts whereas a repulsive potential to negative phase shifts.

Solution. From Problem 12.19, the equation for phase shift δ_l is given by

$$\sin \delta_l = -\frac{2\mu}{k\hbar^2} \int_0^\infty V(r)\, u_l(kr)\, v_l(kr)\, dr$$

where

$$k^2 = \frac{2mE}{\hbar^2}$$

At high energies, for weak potential, the phase shifts are small and

$$u_l(kr) \cong v_l(kr) \cong kr\, j_l(kr)$$

The spherical Bessel function $j_l(kr)$ is related to ordinary Bessel function by

$$j_l(kr) = \left(\frac{\pi}{2kr}\right)^{1/2} J_{l+(1/2)}(kr)$$

$$\sin\delta_l = \delta_l = -\frac{2\mu k}{\hbar^2}\int_0^\infty V(r)\, j_l^2(kr) r^2 dr$$

$$= -\frac{\pi\mu}{\hbar^2}\int_0^\infty V(r)\left[j_{l+(1/2)}(kr)\right]^2 r\, dr$$

From this equation it is obvious that an attractive potential ($V \le 0$) leads to positive phase shifts, whereas repulsive potential ($V \ge 0$) to negative phase shifts.

12.21 Use the Born approximation to obtain differential scattering cross-section when a particle moves in the potential $V(r) = -V_0 \exp(-r/r_0)$, where V_0 and r_0 are positive constants. Given

$$\int_0^\infty x \exp(-ax) \sin(bx)\, dx = \frac{2ab}{(a^2+b^2)^2}, \qquad a > 0$$

Solution. The scattering amplitude

$$f(\theta) = -\frac{2\mu}{\hbar^2}\int_0^\infty \frac{\sin qr}{qr} V(r) r^2\, dr = \frac{2\mu V_0}{q\hbar^2}\int_0^\infty r e^{-r/r_0} \sin qr\, dr$$

$$\int_0^\infty x e^{-ax} \sin bx\, dx = \frac{2ab}{(a^2+b^2)^2}, \qquad a > 0$$

$$f(\theta) = \frac{2\mu V_0}{q\hbar^2}\,\frac{2q(1/r_0)}{[(1/r_0)^2+q^2]^2} = \frac{4\mu V_0}{\hbar^2 r_0}\left(\frac{r_0^2}{1+q^2r_0^2}\right)^2$$

$$\sigma(\theta) = |f(\theta)|^2 = \frac{16\mu^2V_0^2r_0^6}{\hbar^4(1+q^2r_0^2)^4}$$

12.22 Calculate the scattering amplitude for a particle moving in the potential

$$V(r) = V_0\,\frac{c-r}{r}\exp\left(-\frac{r}{r_0}\right)$$

where V_0 and r_0 are constants.

Solution.

$$f(\theta) = -\frac{2\mu V_0}{q\hbar^2}\int_0^\infty \frac{c-r}{r} e^{-r/r_0}\, r \sin qr\, dr$$

$$= -\frac{2\mu V_0}{q\hbar^2}\left[\int_0^\infty c e^{-r/r_0} \sin qr\, dr - \int_0^\infty r e^{-r/r_0} \sin qr\, dr\right]$$

$$f(\theta) = -\frac{2\mu V_0}{q\hbar^2}\left[c\,\frac{q}{q^2+(1/r_0^2)} - \frac{2q}{r_0}\,\frac{1}{[(1/r_0^2)+q^2]^2}\right]$$

$$= -\frac{2\mu V_0}{\hbar^2}\left[\frac{cr_0^2}{1+q^2r_0^2} - \frac{2r_0^3}{(1+q^2r_0^2)^2}\right]$$

12.23 In scattering from a potential $V(r)$; the wave function $\psi(r)$ is written as an incident plane wave plus an outgoing scattered wave: $\psi = e^{ikz} + f(r)$. Derive a differential equation for $f(r)$ in the first Born approximation.

Solution. The Schrödinger equation that describes the scattering is given by

$$-\frac{\hbar^2}{2\mu}\nabla^2\psi + V(r)\,\psi = E\psi$$

Writing

$$k^2 = \frac{2\mu E}{\hbar^2}, \qquad U(r) = \frac{2\mu V(r)}{\hbar^2}$$

we get

$$(\nabla^2 + k^2)\,\psi = U(r)\,\psi$$

Substituting ψ, we obtain

$$(\nabla^2 + k^2)\,(e^{ikz} + f) = U(e^{ikz} + f)$$

Since $(\nabla^2 + k^2)\,e^{ikz} = 0$,

$$(\nabla^2 + k^2)\,f(r) = U(e^{ikz} + f)$$

In the first Born approximation, $e^{ikz} + f(r) \simeq e^{ikz}$, and hence the differential equation for $f(r)$ becomes

$$(\nabla^2 + k^2)\,f(r) = \frac{2m}{\hbar^2}Ve^{ikz}$$

12.24 Use the Born approximation to calculate the differential scattering cross-section for a particle of mass m moving in the potential $V(r) = A\exp(-r^2/a^2)$, where A and a are constants. Given

$$\int_0^\infty e^{-a^2x^2}\cos bx\,dx = \frac{\sqrt{\pi}}{2a}\exp\left(\frac{-b^2}{4a^2}\right)$$

Solution. In the Born approximation, we have

$$f(\theta) = -\frac{2m}{\hbar^2}\int_0^\infty \frac{\sin(qr')}{qr'}V(r')\,r'^2dr'$$

$$= -\frac{2mA}{q\hbar^2}\int_0^\infty r\sin(qr)\exp\left(\frac{-r^2}{a^2}\right)dr$$

$$|q| = 2k\sin\frac{\theta}{2}$$

Integrating by parts, we get

$$f(\theta) = -\frac{2mA}{q\hbar^2}\left[-\frac{a^2\exp(-r^2/a^2)}{2}\sin qr\right]_0^\infty - \frac{2mA}{q\hbar^2}\frac{a^2q}{2}\int_0^\infty \exp\left(\frac{-r^2}{a^2}\right)\cos qr\,dr$$

As the integrated term vanishes

$$f(\theta) = -\frac{mAa^2}{\hbar^2} \int_0^\infty e^{-r^2/a^2} \cos (qr)\, dr$$

$$= -\frac{mAa^2}{\hbar^2} \frac{\sqrt{\pi}\, a \exp (-q^2a^2/4)}{2} = -\frac{mAa^3\pi^{1/2}}{2\hbar^2} \exp\left(\frac{-q^2a^2}{4}\right)$$

$$\sigma(\theta) = |f(\theta)|^2 = \frac{m^2A^2a^6\pi}{4\hbar^4} \exp\left(\frac{-q^2a^2}{2}\right)$$

12.25 A particle of mass m and energy E is scattered by a spherically symmetric potential $A\delta(r - a)$, where A and a are constants. Calculate the differential scattering cross-section when the energy is very high.

Solution. At high energies, the Born approximation is more appropriate. From Eq. (12.15a), the scattering amplitude

$$f(\theta) = -\frac{2m}{\hbar^2} \int_0^\infty \frac{\sin qr}{qr} V(r)\, r^2\, dr$$

Substituting the value of $V(r)$, we get

$$f(\theta) = -\frac{2m}{\hbar^2} A \int_0^\infty \frac{\sin qr}{qr} \delta(r - a)\, r^2\, dr$$

$$= -\frac{2mA}{\hbar^2} \frac{a \sin qa}{q}$$

The differential scattering cross-section

$$\sigma(\theta) = |f(\theta)|^2 = \frac{4m^2A^2a^2 \sin^2 qa}{q^2\hbar^4}$$

12.26 For the attractive square well potential,

$$V = -V_0 \quad \text{for } 0 \le r < a,$$

$$V = 0 \quad \text{for } r > a$$

Calculate the scattering cross-section for a low energy particle by the method of partial wave analysis. Compare the result with the Born approximation result. Given

$$\int_0^\infty \exp (-ax) \sin bx\, dx = \frac{b}{a^2 + b^2}$$

Solution. The scattering of a particle by an attractive square well potential of the same type by the method of partial wave analysis has been discussed in Problem 12.5. The phase shift δ_0 is given by

$$\tan \delta_0 = \frac{k}{k_1} \tan (k_1a) - ka$$

where

$$k^2 = \frac{2\mu E}{\hbar^2}, \qquad k_1^2 = \frac{2\mu(E + V_0)}{\hbar^2}$$

For low energy particles,

$$k \to 0, \quad k_1 \to k_0 = \frac{2\mu V_0}{\hbar^2}$$

Consequently, the above relations reduce to

$$\delta_0 \simeq ka\left[\frac{\tan(k_0 a)}{k_0 a} - 1\right]$$

The total scattering cross-section

$$\sigma \simeq \frac{4\pi}{k^2}\sin^2\delta_0 \simeq \frac{4\pi}{k^2}\delta_0^2$$

$$= 4\pi a^2\left(\frac{\tan(k_0 a)}{k_0 a} - 1\right)^2$$

If $k_0 a \ll 1$,

$$\sigma \simeq 4\pi a^2\left(\frac{k_0 a}{k_0 a} + \frac{(k_0 a)^3}{3k_0 a} - 1\right)^2 = 4\pi a^2\left[\frac{(k_0 a)^3}{3k_0 a}\right]^2$$

$$= \frac{16\pi a^6\mu^2 V_0^2}{9\hbar^4}$$

In the Born approximation, the scattering amplitude (refer Problem 12.14)

$$f(\theta) = \frac{2\mu V_0}{\hbar^2 q^3}[\sin(qa) - qa\cos(qa)]$$

$$\sigma(\theta) = |f(\theta)|^2 = \frac{4\mu^2 V_0^2}{\hbar^4 q^6}[\sin(qa) - qa\cos(qa)]^2$$

where

$$q = 2k\sin\frac{\theta}{2}, \qquad k^2 = \frac{2\mu E}{\hbar^2}$$

where θ is the scattering angle. At low energies, $k \to 0$, $q \to 0$, and hence

$$\sin(qa) \simeq qa - \frac{1}{3!}(qa)^3, \qquad \cos(qa) \simeq 1 - \frac{1}{2!}(qa)^2$$

Hence,

$$\sigma(\theta) = \frac{4\mu^2 V_0^2 a^6}{9\hbar^4}$$

The total cross-section for scattering is

$$\sigma = \int \sigma(\theta)\, d\Omega = \int_0^{\pi} \int_0^{2\pi} \sigma(\theta) \sin\theta\, d\theta\, d\phi$$

$$= \frac{16\pi\mu^2 V_0^2 a^6}{9\hbar^4}$$

At low energies the two methods give the same result.

12.27 In partial wave analysis of scattering, one has to consider waves with $l = 0, 1, 2, 3, \ldots$. For a given energy, for spherically symmetric potentials having range r_0, up to what value of l should one consider?

Solution. The wave vector $k = \sqrt{2\mu E}/\hbar$, where E is the energy and μ is the reduced mass.

The linear momentum of the particle $p = k\hbar$

Angular momentum $= l\hbar$

If b is the impact parameter, classically, then

Angular momentum $= pb = k\hbar b$

Equating the two expressions for angular momentum, we get

$$k\hbar b = l\hbar \quad \text{or} \quad l = kb$$

When the impact parameter $b > r_0$, the particle will not see the potential region and a classical particle will not get scattered if $l > kr_0$. Hence we need to consider partial waves up to $l = kr_0$.

12.28 (i) Write the asymptotic form of the wave function in the case of scattering by a fixed potential and explain.

(ii) What is Born approximation?

(iii) What is the formula for the first Born approximation for scattering amplitude $f(\theta)$?

(iv) Under what condition is the Born approximation valid?

Solution.

(i) The general asymptotic solution is

$$\psi \xrightarrow[r\to\infty]{} A\left[e^{ikz} + f(\theta, \phi)\frac{e^{ikz}}{r}\right] \tag{i}$$

where A is a constant.

In this, the part e^{ikz} represents the incident plane wave along the z-axis. The wave vector k is given by

$$k^2 = \frac{2mE}{\hbar^2},$$

where E is the energy.

The second term of Eq. (i) represents the spherically diverging scattered wave. The amplitude factor $f(\theta, \phi)$ is called the *scattering amplitude*.

(ii) A general analysis of the scattering problem requires expressing the wave function in the form of an integral equation. In this expression for the wave function, the wave function appears under the integral sign. In the first Born approxiamtion, $\psi(r)$ in the integrand is replaced by the incoming plane wave exp $(i\boldsymbol{k} \cdot \boldsymbol{r})$. This leads to an improved value for the wave function which is used in the integral in the second Born approximation. This iterative procedure is continued till the input and output ψ's are almost equal.

(iii) In the first Born approximation, the scattering amplitude

$$f(\theta) = -\frac{2\mu}{\hbar^2}\int_0^\infty \frac{\sin(qr')}{qr'} V(r')\, r'^2\, dr'$$

where $q\hbar$ is the momentum transfer from the incident particle to the scattering potential and

$$|q| = 2|k| \sin\frac{\theta}{2}$$

with angle θ being the scattering angle, $V(r)$ the potential, and μ the reduced mass.

(iv) The Born approximation is valid for weak potentials at high energies.

12.29 In the scattering experiment, the measurement is done in the laboratory system. Discuss its motion in the centre of mass system and illustrate it with a diagram.

Solution. Consider a particle of mass m moving in the positive z-direction with velocity v_L and encountering a scattering centre of mass M which is at rest at O. After scattering, it gets scattered in the direction (θ_L, ϕ_L). The velocity of the centre of mass

$$v_{cm} = \frac{mv_L}{m + M}$$

We shall now examine the situation with respect to an observer located at the centre of mass. The observer sees the particle M approaching him from the right with velocity $-mv_L/(m + M)$, the particle m approaching him from left with velocity

$$v_c = v_L - v_{cm} = v_L - \frac{mv_L}{m + M} = \frac{mv_L}{m + M}$$

After encounter to keep the centre of mass at rest, the two particles must be scattered in the opposite directions with speeds unchanged (elastic scattering). The collision process in the centre of mass system is illustrated in Fig. 12.1.

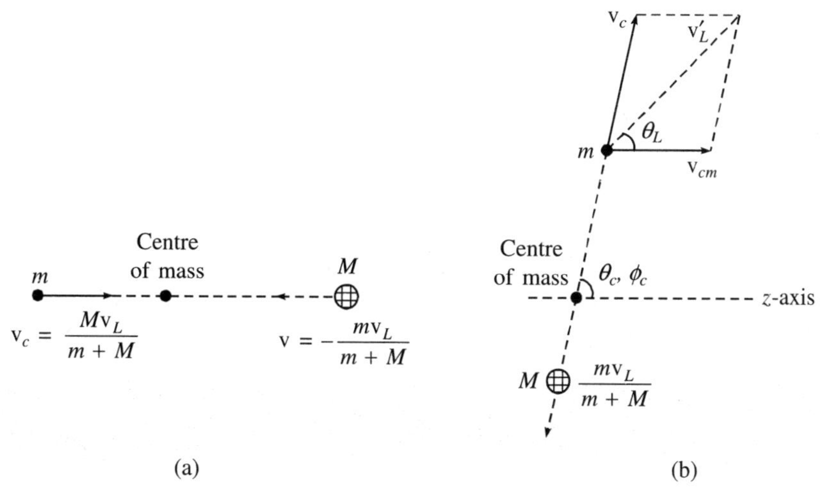

Fig. 12.1 Motion of the particles in the centre of mass system: (a) before collision; (b) after collision.

CHAPTER 13

Relativistic Equations

The quantum mechanics discussed so far does not satisfy the requirements of the Special Theory of Relativity as it is based on a nonrelativistic Hamiltonian. Based on the relativistic Hamiltonian, two relativistic wave equations were developed, one by Klein and Gordon and the other by P.A.M. Dirac.

13.1 Klein-Gordon Equation

The Klein-Gordon equation is based on the relativistic energy expression

$$E^2 = c^2p^2 + m^2c^4 \tag{13.1}$$

where m is the rest mass of the particle and p its momentum. Replacing $\boldsymbol{p}$ by $-i\hbar\nabla$ and E by $i\hbar(\partial/\partial t)$, we get

$$\left(\nabla^2 - \frac{1}{c^2}\frac{\partial^2}{\partial t^2}\right)\Psi(\boldsymbol{r}, t) = \frac{m^2c^2}{\hbar^2}\,\Psi(\boldsymbol{r}, t) \tag{13.2}$$

which is the Klein-Gordon equation.

To get the equation of continuity (2.15) in the relativistic theory, we have to define the position probability density by

$$\boldsymbol{P}(\boldsymbol{r}, t) = \frac{i\hbar}{2mc^2}\left(\Psi^*\frac{\partial\Psi}{\partial t} - \Psi\frac{\partial\Psi^*}{\partial t}\right) \tag{13.3}$$

and the probability current density by the same definition, Eq. (2.14). This definition of $P(\boldsymbol{r}, t)$ leads to both positive and negative values for it. By interpreting eP as the electrical charge density and ej as the corresponding electric current, the Klein-Gordon equation is used for a system of particles having both positive and negative charges.

13.2 Dirac's Equation for a Free Particle

To get a first derivative equation in both time and space coordinates, Dirac unambignously wrote the Hamiltonian as

$$E = H = c(\alpha_x p_x + \alpha_y p_y + \alpha_z p_z) + \beta mc^2 \tag{13.4}$$

$$E = H = c\boldsymbol{\alpha} \cdot \boldsymbol{p} + \beta mc^2$$

where α_x, α_y, α_z and β are matrices. Replacing E and $\boldsymbol{p}$ by their operators and allowing the resulting operator equation to operate on $\Psi(\boldsymbol{r}, t)$, we obtain

$$i\hbar \frac{\partial \Psi(\boldsymbol{r}, t)}{\partial t} = -ic\hbar \left(\alpha_x \frac{\partial}{\partial x} + \alpha_x \frac{\partial}{\partial y} + \alpha_x \frac{\partial}{\partial z} \right) \Psi(\boldsymbol{r}, t) + \beta mc^2 \Psi(\boldsymbol{r}, t) \tag{13.5}$$

which is Dirac's relativistic equation for a free particle. The $\boldsymbol{\alpha}$ and β matrices are given by

$$\alpha_x = \begin{pmatrix} 0 & \sigma_x \\ \sigma_x & 0 \end{pmatrix}, \qquad \alpha_y = \begin{pmatrix} 0 & \sigma_y \\ \sigma_y & 0 \end{pmatrix} \tag{13.6}$$

$$\alpha_z = \begin{pmatrix} 0 & \sigma_z \\ \sigma_z & 0 \end{pmatrix}, \qquad \beta = \begin{pmatrix} I & 0 \\ 0 & -I \end{pmatrix}$$

where σ_x, σ_y and σ_z are Pauli's spin matrices and I is a unit 2×2 matrix. Since α_x, α_y, α_z and β are 4×4 matrices, the Dirac wave function $\Psi(\boldsymbol{r}, t)$ must be a 4-coulumn vector

$$\Psi(\boldsymbol{r}, t) = \begin{pmatrix} \Psi_1 \\ \Psi_2 \\ \Psi_3 \\ \Psi_4 \end{pmatrix}, \qquad \Psi^\dagger = (\Psi_1^*, \Psi_2^*, \Psi_3^*, \Psi_4^*) \tag{13.7}$$

The probability density $P(\boldsymbol{r}, t)$ and the probability current density $\boldsymbol{j}(\boldsymbol{r}, t)$ are defined by the relations

$$P(\boldsymbol{r}, t) = \Psi^\dagger \Psi, \qquad \boldsymbol{j}(\boldsymbol{r}, t) = c\Psi^\dagger \boldsymbol{\alpha} \Psi \tag{13.8}$$

PROBLEMS

13.1 Starting from the Klein-Gordon equation, obtain the equation of continuity.

Solution. The Klein-Gordon equation and its complex conjugate are

$$-\hbar^2 \frac{\partial^2 \Psi}{\partial t^2} = -c^2\hbar^2\nabla^2\psi(\mathbf{r}, t) + m^2c^4\psi(\mathbf{r}, t)$$

$$-\hbar^2 \frac{\partial^2 \Psi^*}{\partial t^2} = -c^2\hbar^2\nabla^2\Psi^* + m^2c^4\Psi^*$$

Multiplying the first equation from the LHS by Ψ^* and the second equation from the LHS by Ψ and subtracting, we get

$$\Psi^* \frac{\partial^2 \Psi}{\partial t^2} - \Psi \frac{\partial^2 \Psi^*}{\partial t^2} = c^2(\Psi^*\nabla^2\Psi - \Psi\nabla^2\Psi^*)$$

$$\frac{\partial}{\partial t}\left(\Psi^* \frac{\partial \Psi}{\partial t} - \Psi \frac{\partial \Psi^*}{\partial t}\right) = -c^2\nabla\,(\Psi\nabla\Psi^* - \Psi^*\nabla\Psi)$$

$$\frac{\partial}{\partial t}P(\mathbf{r}, t) + \nabla \cdot j(\mathbf{r}, t) = 0$$

$$P(r, t) = \frac{i\hbar}{2mc^2}\left(\Psi^* \frac{\partial \Psi}{\partial t} - \Psi \frac{\partial \Psi^*}{\partial t}\right), \qquad j(\mathbf{r}, t) = \frac{i\hbar}{2m}(\Psi\nabla\Psi^* - \Psi^*\nabla\Psi)$$

13.2 Show that the Dirac matrices α_x, α_y, α_z and β anticommute in pairs and their squares are unity.

Solution.

$$\alpha_x = \begin{pmatrix} 0 & \sigma_x \\ \sigma_x & 0 \end{pmatrix}, \quad \alpha_y = \begin{pmatrix} 0 & \sigma_y \\ \sigma_y & 0 \end{pmatrix}, \quad \alpha_z = \begin{pmatrix} 0 & \sigma_z \\ \sigma_z & 0 \end{pmatrix}, \quad \beta = \begin{pmatrix} I & 0 \\ 0 & -I \end{pmatrix}$$

$$\alpha_x\alpha_y + \alpha_y\alpha_x = \begin{pmatrix} 0 & \sigma_x \\ \sigma_x & 0 \end{pmatrix}\begin{pmatrix} 0 & \sigma_y \\ \sigma_y & 0 \end{pmatrix} + \begin{pmatrix} 0 & \sigma_y \\ \sigma_y & 0 \end{pmatrix}\begin{pmatrix} 0 & \sigma_x \\ \sigma_x & 0 \end{pmatrix}$$

$$= \begin{pmatrix} \sigma_x\sigma_y & 0 \\ 0 & \sigma_x\sigma_y \end{pmatrix} + \begin{pmatrix} \sigma_y\sigma_x & 0 \\ 0 & \sigma_y\sigma_x \end{pmatrix}$$

Since $\sigma_x\sigma_y = i\sigma_z$, $\sigma_y\sigma_x = -i\sigma_z$, we have

$$\alpha_x\alpha_y + \alpha_y\alpha_x = \begin{pmatrix} i\sigma_z & 0 \\ 0 & i\sigma_z \end{pmatrix} + \begin{pmatrix} -i\sigma_z & 0 \\ 0 & -i\sigma_z \end{pmatrix} = 0$$

i.e., α_x and α_y anticommute. Similarly,

$$\alpha_y\alpha_z + \alpha_z\alpha_y = \alpha_z\alpha_x + \alpha_x\alpha_z = 0$$

$$\alpha_x\beta + \beta\alpha_x = \begin{pmatrix} 0 & \sigma_x \\ \sigma_x & 0 \end{pmatrix}\begin{pmatrix} I & 0 \\ 0 & -I \end{pmatrix} + \begin{pmatrix} I & 0 \\ 0 & -I \end{pmatrix}\begin{pmatrix} 0 & \sigma_x \\ \sigma_x & 0 \end{pmatrix}$$

$$= \begin{pmatrix} 0 & -\sigma_x I \\ \sigma_x I & 0 \end{pmatrix} + \begin{pmatrix} 0 & I\sigma_x \\ -I\sigma_x & 0 \end{pmatrix}$$

As I commutes with σ_x, RHS of the above vanishes, and hence

$$\alpha_y\beta + \beta\alpha_y = \alpha_z\beta + \beta\alpha_z = 0$$

$$\alpha_x^2 = \begin{pmatrix} 0 & \sigma_x \\ \sigma_x & 0 \end{pmatrix}\begin{pmatrix} 0 & \sigma_x \\ \sigma_x & 0 \end{pmatrix} = \begin{pmatrix} \sigma_x^2 & 0 \\ 0 & \sigma_x^2 \end{pmatrix} = \begin{pmatrix} 1 & 0 \\ 0 & 1 \end{pmatrix}$$

since $\sigma_x^2 = 1$. Similarly, $\alpha_x^2 = a_z^2 = \beta^2 = 1$. Hence, α_x, α_y, α_z and β anticommute in pairs and their squares are unity.

13.3 Write Dirac's equation for a free particle. Find the form of the probability density and the probability current density in Dirac's formalism.

Solution. Dirac's equation for a free particle is

$$i\hbar \frac{\partial}{\partial t}\Psi(\mathbf{r}, t) = -ic\hbar\boldsymbol{a} \cdot \nabla\Psi + \beta mc^2\Psi \tag{i}$$

Here, α and β are 4×4 matrices and $\Psi(\mathbf{r}, t)$ is a four-column vector. The Hermitian conjugate of Eq (i) is

$$-i\hbar \frac{\partial}{\partial t}\Psi^\dagger = ic\hbar\nabla\Psi^\dagger \cdot \boldsymbol{\alpha} + \Psi^\dagger\beta mc^2 \tag{ii}$$

Multiplying Eq (i) by $\Psi^\dagger$ on left, Eq (ii) by Ψ on the RHS, and subtracting one from the other, we get

$$i\hbar\left(\Psi^\dagger \frac{\partial\Psi}{\partial t} - \frac{\partial\Psi^\dagger}{\partial t}\Psi\right) = -ic\hbar(\Psi^\dagger\boldsymbol{\alpha} \cdot \nabla\Psi + \nabla\Psi^\dagger \cdot \boldsymbol{\alpha}\Psi)$$

$$\frac{\partial}{\partial t}(\Psi^\dagger\Psi) + \nabla \cdot (c\Psi^\dagger\boldsymbol{\alpha}\Psi) = 0$$

$$\frac{\partial}{\partial t}P(\mathbf{r}, t) + \nabla \cdot j(\mathbf{r}, t) = 0 \tag{iii}$$

where

$$j(\mathbf{r}, t) = c\Psi^\dagger\boldsymbol{\alpha}\Psi, \qquad P(\mathbf{r}, t) = \Psi^\dagger\Psi \tag{iv}$$

Equation (iii) is the continuity equation and the quantities $P(\mathbf{r}, t)$ and $\mathbf{j}(\mathbf{r}, t)$ are the probability density and probability current density, respectively.

13.4 In Dirac's theory, the probability current density is defined by the relation $\mathbf{j}(\mathbf{r}, t) = c\Psi^\dagger\alpha\,\Psi$, where Ψ is the four-component wave vector. Write the relations for j_x, j_y and j_z in terms of the components of Ψ, i.e.,

$$j(\boldsymbol{r}, t) = c\Psi^\dagger \alpha \Psi, \qquad j_x = c\Psi^\dagger \alpha_x \Psi$$

$$j_x = c\,(\Psi_1^* \;\Psi_2^* \;\Psi_3^* \;\Psi_4^*)\begin{pmatrix} 0 & 0 & 0 & 1 \\ 0 & 0 & 1 & 0 \\ 0 & 1 & 0 & 0 \\ 1 & 0 & 0 & 0 \end{pmatrix}\begin{pmatrix} \Psi_1 \\ \Psi_2 \\ \Psi_3 \\ \Psi_4 \end{pmatrix}$$

$$= c\,(\Psi_1^* \;\Psi_2^* \;\Psi_3^* \;\Psi_4^*)\begin{pmatrix} \Psi_4 \\ \Psi_3 \\ \Psi_2 \\ \Psi_1 \end{pmatrix}$$

$$= c\,(\Psi_1^* \Psi_4 + \Psi_2^* \Psi_3 + \Psi_3^* \Psi_2 + \Psi_4^* \Psi_1)$$

Proceeding on a similar line, we have

$$j_y = ic\,(-\Psi_1^* \Psi_4 + \Psi_2^* \Psi_3 - \Psi_3^* \Psi_2 + \Psi_4^* \Psi_1)$$

$$j_z = c\,(\Psi_1^* \Psi_3 - \Psi_2^* \Psi_4 + \Psi_3^* \Psi_1 - \Psi_4^* \Psi_2)$$

13.5 Prove that the operator $c\alpha$, where α stands for Dirac matrix, can be interpreted as the velocity operator.

Solution. In the Heisenberg picture, the equation of motion of the position vector $\boldsymbol{r}$, which has no explicit time dependence, is given by

$$\frac{d\boldsymbol{r}}{dt} = \frac{1}{i\hbar}[\boldsymbol{r}, H], \qquad H = c\boldsymbol{\alpha}\cdot\boldsymbol{p} + \beta mc^2$$

Since α commutes with x, the x-component of the above equation reduces to

$$\frac{dx}{dt} = \frac{1}{i\hbar}[x, H] = \frac{1}{i\hbar}(xH - Hx) = \frac{c}{i\hbar}(x\alpha_x p_x - \alpha_x p_x x)$$

$$= \frac{c}{i\hbar}\alpha_x(xp_x - p_x x) = c\alpha_x$$

Similarly, $$\frac{dy}{dt} = c\alpha_y, \qquad \frac{dz}{dt} = c\alpha_z$$

Thus, $c\alpha$ is the velocity vector.

13.6 Show that $(\boldsymbol{\alpha}\cdot\mathbf{A})\,(\boldsymbol{\alpha}\cdot\mathbf{B}) = (\mathbf{A}\cdot\mathbf{B}) + i\sigma'\cdot(\mathbf{A}\times\mathbf{B})$, where **A** and **B** commute with α and $\sigma' = \begin{pmatrix} \sigma & 0 \\ 0 & \sigma \end{pmatrix}$.

Solution.

$$(\boldsymbol{\alpha}\cdot\mathbf{A})\,(\boldsymbol{\alpha}\cdot\mathbf{B}) = (\alpha_x A_x + \alpha_y A_y + \alpha_z A_z)(\alpha_x B_x + \alpha_y B_y + \alpha_z B_z)$$

$$= \alpha_x^2 A_x B_x + \alpha_y^2 A_y B_y + \alpha_z^2 A_z B_z + a_x a_y A_x B_y + \alpha_x \alpha_z A_x B_z$$

$$+ \alpha_y \alpha_x A_y B_x + \alpha_y \alpha_z A_y B_z + \alpha_z \alpha_x A_z B_x + \alpha_z \alpha_y A_z B_y$$

Since $\alpha_x^2 = \alpha_y^2 = \alpha_z^2 = 1$, $\alpha_x\alpha_y = -\alpha_y\alpha_x$ and the cyclic relations

$$(\alpha \cdot A)(\alpha \cdot B) = (A \cdot B) + \alpha_x\alpha_y(A_xB_y - A_yB_x) + \alpha_y\alpha_z(A_yB_z - A_zB_y)$$
$$+ a_z a_x (A_zB_x - A_xB_z)$$

$$\alpha_x\alpha_y = \begin{pmatrix} 0 & \sigma_x \\ \sigma_x & 0 \end{pmatrix}\begin{pmatrix} 0 & \sigma_y \\ \sigma_y & 0 \end{pmatrix} = \begin{pmatrix} \sigma_x\sigma_y & 0 \\ 0 & \sigma_x\sigma_y \end{pmatrix} = i\begin{pmatrix} \sigma_z & 0 \\ 0 & \sigma_z \end{pmatrix} = i\sigma_z'$$

Using this results and the cyclic relations, we get

$$(\alpha \cdot \mathbf{A})(\alpha \cdot \mathbf{B}) = (\mathbf{A} \cdot \mathbf{B}) + i\sigma' \cdot (\mathbf{A} \times \mathbf{B})$$

13.7 Consider the one-dimensional Dirac equation

$$i\hbar\frac{\partial \psi}{\partial t} = [c\alpha p_z + \beta mc^2 + V(z)]\,\psi, \qquad p_z = -i\hbar\frac{\partial}{\partial z}$$

$$\alpha = \begin{pmatrix} 0 & \sigma_z \\ \sigma_z & 0 \end{pmatrix}, \qquad \sigma_z = \begin{pmatrix} 1 & 0 \\ 0 & -1 \end{pmatrix}, \qquad \beta = \begin{pmatrix} I & 0 \\ 0 & -I \end{pmatrix}$$

Show that

(i) $\sigma = \begin{pmatrix} \sigma_z & 0 \\ 0 & \sigma_z \end{pmatrix}$

commutes with H; (ii) The one-dimensional Dirac equation can be written as two coupled first order differential equations.

Solution. The Hamiltonian

$$H = c\alpha\left(-i\hbar\frac{\partial}{\partial z}\right) + \beta mc^2 + V(z)$$

The commutator

$$[\sigma, \alpha] = \left[\begin{pmatrix} \sigma_z & 0 \\ 0 & \sigma_z \end{pmatrix}\begin{pmatrix} 0 & \sigma_z \\ \sigma_z & 0 \end{pmatrix}\right] = \begin{pmatrix} \sigma_z & 0 \\ 0 & \sigma_z \end{pmatrix}\begin{pmatrix} 0 & \sigma_z \\ \sigma_z & 0 \end{pmatrix} - \begin{pmatrix} 0 & \sigma_z \\ \sigma_z & 0 \end{pmatrix}\begin{pmatrix} \sigma_z & 0 \\ 0 & \sigma_z \end{pmatrix}$$

$$= \begin{pmatrix} 0 & \sigma_z^2 \\ \sigma_z^2 & 0 \end{pmatrix} - \begin{pmatrix} 0 & \sigma_z^2 \\ \sigma_z^2 & 0 \end{pmatrix} = 0$$

Similarly,

$$[\sigma, \beta] = \left[\begin{pmatrix} \sigma_z & 0 \\ 0 & \sigma_z \end{pmatrix}, \begin{pmatrix} I & 0 \\ 0 & -I \end{pmatrix}\right] = 0$$

Hence,

$$[\sigma, H] = c[\sigma, \alpha]\, p_z + [\sigma, \beta]\, mc^2 = 0$$

As $[\sigma, H] = 0$, the two operators σ and H have common eigenfunctions σ is a diagonal matrix whose eigenfunction is

$$\begin{pmatrix} \psi_1 \\ \psi_2 \\ \psi_3 \\ \psi_4 \end{pmatrix}$$

$$\sigma \begin{pmatrix} \psi_1 \\ \psi_2 \\ \psi_3 \\ \psi_4 \end{pmatrix} = \begin{pmatrix} 1 & 0 & 0 & 0 \\ 0 & -1 & 0 & 0 \\ 0 & 0 & 1 & 0 \\ 0 & 0 & 0 & -1 \end{pmatrix} \begin{pmatrix} \psi_1 \\ \psi_2 \\ \psi_3 \\ \psi_4 \end{pmatrix} = \begin{pmatrix} \psi_1 \\ -\psi_2 \\ \psi_3 \\ -\psi_4 \end{pmatrix} = \begin{pmatrix} \psi_1 \\ 0 \\ \psi_3 \\ 0 \end{pmatrix} - \begin{pmatrix} 0 \\ \psi_2 \\ 0 \\ \psi_4 \end{pmatrix}$$

From the form of σ, it is obvious that

$$\begin{pmatrix} \psi_1 \\ 0 \\ \psi_3 \\ 0 \end{pmatrix} \quad \text{and} \quad \begin{pmatrix} 0 \\ \psi_2 \\ 0 \\ \psi_4 \end{pmatrix}$$

are the eigenfunctions of σ with the eigenvalues +1 and –1, respectively. Substituting these functions in the Dirac equation, we get

$$i\hbar \frac{\partial}{\partial t} \begin{pmatrix} \psi_1 \\ 0 \\ \psi_3 \\ 0 \end{pmatrix} = \left(-i\hbar c\alpha \frac{\partial}{\partial z} + \beta mc^2 + V \right) \begin{pmatrix} \psi_1 \\ 0 \\ \psi_3 \\ 0 \end{pmatrix}$$

$$i\hbar \frac{\partial}{\partial t} \begin{pmatrix} 0 \\ \psi_2 \\ 0 \\ \psi_4 \end{pmatrix} = \left(-i\hbar c\alpha \frac{\partial}{\partial z} + \beta mc^2 + V \right) \begin{pmatrix} 0 \\ \psi_2 \\ 0 \\ \psi_4 \end{pmatrix}$$

$$\alpha \begin{pmatrix} \partial\psi_1/\partial z \\ 0 \\ \partial\psi_3/\partial z \\ 0 \end{pmatrix} = \begin{pmatrix} 0 & 0 & 1 & 0 \\ 0 & 0 & 0 & -1 \\ 1 & 0 & 0 & 0 \\ 0 & -1 & 0 & 0 \end{pmatrix} \begin{pmatrix} \partial\psi_1/\partial z \\ 0 \\ \partial\psi_3/\partial z \\ 0 \end{pmatrix} = \frac{\partial}{\partial z} \begin{pmatrix} \psi_3 \\ 0 \\ \psi_1 \\ 0 \end{pmatrix}$$

$$\beta \begin{pmatrix} \psi_1 \\ 0 \\ \psi_3 \\ 0 \end{pmatrix} = \begin{pmatrix} 1 & 0 & 0 & 0 \\ 0 & 1 & 0 & 0 \\ 0 & 0 & -1 & 0 \\ 0 & 0 & 0 & -1 \end{pmatrix} \begin{pmatrix} \psi_1 \\ 0 \\ \psi_3 \\ 0 \end{pmatrix} = \begin{pmatrix} \psi_1 \\ 0 \\ -\psi_3 \\ 0 \end{pmatrix}$$

Similarly,

$$\alpha \begin{pmatrix} 0 \\ \partial\psi_2/\partial z \\ 0 \\ \partial\psi_4/\partial z \end{pmatrix} = \begin{pmatrix} 0 & 0 & 1 & 0 \\ 0 & 0 & 0 & -1 \\ 1 & 0 & 0 & 0 \\ 0 & -1 & 0 & 0 \end{pmatrix} \begin{pmatrix} 0 \\ \partial\psi_2/\partial z \\ 0 \\ \partial\psi_4/\partial z \end{pmatrix} = \frac{\partial}{\partial z} \begin{pmatrix} 0 \\ -\psi_4 \\ 0 \\ -\psi_2 \end{pmatrix}$$

Substituting this equation in the Dirac equations, we have

$$i\hbar \frac{\partial}{\partial t} \begin{pmatrix} \psi_1 \\ 0 \\ \psi_3 \\ 0 \end{pmatrix} = -i\hbar c \frac{\partial}{\partial z} \begin{pmatrix} \psi_3 \\ 0 \\ \psi_1 \\ 0 \end{pmatrix} + mc^2 \begin{pmatrix} \psi_1 \\ 0 \\ -\psi_3 \\ 0 \end{pmatrix} + V(z) \begin{pmatrix} \psi_1 \\ 0 \\ \psi_3 \\ 0 \end{pmatrix}$$

$$i\hbar \frac{\partial}{\partial t} \begin{pmatrix} 0 \\ \psi_2 \\ 0 \\ \psi_4 \end{pmatrix} = -i\hbar c \frac{\partial}{\partial z} \begin{pmatrix} 0 \\ -\psi_4 \\ 0 \\ -\psi_2 \end{pmatrix} + mc^2 \begin{pmatrix} 0 \\ -\psi_2 \\ 0 \\ -\psi_4 \end{pmatrix} + V(z) \begin{pmatrix} 0 \\ \psi_2 \\ 0 \\ \psi_4 \end{pmatrix}$$

Each of these two equations represents two coupled differential equations.

13.8 For a Dirac particle moving in a central potential, show that the orbital angular momentum is not a constant of motion.

Solution. In the Heisenberg picture, the time rate of change of the $\boldsymbol{L} = \boldsymbol{r} \times \boldsymbol{p}$ is given by

$$i\hbar \frac{d\boldsymbol{L}}{dt} = [L, H]$$

Its x-component is

$$i\hbar \frac{d}{dt} L_x = [L_x, H] = [yp_z - zp_y, c\boldsymbol{\alpha} \cdot \boldsymbol{p} + \beta mc^2]$$

Since $\boldsymbol{\alpha}$ and β commute with $\boldsymbol{r}$ and $\boldsymbol{p}$,

$$\begin{aligned} i\hbar \frac{d}{dt} L_x &= [yp_z, c\alpha_y \boldsymbol{p}_y] - [zp_y, c\alpha_z \boldsymbol{p}_z] \\ &= c[y, p_y]\, p_z\alpha_y - c[z, p_z]\, p_y\alpha_z \\ &= ci\hbar p_z\alpha_y - ci\hbar p_y\alpha_z \\ &= ic\hbar (p_z\alpha - p_y\alpha_z) \end{aligned}$$

which shows that L_x is not a constant of motion. Similar relations hold good for L_y and L_z components. Hence the orbital angular momentum $\boldsymbol{L}$ is not a constant of motion.

13.9 Prove that the quatity $\boldsymbol{L} + (1/2)\hbar\boldsymbol{\sigma}'$, where L is the orbital angular momentum of a particle, and $\boldsymbol{\sigma}' = \begin{pmatrix} \sigma' & 0 \\ 0 & \sigma' \end{pmatrix}$ is a constant of motion for the particle in Dirac's formalism. Hence give an interpretation for the additional angular momentum $1/2\ \hbar\boldsymbol{\sigma}'$.

Solution. In Dirac's formalism, the Hamiltonian of a free particle is

$$H = c\boldsymbol{\alpha} \cdot \boldsymbol{p} + \beta mc^2 \tag{i}$$

In the Heisenberg picture, the equation of motion for an operator $\boldsymbol{M}$ is given by

$$i\hbar \frac{d}{dt} \boldsymbol{M} = [\boldsymbol{M}, H] \tag{ii}$$

Hence, for a dynamical variable to be a constant of motion, it should commute with its Hamiltonian. Writing

$$\boldsymbol{M} = \boldsymbol{L} + \frac{1}{2}\hbar\boldsymbol{\sigma}' \tag{iii}$$

where equation of motion is

$$i\hbar \frac{d}{dt}\left(\boldsymbol{L} + \frac{1}{2}\hbar\boldsymbol{\sigma}'\right) = \left[\boldsymbol{L} + \frac{1}{2}\hbar\boldsymbol{\sigma}', c\boldsymbol{\alpha} \cdot \boldsymbol{p} + \beta mc^2\right] \tag{iv}$$

The x-component of Eq. (iv) is

$$i\hbar \frac{d}{dt}\left(L_x + \frac{1}{2}\hbar\sigma'_x\right) = \left[L_x + \frac{1}{2}\hbar\sigma'_x, c\boldsymbol{\alpha} \cdot \boldsymbol{p} + \beta mc^2\right]$$

$$= [L_x, c\boldsymbol{\alpha} \cdot \boldsymbol{p} + \beta mc^2] + \frac{1}{2}\hbar\,[\sigma'_x, c\boldsymbol{\alpha} \cdot \boldsymbol{p} + \beta mc^2] \tag{v}$$

Let us now evaluate the commutators on the right side of (v) one by one

$$[L_x, c\boldsymbol{\alpha} \cdot \boldsymbol{p} + \beta mc^2] = [yp_z - zp_y, c\alpha_x p_x + c\alpha_y p_y + c\alpha_z p_z + \beta mc^2]$$

Since α and β commute with $\boldsymbol{r}$ and $\boldsymbol{p}$,

$$[L_x, c\boldsymbol{\alpha} \cdot \boldsymbol{p} + \beta mc^2] = [yp_z, c\alpha_y p_y] - [zp_z, c\alpha_z p_z]$$

$$= c[y, p_y]p_z\alpha_y - c[z, p_z]p_y\alpha_z$$

$$= ic\hbar\,(\alpha_y p_z - \alpha_z p_y) \tag{vi}$$

The second commutator in Eq. (v) is

$$[\sigma'_x, c\boldsymbol{\alpha} \cdot \boldsymbol{p} + \beta mc^2] = [\sigma'_x, c\alpha_x p_x + c\alpha_y p_y + c\alpha_z p_z + \beta mc^2]$$

$$= [\sigma'_x, c\alpha_x p_x] + [\sigma'_x, c\alpha_y p_y] + [\sigma'_x, c\alpha_z p_z] + [\sigma'_x, \beta mc^2]$$

From Problem 13, we have

$$[\sigma'_x, \beta] = 0, \qquad [\sigma'_x, \alpha_x] = 0, \qquad [\sigma'_x, \alpha_y] = 2i\alpha_z, \qquad [\sigma'_x, \alpha_z] = -2i\alpha_y$$

Substituting these commutators in the above equation, we get

$$[\sigma'_x, c\boldsymbol{\alpha} \cdot \boldsymbol{p} + \beta mc^2] = c\,[\sigma'_x, \alpha_y]\,p_y + c\,[\sigma'_x, \alpha_z]\,p_z$$

$$= 2ic\alpha_z p_y - 2ic\alpha_y p_z \tag{vii}$$

From Eqs. (v)–(vii),

$$i\hbar \frac{d}{dt}\left(L_x + \frac{1}{2}\hbar\sigma'_x\right) = ic\hbar(\alpha_y p_z - \alpha_z p_y) + \frac{1}{2}\hbar \times 2ic(\alpha_z p_y - \alpha_y p_z) \tag{viii}$$

$$= 0$$

$$L_x + \frac{1}{2}\hbar\sigma'_x = \text{constant} \tag{ix}$$

Similar relations are obtained for the y- and z-components. Hence,

$$\boldsymbol{L} + \frac{1}{2}\hbar\boldsymbol{\sigma}' = \text{constant} \tag{x}$$

From the structure of the σ' matrix, we can write

$$\sigma_x'^2 = \sigma_y'^2 = \sigma_z'^2 = 1$$

This gives the eigenvalues of $\frac{1}{2}\hbar\boldsymbol{\sigma}'$ as $+\frac{1}{2}\hbar$ or $-\frac{1}{2}\hbar$. Thus, the additional angular momentum $\frac{1}{2}\hbar\boldsymbol{\sigma}'$ can be interpreted as the spin angular momentum, i.e.,

$$\boldsymbol{S} = \frac{1}{2}\hbar\begin{pmatrix} \boldsymbol{\sigma} & 0 \\ 0 & \boldsymbol{\sigma} \end{pmatrix}$$

13.10 If the radial momentum p_r and radial velocity α_r for an electron in a central potential are defined by

$$p_r = \frac{\boldsymbol{r}\cdot\boldsymbol{p} - i\hbar}{r}, \qquad \alpha_r = \frac{\boldsymbol{\alpha}\cdot\boldsymbol{r}}{r}$$

show that

$$(\boldsymbol{\alpha}\cdot\boldsymbol{p}) = \alpha_r\, p_r + \frac{i\hbar k\beta\alpha_r}{r}$$

where $k = \dfrac{\beta(\boldsymbol{\sigma}'\cdot\boldsymbol{L} + \hbar)}{\hbar}$.

Solution. The relativistic Hamiltonian of an electron in a central potential $V(r)$ is given by

$$H = c(\boldsymbol{\alpha}\cdot\boldsymbol{p}) + \beta mc^2 + V(r)$$

If $\boldsymbol{A}$ and $\mathbf{B}$ are operators, then

$$(\boldsymbol{a}\cdot\boldsymbol{A})(\boldsymbol{a}\cdot\boldsymbol{B}) = (\boldsymbol{A}\cdot\boldsymbol{B}) + \boldsymbol{\sigma}'\cdot(\boldsymbol{A}\times\boldsymbol{B})$$

Setting $\boldsymbol{A} = \boldsymbol{B} = \boldsymbol{r}$, we have $(\boldsymbol{a}\cdot\boldsymbol{r})^2 = r^2$. Taking $\boldsymbol{A} = \boldsymbol{r}$ and $\boldsymbol{B} = \boldsymbol{p}$, we get

$$(\boldsymbol{a}\cdot\boldsymbol{r})(\boldsymbol{a}\cdot\boldsymbol{p}) = (\boldsymbol{r}\cdot\boldsymbol{p}) + i\boldsymbol{\sigma}'\cdot\boldsymbol{L}$$

Given

$$k = \frac{\beta[(\boldsymbol{\sigma}'\cdot\boldsymbol{L}) + \hbar]}{\hbar} \quad \text{or} \quad \boldsymbol{\sigma}'\cdot\boldsymbol{L} = \frac{k\hbar}{\beta} - \hbar$$

Substituting this value of $\sigma' \cdot L$ and multiplying by $\alpha \cdot r$, we obtain

$$(\boldsymbol{a} \cdot \boldsymbol{r})^2 (\boldsymbol{a} \cdot \boldsymbol{p}) = (\boldsymbol{a} \cdot \boldsymbol{r}) \left[(\boldsymbol{r} \cdot \boldsymbol{p}) + i\left(\frac{k\hbar}{\beta} - \hbar \right) \right]$$

Since

$$(\boldsymbol{a} \cdot \boldsymbol{r})^2 = r^2$$

we have

$$\boldsymbol{a} \cdot \boldsymbol{p} = \frac{\boldsymbol{a} \cdot \boldsymbol{r}}{r^2} \left[(\boldsymbol{r} \cdot \boldsymbol{p}) - i\hbar + \frac{ik\hbar}{\beta} \right] = \frac{\boldsymbol{a} \cdot \boldsymbol{r}}{r} \left[\frac{(\boldsymbol{r} \cdot \boldsymbol{p}) - i\hbar}{r} + \frac{ik\hbar}{\beta r} \right]$$

Using the definitions of p_r and α_r, we get

$$\alpha \cdot \boldsymbol{p} = \alpha_r p_r + \frac{i\hbar k \alpha_r}{\beta r} = \alpha_r p_r + \frac{i\hbar k \beta^2 \alpha_r}{\beta r}$$

$$= \alpha_r p_r + \frac{i\hbar k \beta \alpha_r}{r}$$

13.11 If one wants to write the relativistic energy E of a free particle as

$$\frac{E^2}{c^2} = (\alpha \cdot \boldsymbol{p} + \beta mc)^2,$$

show that α's and β's have to be matrices and establish that they are nonsingular and Hermitian.

Solution. The relativistic energy (E) of a free particle is given by

$$E^2 = c^2p^2 + m^2c^4 = c^2(p^2 + m^2c^2)$$

When E^2/c^2 is written as given in the problem,

$$p^2 + m^2c^2 = (\alpha \cdot \boldsymbol{p} + \beta mc)^2 = \alpha_x^2 p_x^2 + \alpha_y^2 p_y^2 + \alpha_z^2 p_z^2$$

$$+ \beta^2 m^2 c^2 + (\alpha_x \alpha_y + \alpha_y \alpha_x)\, p_x p_y + (\alpha_x \alpha_z + \alpha_z \alpha_x)\, p_x p_z$$

$$+ (\alpha_y \alpha_z + \alpha_z \alpha_y)\, p_y p_z + (\alpha_x \beta + \beta \alpha_x)\, mcp_x$$

$$+ (\alpha_y \beta + \beta \alpha_y)\, mcp_y + (\alpha_z \beta + \beta \alpha_z)\, mcp_z$$

For this equation to be valid, it is necessary that

$$\alpha_x^2 = \alpha_y^2 = \alpha_z^2 = \beta^2 = 1, \qquad [\alpha_x, \alpha_y]_+ = 0, [\alpha_y, \alpha_z]_+ = 0$$

$$[\alpha_x, \alpha_z]_+ = 0, [\alpha_x, \beta]_+ = 0, \qquad [\alpha_y, \beta]_+ = 0, [\alpha_z, \beta]_+ = 0$$

It is obvious that the α's and β cannot be ordinary numbers. The anticommuting nature of the α's and β suggests that they have to be matrices. Since the squares of these matrices are unit matrices, they are nonsingular. As the α's and β determine the Hamiltonian, they must be Hermitian.

13.12 If $\sigma' = \begin{pmatrix} \sigma & 0 \\ 0 & \sigma \end{pmatrix}$, show that

(i) $\sigma_x'^2 = \sigma_y'^2 = \sigma_z'^2 = 1.$

(ii) $[\sigma_x', \alpha_x] = 0, \quad [\sigma_x', \alpha_y] = 2i\alpha_z, \quad [\sigma_x', \alpha_z] = -2i\alpha_y,$

where σ is the Pauli matrix and α_x, α_y, α_z are the Dirac matrices.

Solution.

$$\sigma' = \begin{pmatrix} \sigma & 0 \\ 0 & \sigma \end{pmatrix}, \qquad \sigma_x' = \begin{pmatrix} \sigma_x & 0 \\ 0 & \sigma_x \end{pmatrix}$$

(i) $$\sigma_x'^2 = \begin{pmatrix} \sigma_x & 0 \\ 0 & \sigma_x \end{pmatrix}\begin{pmatrix} \sigma_x & 0 \\ 0 & \sigma_x \end{pmatrix} = \begin{pmatrix} \sigma_x^2 & 0 \\ 0 & \sigma_x^2 \end{pmatrix} = \begin{pmatrix} 1 & 0 \\ 0 & 1 \end{pmatrix}$$

A similar procedure gives the values of $\sigma_y'^2$ and $\sigma_z'^2$. Hence the result

(ii) $$[\sigma_x', \alpha_x] = \begin{pmatrix} \sigma_x & 0 \\ 0 & \sigma_x \end{pmatrix}\begin{pmatrix} 0 & \sigma_x \\ \sigma_x & 0 \end{pmatrix} - \begin{pmatrix} 0 & \sigma_x \\ \sigma_x & 0 \end{pmatrix}\begin{pmatrix} \sigma_x & 0 \\ 0 & \sigma_x \end{pmatrix}$$

$$= \begin{pmatrix} 0 & \sigma_x^2 \\ \sigma_x^2 & 0 \end{pmatrix} - \begin{pmatrix} 0 & \sigma_x^2 \\ \sigma_x^2 & 0 \end{pmatrix} = 0$$

$$[\sigma_x', \alpha_y] = \begin{pmatrix} \sigma_x & 0 \\ 0 & \sigma_x \end{pmatrix}\begin{pmatrix} 0 & \sigma_y \\ \sigma_y & 0 \end{pmatrix} - \begin{pmatrix} 0 & \sigma_y \\ \sigma_y & 0 \end{pmatrix}\begin{pmatrix} \sigma_x & 0 \\ 0 & \sigma_x \end{pmatrix}$$

$$\begin{pmatrix} 0 & \sigma_x\sigma_y \\ \sigma_x\sigma_y & 0 \end{pmatrix} - \begin{pmatrix} 0 & \sigma_y\sigma_x \\ \sigma_y\sigma_x & 0 \end{pmatrix} = \begin{pmatrix} 0 & \sigma_x\sigma_y - \sigma_y\sigma_x \\ \sigma_x\sigma_y - \sigma_y\sigma_x & 0 \end{pmatrix}$$

$$= \begin{pmatrix} 0 & 2i\sigma_z \\ 2i\sigma_z & 0 \end{pmatrix} = 2i\alpha_z$$

Proof of the other relation is straightforward.

13.13 Show that matrix $\sigma' = \begin{pmatrix} \sigma & 0 \\ 0 & \sigma \end{pmatrix}$ is not a constant of motion.

Solution. The equation of motion of σ' in the Heisenberg picture is

$$i\hbar \frac{d\sigma'}{dt} = [\sigma', H]$$

Hence for σ' to be a constant of motion, σ_x', σ_y' and σ_z' should commute with the Hamiltonian. Thus,

$$[\sigma_x', H] = [\sigma_x', c\boldsymbol{\alpha} \cdot \boldsymbol{p} + \beta mc^2]$$

Since σ'_x commutes with β,

$$[\sigma'_x, H] = [\sigma'_x, c\alpha_x p_x] + [\sigma'_x, c\alpha_y p_y] + [\sigma'_x, c\alpha_z p_z]$$

From Problem 13.12,

$$[\sigma'_x, \alpha_x] = 0, \qquad [\sigma'_x, \alpha_y] = 2i\alpha_z, \qquad [\sigma'_x, \alpha_z] = -2i\alpha_y$$

$$[\sigma'_x, H] = 2ic\,(\alpha_z p_y - \alpha_y p_z) \neq 0$$

Hence the result.

13.14 Show that Dirac's Hamiltonian for a free particle commutes with the operator $\boldsymbol{\sigma} \cdot \boldsymbol{p}$, where $\boldsymbol{p}$ is the momentum operator and $\boldsymbol{\sigma}$ is the Pauli spin operator in the space of four component spinors.

Solution. Dirac's Hamiltonian for a free particle is

$$H = c\,(\boldsymbol{\alpha} \cdot \boldsymbol{p}) + \beta mc^2$$

where

$$\alpha = \begin{pmatrix} 0 & \sigma \\ \sigma & 0 \end{pmatrix}, \qquad \beta = \begin{pmatrix} I & 0 \\ 0 & -I \end{pmatrix}$$

$$\alpha \cdot \boldsymbol{p} = \begin{pmatrix} 0 & \sigma \\ \sigma & 0 \end{pmatrix} \cdot \boldsymbol{p} = \begin{pmatrix} 0 & \sigma \cdot \boldsymbol{p} \\ \sigma \cdot \boldsymbol{p} & 0 \end{pmatrix}$$

$$\sigma \cdot \boldsymbol{p} = \sigma \cdot \boldsymbol{p} I = \begin{pmatrix} \sigma \cdot \boldsymbol{p} & 0 \\ 0 & \sigma \cdot \boldsymbol{p} \end{pmatrix}$$

$$\begin{aligned}
[\sigma \cdot \boldsymbol{p}, H] &= [\sigma \cdot \boldsymbol{p},\, c\alpha \cdot \boldsymbol{p} + \beta mc^2] \\
&= c\,[(\sigma \cdot \boldsymbol{p}),\, \alpha \cdot \boldsymbol{p}] + [\sigma \cdot \boldsymbol{p},\, \beta mc^2] \\
&= c\left[\begin{pmatrix} \sigma \cdot \boldsymbol{p} & 0 \\ 0 & \sigma \cdot \boldsymbol{p} \end{pmatrix}, \begin{pmatrix} 0 & \sigma \cdot \boldsymbol{p} \\ \sigma \cdot \boldsymbol{p} & 0 \end{pmatrix}\right] + mc^2\left[\begin{pmatrix} \sigma \cdot \boldsymbol{p} & 0 \\ 0 & \sigma \cdot \boldsymbol{p} \end{pmatrix}, \begin{pmatrix} 1 & 0 \\ 0 & -1 \end{pmatrix}\right] \\
&= 0 + 0 = 0
\end{aligned}$$

Hence the result.

CHAPTER 14

Chemical Bonding

With the advent of quantum mechanics, elegant methods were developed to study the mechanism that holds the atoms together in molecules. The molecular orbital (MO) and valence bond (VB) methods are the two commonly used methods. Recent computational works mainly use the MO methods.

14.1 Born–Oppenheimer Approximation

In molecules, one has to deal with not only the moving electrons but also the moving nuclei. Born and Oppenheimer assumed the nuclei as stationary and in such a case, the Hamiltonian representing the electronic motion is

$$H = -\frac{\hbar^2}{2m}\sum_i \nabla_i^2 - \sum_\alpha \sum_i \frac{kz_\alpha e^2}{r_{i\alpha}} + \sum_i \sum_{j>i} \frac{ke^2}{r_{ij}} + \sum_\alpha \sum_{\beta>\alpha} \frac{kz_\alpha z_\beta e^2}{r_{\alpha\beta}} \tag{14.1}$$

where i, j refer to electrons, α, β to nuclei and $k = 1/(4\pi\varepsilon_0)$.

14.2 Molecular Orbital and Valence Bond Methods

In the molecular orbital method, developed by Mulliken, molecular wavefunctions, called **molecular orbitals**, are derived first. In the commonly used approach, the molecular orbital ψ is written as a linear combination of the atomic orbitals (LCAO) as

$$\psi = c_1\psi_1 + c_2\psi_2 + \ldots \tag{14.2}$$

where ψ_1, ψ_2, ... are the individual atomic orbitals. The constants c_1, c_2, ... are to be selected in such a way that the energy given by ψ is minimum.

In the valence bond approach, atoms are assumed to maintain their individual identity in a molecule and the bond arises due to the interaction of the valence electrons. That is, a bond is formed when a valence electron in an atomic orbital pairs its spin with that of another valence electron in the other atomic orbital.

14.3 Hydrogen Molecule-ion

Hydrogen molecule-ion consists of an electron of charge $-e$ associated with two protons a and b separated by a distance R (see Fig. 14.1). The electron's atomic orbital, when it is in the neighbourhood of a is

$$\psi_a = \left(\frac{1}{\pi a_0^3}\right)^{1/2} \exp\left(\frac{-r_a}{a_0}\right) \tag{14.3}$$

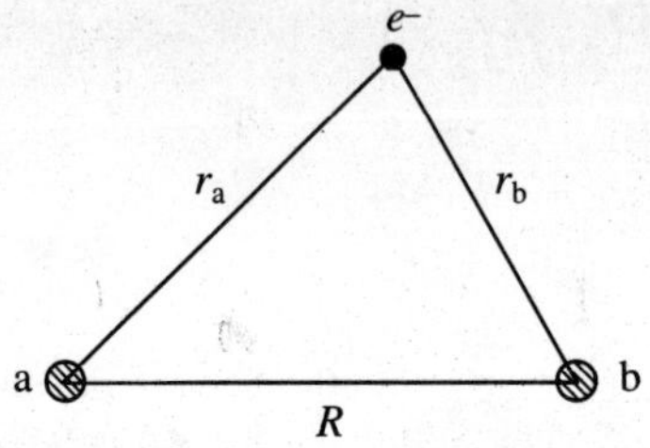

Fig. 14.1 The H_2^+ molecule.

and when it is in the neighbourhood of b, it is

$$\psi_b = \left(\frac{1}{\pi a_0^3}\right)^{1/2} \exp\left(\frac{-r_b}{a_0}\right) \tag{14.4}$$

A reasonable MO will be

$$\psi = c_1\psi_a + c_2\psi_a \tag{14.5}$$

where c_1 and c_2 are constants. Then the energy E of the system is given by

$$E = \frac{\langle\psi|H|\psi\rangle}{\langle\psi|\psi\rangle} \tag{14.6}$$

Substituting the value of ψ and simplifying, we get the energies as

$$E_1 = E_H - \frac{V_{aa} + V_{ab}}{1 + S} + \frac{ke^2}{R} \tag{14.7}$$

$$E_2 = E_H - \frac{V_{aa} - V_{ab}}{1 - S} + \frac{ke^2}{R} \tag{14.8}$$

where

$$V_{aa} = \left\langle \psi_a \left| \frac{ke^2}{r_b} \right| \psi_a \right\rangle, \qquad V_{ab} = \left\langle \psi_a \left| \frac{ke^2}{r_a} \right| \psi_b \right\rangle \tag{14.9}$$

$$S = \langle\psi_a|\psi_b\rangle = \langle\psi_b|\psi_a\rangle \tag{14.10}$$

The normalized wavefunctions corresponding to these energies are

$$\psi_1 = \frac{\psi_a + \psi_b}{\sqrt{2 + 2S}}, \qquad \psi_2 = \frac{\psi_a - \psi_b}{\sqrt{2 - 2S}} \tag{14.11}$$

The wavefunction ψ_1 corresponds to a build-up of electron density between the two nuclei and is therefore called a **bonding molecular orbital**. The wavefunction ψ_2 is called an **antibonding orbital** since it corresponds to a depletion of charge between the nuclei.

14.4 MO Treatment of Hydrogen Molecule

In MO theory the treatment of hydrogen molecule is essentially the same as that of H_2^+ molecule. One can reasonably take that in the ground state both the electrons occupy the bonding orbital ψ_1 (Eq. 14.1) of H_2^+ which is symmetric with respect to interchange of nuclei. The trial wave function of H_2 molecule can then be taken as

$$\psi_{mo} = \psi_1(1)\,\psi_1(2) = \frac{[\psi_a(1) + \psi_b(1)]\,[\psi_a(2) + \psi_b(2)]}{2(1 + S)} \tag{14.12}$$

With this wave function, the energy is calculated.

14.5 Diatomic Molecular Orbitals

Figure 14.2 illustrates the formation of bonding and antibonding orbitals from two 1s atomic orbitals. Both are symmetrical about the internuclear axis. Molecular orbitals which are symmetrical about the internuclear axis are designated by σ (sigma) bond, and those which are not symmetrical about the internuclear axis are designated by π (pi) bond. The bonding orbital discussed is represented by the symbol 1sσ since it is produced from two 1s atomic orbitals. The antibonding state is represented by the symbol 1sσ*, the asterisk representing higher energy.

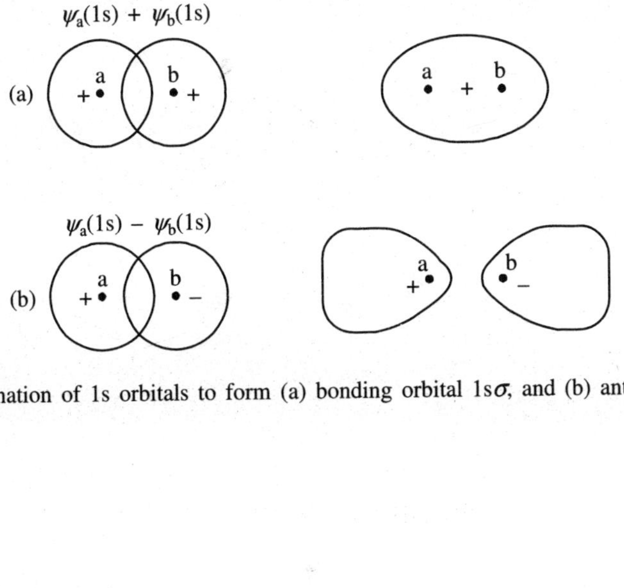

Fig. 14.2 Combination of 1s orbitals to form (a) bonding orbital 1sσ, and (b) antibonding orbital 1sσ*.

If an inversion of a molecular orbital about the centre of symmetry does not change the sign of ψ, it is said to be even and is denoted by the symbol g as a subscript. If the sign changes, the orbital is said to be odd and a subscript u is assigned to the symbol. In this notation, the bonding and antibonding orbitals are respectively denoted by $1s\sigma_g$ and $1s\sigma_u^*$. Two 2s atomic orbitals combine to form again a bonding $2s\sigma_g$ and an antibonding $2s\sigma_u^*$ molecular orbitals. The terminology followed for labelling MOs in the increasing order of energy is

$$1s\sigma < 1s\sigma^* < 2s\sigma < 2s\sigma^* < 2p_x\sigma < (2p_y\pi = 2p_z\pi) < (2p_y\pi^* = 2p_z\pi^*) < 2p_x\sigma^* \quad (14.13)$$

PROBLEMS

14.1 Illustrate, with the help of diagrams the combination of two *p*-orbitals, bringing out the formation of bonding σ_g, antibonding σ_u^*, bonding π_u and antibonding π_u^* orbitals.

Solution. The two lobes of each of the p-orbitals have opposite signs. If the internuclear axis is taken as the *x*-direction, two p_x atomic orbitals combine to give the molecular orbitals $2p_x\sigma_g$ and $2p_x\sigma_u^*$, which is illustrated in Fig. 14.3 Both have symmetry about the bond axis. The combination

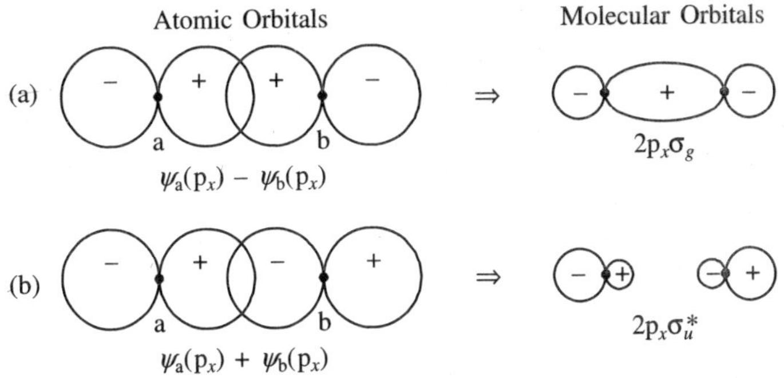

Fig. 14.3 Formation of (a) bonding orbital $2p_x\sigma_g$, and (b) antibonding $2p_x\sigma_u^*$ molecular orbitals from two p_x orbitals.

of two p_y orbitals gives the molecular orbitals $2p_y\pi_u$ and $2p_y\pi_g^*$, see Fig. 14.4. The $p_y\pi_u$ MO consists of two streamers, one above and one below the nuclei. In this case, the bonding orbital is odd and the antibonding orbital is even, unlike the earlier ones. Formation of π molecular orbitals from atomic p_z orbitals is similar to the one from atomic p_y orbitals.

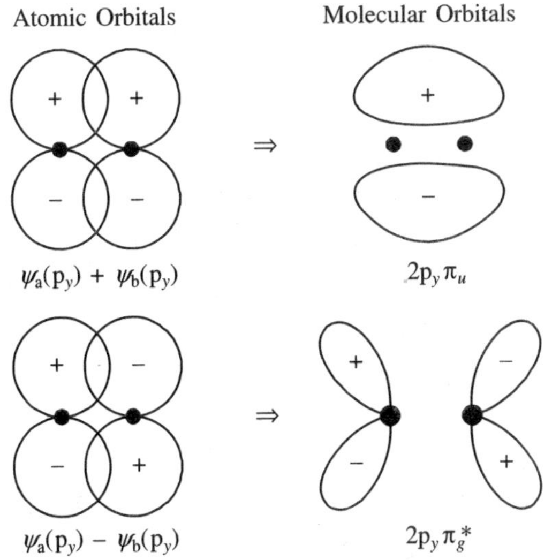

Fig. 14.4 The formation of (a) bonding orbital $2p_y\pi_u$, and (b) antibonding $2p_y\pi_g^*$ from two $2p_y$ orbitals.

14.2 Outline the Heitler-London wavefunctions for hydrogen molecule. What are singlet and triplet states of hydrogen?

Solution. Hydrogen molecule is a system of two hydrogen atoms and, therefore, can be described by the wave function

$$\psi(1, 2) = \psi_a(1)\ \psi_b(2) \tag{i}$$

where a and b refer to the two nuclei, 1 and 2 to the two electrons. The function $\psi_a(1)\ \psi_b(2)$ means electron 1 is associated with the atom whose nucleus is a and electron 2 is associated with the atom whose nucleus is b. The electrons are indistinguishable. Hence,

$$\psi(2, 1) = \psi_a(2)\ \psi_b(1) \tag{ii}$$

is also a wave function. The wave function of the two-electron system is a linear combination of the two.

Since an exchange of electron 1 and electron 2 leaves the Hamiltonian of the system unchanged, the wavefunctions must either be symmetric or antisymmetric with respect to such an exchange. The symmetric ψ_s and antisymmetric ψ_{as} combinations are

$$\psi_s = N_s\,[\psi_a(1)\ \psi_b(2) + \psi_a(2)\ \psi_b(1)] \tag{iii}$$

$$\psi_{as} = N_{as}\,[\psi_a(1)\ \psi_b(2) - \psi_a(2)\ \psi_b(1)] \tag{iv}$$

where N_s and N_{as} are normalization constants. The spin functions of a two-spin half system is given by

$$\chi_{as} = \frac{1}{\sqrt{2}}\,[\alpha(1)\ \beta(2) - \beta(1)\ \alpha(2)] \tag{v}$$

$$\chi_s = \begin{cases} \alpha(1)\ \alpha(2) \\ \dfrac{1}{\sqrt{2}}\,[\alpha(1)\ \beta(2) + \beta(1)\ \alpha(2)] \\ \beta(1)\ \beta(2) \end{cases} \tag{vi}$$

As the total wave function has to be antisymmetric, the symmetric space part combines with the antisymmetric spin part and vice versa. Hence, the inclusion of electron spin leads to the Heitler-London wave functions

$$N_s\,[\psi_a(1)\,\psi_b(2) + \psi_a(2)\,\psi_b(1)]\frac{1}{\sqrt{2}}\,[\alpha(1)\ \beta(2) - \beta(1)\ \alpha(2)] \tag{vii}$$

$$N_{as}\,[\psi_a(1)\,\psi_b(2) - \psi_a(2)\,\psi_b(1)] = \begin{cases} \alpha(1)\ \alpha(2) \\ \dfrac{1}{\sqrt{2}}\,[\alpha(1)\ \beta(2) + \beta(1)\ \alpha(2)] \\ \beta(1)\ \beta(2) \end{cases} \tag{viii}$$

Equation (vii) corresponds to a singlet state since $S = 0$, whereas Eq. (viii) is a triplet state as $S = 1$.

14.3 In the hydrogen molecule ion, the wave functions corresponding to energy E_1 and E_2 are $\psi_1 = c_1(\psi_a + \psi_b)$ and $\psi_2 = c_2(\psi_a - \psi_b)$, where ψ_a and ψ_b are hydrogenic wave functions. Normalize the functions. What will be the normalization factor if the two nuclei are at infinite distance?

Solution. Given

$$\psi_1 = c_1(\psi_a + \psi_b), \qquad \psi_2 = c_2(\psi_a - \psi_b)$$

The normalization of ψ_1 gives

$$|c_1|^2 \langle(\psi_a + \psi_b)|(\psi_a + \psi_b)\rangle = 1$$

$$|c_1|^2 [\langle\psi_a|\psi_a\rangle + \langle\psi_b|\psi_b\rangle + \langle\psi_a|\psi_b\rangle + \langle\psi_b|\psi_a\rangle] = 1$$

Writing $\langle\psi_a|\psi_b\rangle = \langle\psi_b|\psi_a\rangle$, refer Eq. (14.10), we get

$$c_1^2 [1 + 1 + S + S] = 1$$

$$c_1 = \frac{1}{\sqrt{2 + 2S}}, \qquad \psi_1 = \frac{\psi_a + \psi_b}{\sqrt{2 + 2S}}$$

Normalization of ψ_2 gives

$$|c_2|^2 [\langle\psi_a|\psi_a\rangle + \langle\psi_b|\psi_b\rangle - \langle\psi_a|\psi_b\rangle - \langle\psi_b|\psi_a\rangle] = 1$$

$$c_2 = \frac{1}{\sqrt{2 - 2S}}, \qquad \psi_2 = \frac{\psi_a - \psi_b}{\sqrt{2 - 2S}}$$

When the two nuclei are at infinite distance, the overlap integral $\langle\psi_a|\psi_b\rangle = \langle\psi_b|\psi_a\rangle = 0$. Hence the normalization factor for both ψ_1 and ψ_2 is $1/\sqrt{2}$.

14.4 The Heitler-London wave functions for hydrogen molecule are

$$\psi_s = N_s [\psi_a(1)\, \psi_b(2) + \psi_a(2)\, \psi_b(1)]$$

$$\psi_{as} = N_a [\psi_a(1)\, \psi_b(2) - \psi_a(2)\, \psi_b(1)]$$

Evaluate the normalization constants N_s and N_a. What will be the normalization factor if the nuclear separation is infinite.

Solution. The normalization condition of the symmetric Heitler-London trial function gives

$$|N_s|^2 \langle[\psi_a(1)\, \psi_b(2) + \psi_a(2)\, \psi_b(1)]|[\psi_a(1)\, \psi_b(2) + \psi_a(2)\, \psi_b(1)]\rangle = 1$$

$$|N_s|^2 [\langle\psi_a(1)\, \psi_b(2)|\psi_a(1)\, \psi_b(2)\rangle + \langle\psi_a(1)\, \psi_b(2)|\psi_a(2)\, \psi_b(1)\rangle$$

$$+ \langle\psi_a(2)\, \psi_b(1)|\psi_a(1)\, \psi_b(2)\rangle + \langle\psi_a(2)\, \psi_b(1)|\psi_a(2)\, \psi_b(1)\rangle] = 1$$

$$|N_s|^2 [1 + S^2 + S^2 + 1] = 1, \qquad N_s = \frac{1}{\sqrt{2 + 2S^2}}$$

since

$$\langle\psi_a(1)|\,\psi_a(1)\rangle = \langle\psi_b(2)|\,\psi_b(2)\rangle = \langle\psi_a(2)|\,\psi_a(2)\rangle = \langle\psi_b(1)|\,\psi_b(1)\rangle = 1$$

$$\langle\psi_a(1)\psi_b(2)|\,\psi_a(2)\psi_b(1)\rangle = \langle\psi_a(1)|\,\psi_b(1)\rangle\langle\psi_b(2)|\,\psi_a(2)\rangle = S \cdot S = S^2$$

Similarly,

$$N_a = \frac{1}{\sqrt{2 - 2S^2}}$$

For infinite nuclear separation, $S = 0$, $N_s = N_a = 1/\sqrt{2}$.

14.5 Write the electronic configuration of N_2 molecule in the MO concept and explain the formation of the triple bond $N \equiv N$.

Solution. The 14 electrons in the nitrogen molecule are distributed as

$$\text{KK}\,(2\text{s}\sigma_g)^2(2\text{s}\sigma_u^*)^2(2\text{p}_x\sigma_g)^2(2\text{p}_y\pi = 2\text{p}_z\pi)^4$$

The presence of two electrons in the bonding orbital $2s\sigma_g$ and two electrons in the antibondiong $2s\sigma_u^*$ leads to no bonding. The remaining bonding orbitals $(2p_x\sigma_g)^2$ $(2p_y\pi = 2p_z\pi)^4$ are not cancelled by the corresponding antibonding orbitals. These six bonding electrons give the triple bond $N \equiv N$, one bond being σ and the other two are π bonds.

14.6 Write the electronic configuration of O_2 and S_2 and account for their paramagnetism.

Solution. The sixteen electrons in the O_2 molecule are distributed as

$$\text{KK}\,(2\text{s}\sigma_g)^2(2\text{s}\sigma_u^*)^2(2\text{p}\sigma_g)^2(2\text{p}\pi_u)^4(2\text{p}\pi_g^*)^2$$

where KK stands for $(1\text{s}\sigma_g)^2(1\text{s}\sigma_u^*)^2$. The orbital $2\text{p}\pi_g^*$ is degenerate. Hence the two electrons in that antibonding orbital will go one each with parallel spins (Hund's rule). Since the last two electrons are with parallel spins, the net spin is one and the molecule is paramagnetic.

The electronic configuration of S = $1s^2\ 2s^2\ 2p^6\ 3s^2\ 3p^4$ and, therefore, the electronic configuration of S_2 is

$$\text{KKLL}\,(3\text{s}\sigma)^2\ (3\text{s}\sigma^*)^2\ (3\text{p}_x\sigma)^2\ (3\text{p}_y = 3\text{p}_z\pi)^4\ (3\text{p}_y\pi^* = 3\text{p}_z\pi^*)^2$$

where LL stands for the $n = 2$ electrons. The orbitals $3p_y\pi^* = 3p_z\pi^*$ can accommodate four electrons. By Hund's rule, the two available electrons will enter each of these with their spins parallel, giving a paramagnetic molecule.

14.7 The removal of an electron from the O_2 molecule increases the dissociation energy from 5.08 to 6.48 eV, whereas in N_2, the removal of the electron decreases the energy from 9.91 to 8.85 eV. Substantiate.

Solution. The bonding MOs produce charge building between the nuclei, and the antibondig MOs charge depletion between the nuclei. Hence, removal of an electron from an antibonding MO increases the dissociation energy D_e or decreases the bond length of the bond, whereas removal of an electron from a bonding MO decreases D_e or increases the bond length. The electronic configuration of O_2 is

$$\text{KK}\,(2\text{s}\sigma_g)^2(2\text{s}\sigma_u^*)^2(2\text{p}_x\sigma_g)^2(2\text{p}_y\pi_u = 2\text{p}_z\pi_u)^4(2\text{p}\pi_g^*)^2$$

The highest filled MO is antibonding. Hence removal of an electron increases the D_e from 5.08 to 6.48 eV. The electronic configuration of N_2 is

$$\text{KK}\,(2\text{s}\sigma_g)^2(2\text{s}\sigma_u^*)^2(2\text{p}_x\sigma_g)^2(2\text{p}_y\pi_u = 2\text{p}_z\pi_u)^4$$

Removal of an electron from the highest filled bonding orbital decreases the dissociation energy from 9.91 to 8.85 eV.

14.8 Discuss the type of bonding in the heteronuclear diatomic molecule NO. Why is the bond in NO^+ expected to be shorter and stronger than that of NO?

Solution. Nitrogen and oxygen are close to each other in the periodic table and, therefore, their AOs are of similar energy. The nitrogen atom has seven electrons and the oxygen atom eight. The energy levels of the various MOs are the same as those for homonuclear diatomic molecules. Therefore, the electronic configuration of NO molecule is

$$\mathrm{KK}\,(2s\sigma_g)^2(2s\sigma_u^*)^2(2p_x\sigma_g)^2(2p_y\pi_u = 2p_z\pi_u)^4(2p\pi_g^*)^1$$

The inner shell is nonbonding, the bonding and antibonding $(2s\sigma_g)$ and $(2s\sigma_u^*)$ orbitals cancel. Though the four electrons in $(2p_y\pi_u = 2p_z\pi_u)^4$ orbital can give two π bonds, a half-bond is cancelled by the presence of one electron in the antibonding $2p\pi_g^*$ orbital. This leads to a σ-bond $(2p_x\sigma_g)^2$ a full π-bond and a half π-bond form 2p electrons. The molecule is paramagnetic since it has an unpaired electron. Removal of an electron from the system means the removal of an electron from the antibonding orbital. Hence, the bond in NO^+ is expected to be shorter and stronger.

14.9 Compare the MO wavefunction of hydrogen molecule with that of the valence bond theory.

Solution. Equation (14.12) gives the MO wavefunction and the Heitler-London function for hydrogen molecule is given in Problem 14.4. So,

$$\psi_{\text{mo}} = \text{constant}\ [\psi_a(1)\psi_a(2) + \psi_b(1)\psi_b(2) + \psi_a(1)\psi_b(2) + \psi_b(1)\psi_a(2)]$$

$$\psi_{\text{HL}} = \text{constant}\ [\psi_a(1)\psi_b(2) \pm \psi_a(2)\psi_b(1)]$$

The first two terms in ψ_{mo} represent the possibility of both the electrons being on the same proton at the same time.These represent the ionic structures $H_a^- H_b^+$ and $H_a^+ H_b^-$. The third and the fourth terms represent the possibility in which the electrons are shared equally by both the protons, and hence they correspond to covalent structures. Both the terms in the valance bond wavefunction correspond to covalent structures as one electron is associated with one nucleus and the second electron is associated with the other nucleus.

14.10 Write the electronic configuration of Na_2 and S_2 molecules in the MO concept.

Solution. The electronic configuration of Na: $1s^2\ 2s^2\ 2p^6\ 3s^1$.
The electronic configuration of Na_2 molecule is

$$\mathrm{Na_2}\ [\mathrm{KK}\ (2s\sigma)^2\ (2s\sigma^*)^2\ (2p_y\pi = 2p_z\pi)^4\ (2p_x\sigma)^2\ (2p_y\pi^* = 2p_z\pi^*)^4\ (2p_x\sigma^*)^2\ (3s\sigma)^2]$$

$$= \mathrm{Na_2}\ [\mathrm{KK\ LL}\ (3s\sigma)^2]$$

This result may be compared with the electronic configuration of Li_2, another alkali metal.
The electronic configuration of S: $1s^2\ 2s^2\ 2p^6\ 3s^2\ 3p^4$. The electronic configuration of S_2 molecule is

$$\mathrm{S_2}\ [\mathrm{KK\ LL}\ (3s\sigma)^2\ (3s\sigma^*)^2\ (3p_x\sigma)^2\ (3p_y\pi = 3p_z\pi)^4\ (3p_y\pi^* = 3p_z\pi^*)^2]$$

Though the orbitals $3p_y\pi^* = 3p_z\pi^*$ can accomodate four electrons, there are only two. Hence by Hund's rule, one electron will enter each of these with their spins parallel giving a paramagnetic molecule.

14.11 (i) Write the electronic configuration of N_2 molecule and N_2^+ ion
(ii) explain the type of bonding in them.
(iii) which one has the longer equilibrium bond length?
(iv) which one has larger dissociation energy.

Solution. Nitrogen molecule has 14 electrons. They are distributed among the MOs as

$$N_2[KK\ (2s\sigma_g)^2(2s\sigma_u^*)^2(2p\sigma_g)^2(2p\pi_u)^4]$$

The electron configuration of N_2^+ is

$$N_2^+[KK\ (2s\sigma_g)^2(2s\sigma_u^*)^2(2p\sigma_g)^2(2p\pi_u)^3]$$

The two electrons in $2s\sigma_g$ and the two in $2s\sigma_u^*$ antibonding orbital together leads to no bonding. The $(2p\sigma_g)^2$ and $(2p\pi_u)^4$ bonding orbitals together give a triple $N \equiv N$ bond, one bond being σ and the other two being π-bonds, in N_2 molecule. In N_2^+ ion the two electrons in $2p\sigma_g$ gives rise to a single σ-bond, two electrons in $2p\pi_u$ gives a π-bond, and the third electron in $2p\pi_u$ makes a half-bond.

Bonding MOs produce charge building. Hence removal of an electron from $2p\pi_u$ orbital decreases the charge building . Hence, N_2^+ has larger equilibrium bond length. Since charge density is less in N_2^+, the dissociation energy in it is less, or N_2 has larger dissociation energy.

14.12 Using the MO concept of electronic configuration of molecules, show that (i) oxygen is paramagnetic, (ii) the removal of an electron from O_2 decreases the bond length, and (iii) evaluate the bond order of the O_2 molecule.

Solution. The 16 electrons in oxygen molecule gives the electronic configuration

$$O_2\ [KK\ (2s\sigma_g)^2(2s\sigma_u^*)^2(2p\sigma_g)^2(2p\pi_u)^4(2p\pi_g^*)^2]$$

The antibonding MO, $2p\pi_g^*$ is degenerate and can accomodate four electrons. As we have only two electrons in that orbital, the two will align parallel in the two-fold degenerate orbital (Hund's rule). Aligning parallel means, effective spin is 1. Hence the molecule is paramagnetic.

(ii) Removal of an electron from an antibonding orbital increases charge building. Hence, bond length decreases and the equilibrium dissociation energy increases.

(iii) The bond order b is defined as one-half the difference between the number of bonding electrons (n), between the atoms of interest, and the antibonding electrons (n^*):

$$b = \frac{1}{2}(n - n^*)$$

Since $2s\sigma_g$, $2p\sigma_g$ and $2p\pi_u$ are bonding orbitals and $2s\sigma_u^*$ and $2p\pi_g^*$ are anti-bonding orbitals, the bond order

$$b = \frac{1}{2}(8 - 4) = 2$$

14.13 Write the electronic configuration of the F_2 molecule and explain how the configurations of Cl_2 and Br_2 are analogous to those of F_2.

Solution. The electronic configuration of F_2 molecule is

$$F_2\ [KK\ (2s\sigma_g)^2(2s\sigma_u^*)^2(2p\pi_u)^4(2p\sigma_g)^2(2p\pi_g^*)^4]$$

The inner shell is nonbonding and the filled bonding orbitals $(2s\sigma_g)^2$ $(2p\pi_u)^4$ are cancelled by the antibonding orbitals $(2s\sigma_u^*)^2$ $(2p\pi_g^*)^4$. This leaves only the σ-bond provided by the $2p\sigma_g$ orbital. For Cl_2 and Br_2, the electronic configurations are

$$Cl_2\ [\text{KK LL}\,(3s\sigma_g)^2(3s\sigma_u^*)^2(3p\pi_u)^4(3p\sigma_g)^2(3p\pi_g^*)^4]$$

$$Br_2\ [\text{KK LL MM}\ (4s\sigma_g)^2(4s\sigma_u^*)^2(4p\pi_u)^4(4p\sigma_g)^2(4p\pi_g^*)^4]$$

All the three molecules have similar electronic configurations leading to a σ bond.

14.14 On the basis of directed valence, illustrate how the *p*-valence shell orbitals of nitrogen atom combine with the *s*-orbitals of the attached hydrogen atoms to give molecular orbitals for the NH_3 molecule.

Solution. In NH_3, the central nitrogen atom has the electron configuration

$$1s^2\ 2s^2\ 2p_x^1\ 2p_y^1\ 2p_z^1$$

The maximum overlapping of the three *p* orbitals with the 1*s* hydrogen orbitals are possible along the *x*, *y* and *z*-directions (Fig. 14.5). The bond angle in this case is found to be 107.3°, which is again partly due to the mutual repulsion between the hydrogen atoms.

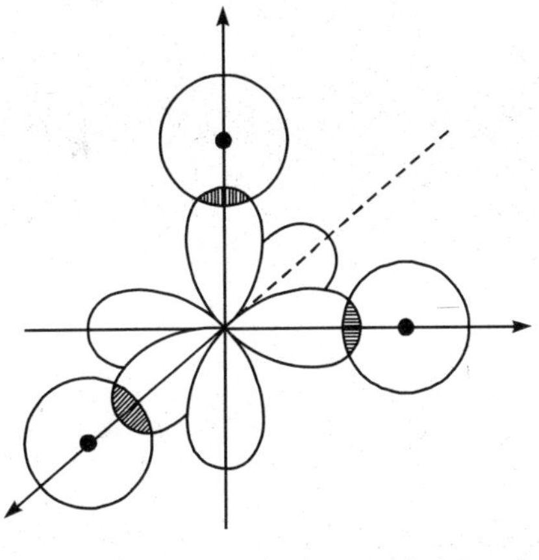

Fig. 14.5 The formation of ammonia molecule. (The singly occupied $2p_x$, $2p_y$ and $2p_z$ orbitals of nitrogen overlap with the hydrogen 1s orbitals).

14.15 A gas consisting of B_2 molecules is found to be paramagnetic. What pattern of molecular orbitals must apply in this case?

Solution. The 10 electrons in this molecule are expected to be distributed as

$$B_2\ [\text{KK}\ (2s\sigma_g)^2(2s\sigma_u^*)^2(2p\sigma_g)^2]$$

The next orbital is $2p\pi_u$ which has nearly the same energy as that of $2p\sigma_g$. Hence, instead of $(2p\sigma_g)^2$, the alternate configuration $(2p\sigma_g)^1$ $(2p\pi_u)^1$, leading to a total spin of one is possible. These two unpaired electrons per molecule lead to the observed paramagnetism of B_2. The molecular orbital pattern of B_2 is, therefore,

$$B_2\ [\text{KK}\ (2s\sigma_g)^2(2s\sigma_u^*)^2(2p\sigma_g)^1\ (2p\pi_u)^1]$$

14.16 Find the relative bond strengths of (i) F_2 molecule and F_2^+ ion; (ii) F_2 and O_2 molecules.

Solution.

(i) The electronic configaration of F_2 is

$$F_2\,[KK\,(2s\sigma_g)^2(2s\sigma_u^*)^2(2p\pi_u)^4(2p\sigma_g)^2(2p\pi_g^*)^4]$$

Removal of an electron means, only three electrons in the antibonding orbital $2p\pi_g^*$. Removal of an electron from an antibonding orbital means an increase in charge building in the bond. Hence bond strength increases in F_2^+. The electronic configuration of O_2 is

$$O_2\,[KK\,(2s\sigma_g)^2(2s\sigma_u^*)^2(2p\sigma_g)^2(2p\pi_u)^4(2p\pi_g^*)^2]$$

(ii) In O_2, there is an excess of four bonding electrons over the antibonding ones, whereas in F_2 there is an excess of only two bonding electrons over the antibonding ones. Hence the bond in O_2 is stronger than that in F_2.

14.17 In *sp* hybridization, show that the angle between the two hybrid bonds is 180°.

Solution. As the two hybrids are equivalent, each must have equal *s* and *p* character. Hence the wave function of the first hybrid is

$$\psi_1 = \frac{1}{\sqrt{2}}s + \frac{1}{\sqrt{2}}p_1$$

and that of the second hybrid is

$$\psi_2 = \frac{1}{\sqrt{2}}s + \frac{1}{\sqrt{2}}p_2$$

Since $\langle\psi_1|\psi_2\rangle = 0$,

$$\left\langle \frac{1}{\sqrt{2}}(s + p_1) \middle| \frac{1}{\sqrt{2}}(s + p_2) \right\rangle = 0$$

$$\frac{1}{2}\langle s|s\rangle + \frac{1}{2}\langle p_1|p_2\rangle + \frac{1}{2}\langle s|p_2\rangle + \frac{1}{2}\langle p_1|s\rangle = 0$$

The last two terms are zero. If θ_{12} is the angle between the hybrids,

$$\frac{1}{2} + \frac{1}{2}\cos\theta_{12} = 0 \quad \text{or} \quad \cos\theta_{12} = -1$$

$$\theta_{12} = 180°$$

14.18 Show that the three hybrid bonds in sp^2 hybridization are inclined to each other by 120°.

Solution. Of the 3p-orbitals we leave one, say the p_z, unmixed and the other two to mix with the s-orbital. Hence, the three hybrid orbitals should be directed in the *xy*-plane. Consider the linear combination of these two p-orbitals

$$\phi = ap_x + bp_y$$

which gives rise to p_1 in the direction of the first hybrid bond. Then the wave function of the first hybrid can be written as

$$\psi_1 = c_1 s + c_2 p_1$$

where c_1 and c_2 are constants. As all the three hybrids are equivalent, each one must have the same amount of s-character and the same amount of p-character. Hence, each bond will have one-third s-character and two-third p-character, i.e., ψ_1^2 must have $(1/3)s^2$ and $(1/3)p^2$. Therefore,

$$c_1^2 = \left(\frac{1}{3}\right) \text{ and } c_2^2 = \left(\frac{2}{3}\right) \quad \text{or} \quad c_1 = \frac{1}{\sqrt{3}} \text{ and } c_2 = \sqrt{\frac{2}{3}}$$

The hybrid orbital of the first bond is

$$\psi_1 = \frac{1}{\sqrt{3}}\text{s} + \sqrt{\frac{2}{3}}\,\text{p}_1$$

The hybrid obrital of the second bond is

$$\psi_2 = \frac{1}{\sqrt{3}}\text{s} + \sqrt{\frac{2}{3}}\,\text{p}_2$$

Since ψ_1 and ψ_2 are orthogonal,

$$\langle\psi_1|\psi_2\rangle = \left\langle\left(\frac{1}{\sqrt{3}}\text{s} + \sqrt{\frac{2}{3}}\,\text{p}_1\right)\middle|\left(\frac{1}{\sqrt{3}}\text{s} + \sqrt{\frac{2}{3}}\,\text{p}_2\right)\right\rangle = 0$$

$$\frac{1}{3}\langle\text{s}|\text{s}\rangle + \frac{2}{3}\langle\text{p}_1|\text{p}_2\rangle + \frac{\sqrt{2}}{3}\langle\text{s}|\text{p}_2\rangle + \frac{\sqrt{2}}{3}\langle\text{p}_1|\text{s}\rangle = 0$$

Since the net overlap between an s and a p orbital centred on the same nucleus is zero, the third and the fourth terms are zero. Writing

$$\text{p}_2 = \text{p}_1 \cos\theta_{12}$$

we have

$$\frac{1}{3} + \frac{2}{3}\langle\text{p}_1|\text{p}_1\rangle\cos\theta_{12} = 0 \quad \text{or} \quad \cos\theta_{12} = -\frac{1}{2}$$

$$\theta_{12} = 120°$$

14.19 Prove that the angle between any two of the sp^3 hybrids is 109° 28′.

Solution. It can be proved that the linear combination of three p-orbitals $\phi = ap_x + bp_y + cp_z$ can give rise to another p-orbital oriented in a direction depending on the values of the constants a, b, and c. Consider an appropriate combination p_1 of the three p-orbitals in the direction of the first bond. Then the wavefunction of the hybrid of the first bond can be written as

$$\psi_1 = c_1\text{s} + c_2\,\text{p}_1$$

where c_1, c_2 are constants.

As all the four hybrids are equivalent, each one must have the same amount of s-character and the same amount of p-character. Hence each bond will have 1/4 s-character and 3/4 p-character, i.e., ψ_1^2 must contain $1/4s^2$ and $3/4p^2$. Therefore, $c_1^2 = 1/4$ and $c_2^2 = 3/4$.

Hybrid orbital of the first bond: $\psi_1 = \dfrac{1}{2}\text{s} + \dfrac{\sqrt{3}}{2}\text{p}_1$

Hybrid orbital of the second bond: $\psi_2 = \dfrac{1}{2}\text{s} + \dfrac{\sqrt{3}}{2}\text{p}_2$

Since ψ_1 and ψ_2 are orthogonal,

$$\langle \psi_1 | \psi_2 \rangle = \left\langle \left(\frac{1}{2}\mathrm{s} + \frac{\sqrt{3}}{2}\mathrm{p}_1 \right) \middle| \frac{1}{2}\mathrm{s} + \frac{\sqrt{3}}{2}\mathrm{p}_2 \right\rangle = 0$$

$$\frac{1}{4}\langle \mathrm{s}|\mathrm{s}\rangle + \frac{3}{4}\langle \mathrm{p}_1|\mathrm{p}_2\rangle + \frac{\sqrt{3}}{4}\langle \mathrm{s}|\mathrm{p}_2\rangle + \frac{\sqrt{3}}{4}\langle \mathrm{p}_1|\mathrm{s}\rangle = 0$$

The net overlap between a s-and a p-orbital centred on the same nucleus is zero, which makes the third and the fourth terms zero. Writing $p_2 = p_1 \cos \theta_{12}$, we have

$$\frac{1}{4} + \frac{3}{4}\langle \mathrm{p}_1|\mathrm{p}_1\rangle \cos\theta_{12} = 0$$

$$\cos\theta_{12} = -\frac{1}{3} \quad \text{or} \quad \theta_{12} = 109^\circ\ 28'$$

14.20 Sketch the molecular orbital formation in ethane and ethylene.

Ethane (C_2H_6): In ethane each atom is sp^3 hybridized. Three of these hybrid orbitals in each carbon atom overlap with the s-orbitals of three hydrogen atoms and the fourth one with the corresponding one of the other carbon atom. All the bonds are of s type. The molecular orbital formation is illustrated in Fig. 14.6.

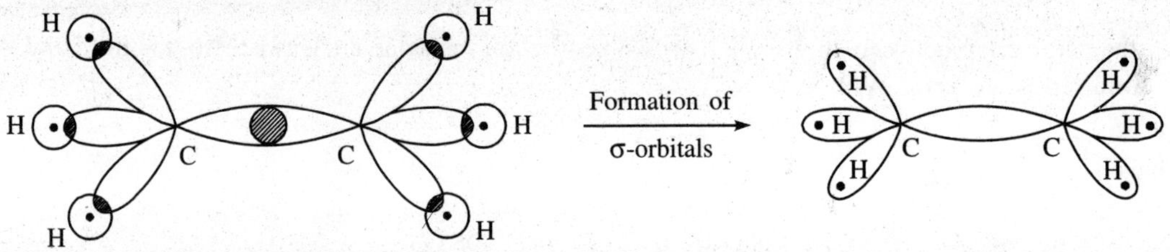

(a) sp^3 hybrids of C and 1s atomic orbitals of H

(b) Molecular orbitals

Fig. 14.6 Molecular orbital formation in ethane.

Ethylene (C_2H_4): Each carbon atom is sp^2 hybridized. Two of these form localized σ-type MO by overlapping with 1s orbital of hydrogen atom and the third overlaps with the second carbon forming another localized σ MO (Fig. 14.7a). These three σ-bonds lie in a plane, the molecular plane. Each carbon atom is left with a singly occupied p-orbital with its axis perpendicular to the plane of the molecule. The lateral overlap of these two p-orbitals give a π-bond (Fig. 14.7b), the second bond between the two carbon atoms. The plane of the molecule is the nodal plane of the π-orbital.

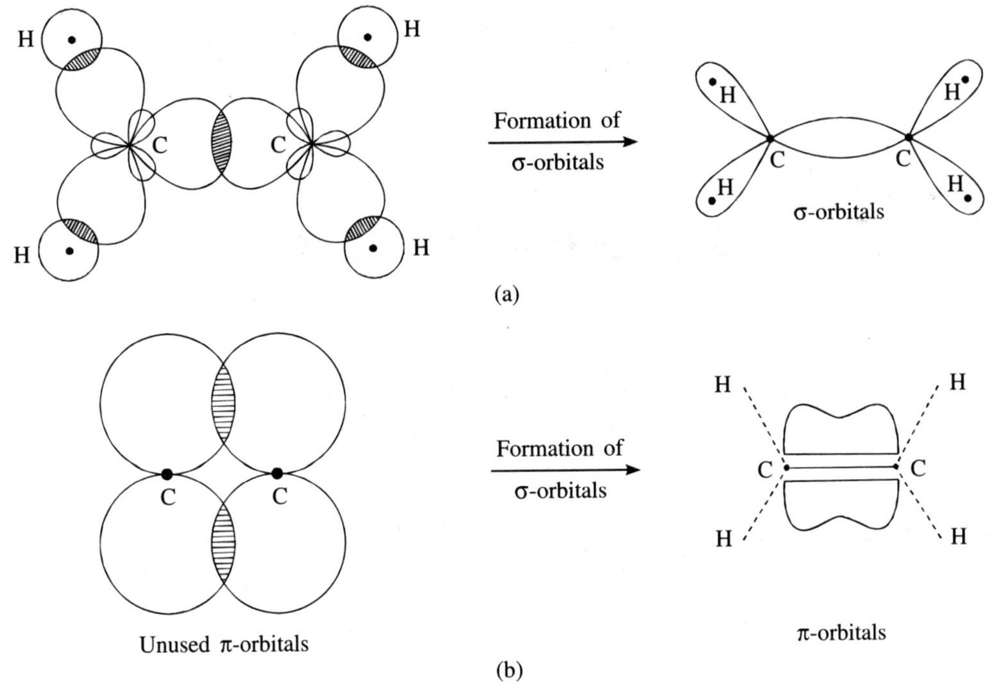

Fig. 14.7 Formation of (a) σ-orbitals (b) π-orbitals in ethylene.

APPENDIX

Some Useful Integrals

1. $\int_0^\infty \exp(-ax^2)\, dx = \frac{1}{2}\left(\frac{\pi}{a}\right)^{1/2}$

2. $\int_0^\infty x^2 \exp(-ax^2)\, dx = \frac{\sqrt{\pi}}{4}\left(\frac{1}{a^{3/2}}\right)$

3. $\int_0^\infty x^4 \exp(-ax^2)\, dx = \frac{3\sqrt{\pi}}{8}\left(\frac{1}{a^{5/2}}\right)$

4. $\int_0^\infty x^6 \exp(-ax^2)\, dx = \frac{15\sqrt{\pi}}{16}\left(\frac{1}{a^{7/2}}\right)$

5. $\int_0^\infty x \exp(-ax^2)\, dx = \frac{1}{2}a$

6. $\int_0^\infty x^3 \exp(-ax^2)\, dx = \frac{1}{2}a^2$

7. $\int_0^\infty x^5 \exp(-ax^2)\, dx = \frac{1}{a^3}$

8. $\int_{-\infty}^\infty x^n \exp(-ax^2)\, dx = 0$ if n is odd

9. $\int_0^\infty x^n \exp(-ax)\, dx = \frac{n!}{a^{n+1}}, \qquad n \geq 0,\ a > 0$

10. $\int_0^\infty \frac{x\, dx}{e^x - 1} = \frac{\pi^2}{6}$

11. $\int_0^\infty \frac{x^3\,dx}{e^x - 1} = \frac{\pi^4}{15}$

12. $\int_0^\infty \cos bx \exp(-ax)\,dx = \frac{a}{(a^2 + b^2)}, \qquad a > 0$

13. $\int_0^\infty \sin bx \exp(-ax)\,dx = \frac{b}{(a^2 + b^2)}, \qquad a > 0$

14. $\int_0^\infty \cos bx \exp(-a^2x^2)\,dx = \frac{\sqrt{\pi}\exp(-b^2/4a^2)}{2a}, \qquad a > 0$

15. $\int xe^{ax}\,dx = (ax - 1)\frac{e^{ax}}{a^2}$

16. $\int x^2e^{ax}\,dx = \left(\frac{x^2}{a} - \frac{2x}{a^2} + \frac{2}{a^3}\right)e^{ax}$

Index